W9-BXO-849

Wallpaper in America

Frontispiece (overleaf) This is the cover of a promotional booklet published by a New York wallpaper manufacturer, Warren Fuller and Company, in 1880. The author, Clarence Cook (1828–1900), was one of New York's leading art critics and arbiters of taste. Cooper-Hewitt Museum library.

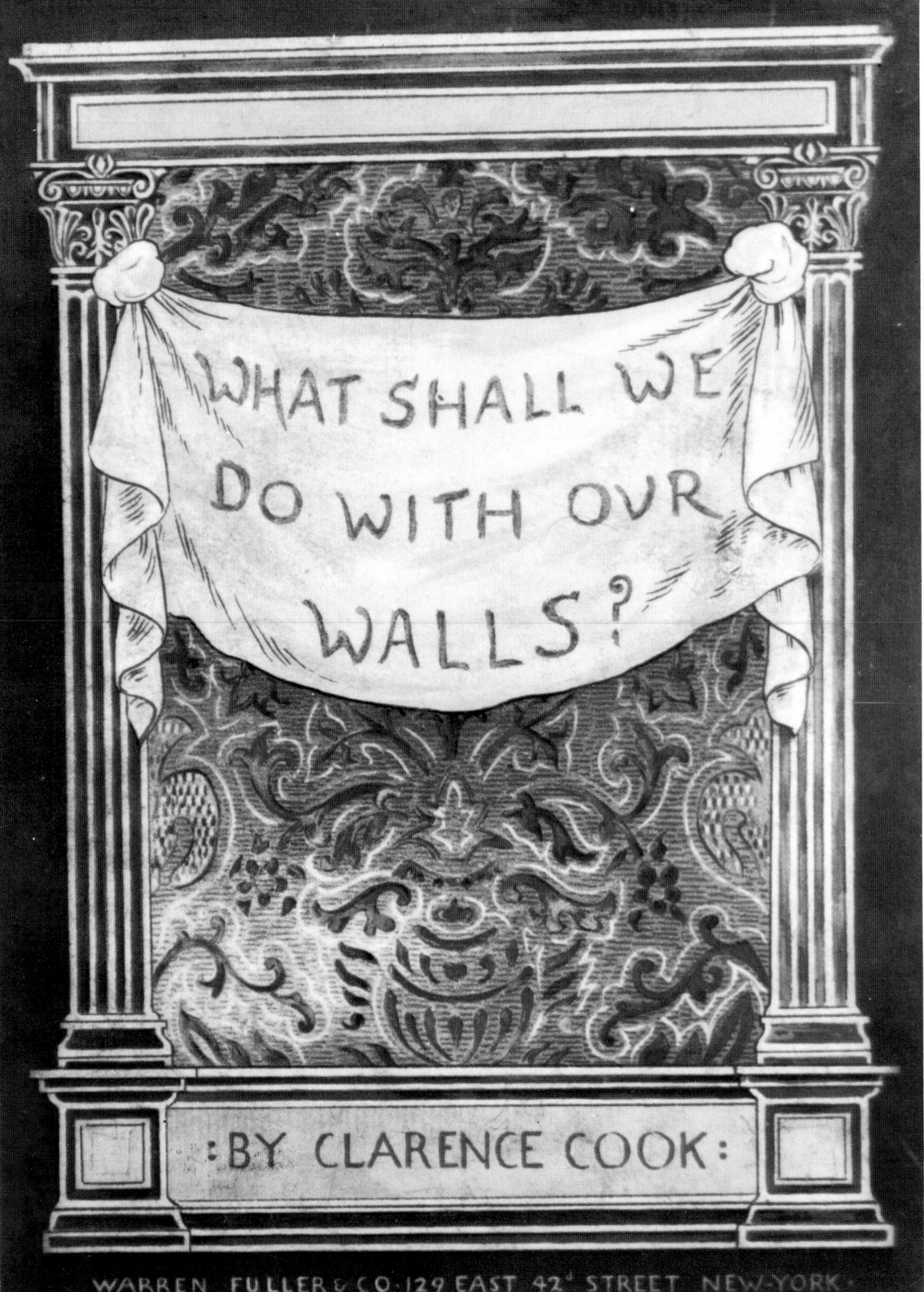
WHAT SHALL WE DO WITH OUR WALLS?
:BY CLARENCE COOK:
WARREN FULLER & CO. 129 EAST 42d STREET NEW-YORK.

Wallpaper in America

From the Seventeenth Century to World War I

Catherine Lynn

With a foreword by Charles van Ravenswaay,
Director Emeritus,
Henry Francis du Pont Winterthur Museum

A Barra Foundation/
Cooper-Hewitt Museum Book
W. W. Norton & Company, Inc. New York

Published simultaneously in Canada by George J. McLeod
Limited, Toronto.

Printed in the United States of America

First Edition

THIS BOOK WAS EDITED BY AND PRODUCED UNDER THE
SUPERVISION OF REGINA RYAN FOR THE BARRA FOUNDATION, INC.

DESIGNER: ULRICH RUCHTI

W. W. Norton & Company, Inc. 500 Fifth Avenue, New York, N.Y., 10110
W. W. Norton & Company Ltd. 25 New Street Square, London EC4A 3NT

ISBN 0 393 01448 7 cloth edition

TABLE OF CONTENTS

FOREWORD

"Why should the Old Time Wall-Papers alone [among the decorative arts] be left unchronicled and forgotten?" Kate Sanborn, a no-nonsense New Hampshire teacher, lecturer, and journalist, asked early in this century. "If a book has ever been written on this subject it has been impossible to discover." She was speaking, of course, about the lack of a history of wallpaper in America and, after pondering her question, she accepted as her mission the task of compiling one. That task also produced a litany of frustrations and disappointments. Friends pronounced her subject "odd"; facts and examples of early papers were very hard to track down. But, undaunted, she persevered and in 1905 her pioneering study, *Old Time Wall-Papers,* was handsomely published. She had achieved her purpose of rescuing antique papers from oblivion, but—perfectionist that she was—she considered her text "scanty and superficial."

In 1924, twenty years after Miss Sanborn's work appeared, came Nancy V. McClelland's *Historic Wall-Papers.* While much more comprehensive than Miss Sanborn's book, it was limited by the data then available. But it had the great advantage of being published at a time of heightened interest in America's early buildings and its decorative arts. The book became the Bible of restorationists and museum curators, exciting interest in antique papers and their preservation. It also influenced interior decoration by directing attention once again to the beauty and versatility of wallpaper.

In addition to these two American studies, a limited number of articles on wallpaper appeared in various periodicals. In the indices of the magazine *Antiques* during its first fifty years (1922–71), there are only seventy-nine references to wallpaper, many of which concern reproductions, bandbox papers, excise marks, and articles about single designs.

Now, some fifty years after the McClelland book appeared, we are fortunate in having Catherine Lynn's *Wallpaper in America.* It represents many years of dedicated study, beginning in the great wallpaper collection at the Cooper-Hewitt Museum, with which she was formerly associated, and continuing in collections both here and abroad. Ms. Lynn has corrected old errors of fact or opinion, and has made good use of the vast amount of research that has been carried on in the field of the decorative arts during the past half-century. The result is a carefully designed and researched study of a neglected subject. It would have greatly pleased the pioneers in her field; it will delight her readers today.

Nineteenth-century writers frequently stressed that much of the "feeling" of a room depended upon the treatment of the walls. William Morris, the English designer and author, was emphatic on that point. "Whatever you have in your rooms," he wrote, "think first of the walls, for they are that which makes your house and home." If you don't, he continued, your rooms will have "a sort of makeshift, lodging-house look about them, however rich your movables may be." By selecting wallpaper with "judgment" as to its "fitness and truthfulness," one could enhance the elegance of mansions and add cheerfulness to humble cottages, A. J. Downing, an American taste maker, advised in 1850. Wallpaper had matured into an independent art form, individual in character and serving many functions, as artists, decorators, and others trained in the intricacies of design and color have always known or sensed. These uses are not always apparent, for they are based upon subtleties of the eye and the mind, including individual emotional responses to visual suggestions. Designs not only reflect changing fashions but with their colors and color tones also express that most indefinable quality, the mood of the times. Because of these and other qualities, wallpaper (as with all the American decorative arts) is a significant record of our national culture.

Among wallpaper's many interesting functions is enhancing the architecture of interiors, or suggesting architectural details where none exist. It can also serve to unify, aesthetically, interior architecture and furnishings while enhancing both, a possibility achieved only when the elements are compatible. Inappropriate backgrounds can insidiously alter the appearance of furnishings in surprising ways, as some restorationists and others—with more enthusiasm than knowledge—have learned to their surprise.

It is never easy to find the real causes of change; often they are not the ones most obvious. When monochromatic, painted wall surfaces began to replace wallpaper after World War I, changing tastes in architecture and interior decoration—and economic factors—appeared to be the chief reasons. This new look seemed fresh and uncluttered, and the functional, ingeniously designed, and impersonal furnishings then coming into vogue looked well against such anonymous walls. But more than the new settings of our lives was becoming impersonal. Some resisted this change in fashion and all that it implied, aided by the growing interest in American antiques, restorations, and the re-creation of period interiors.

Wallpaper continues to be used, and while designers still reproduce or adapt the work of their predecessors, innovations are also being made in the field. It is encouraging to believe that this continuing vitality comes from wallpaper's visual and emotional appeal, the many functions it serves, and—most of all—because it is an expression of the personal and the humane.

CHARLES VAN RAVENSWAAY
Director Emeritus, Henry Francis du Pont
Winterthur Museum.
July 1980

INTRODUCTION

"What Shall We Do With Our Walls?" asks Clarence Cook, the nineteenth-century American art critic. Wallpaper them, he answers, in a promotional booklet published a hundred years ago by Warren Fuller and Company, New York wallpaper manufacturers. Indeed, the publication of *"What Shall We Do With Our Walls?"* in 1880 coincides with the high point in the popularity of wallpaper in America. It was a time when Americans papered the walls of every house, large or small. At the same time, the publication of the booklet marked the beginning of the end of serious interest in wallpaper design. That art has never quite recovered from the reaction against the multi-patterned, multi-bordered wallpapering craze that Cook's publication encouraged.

Wallpaper manufacturers were not alone in promoting the use of their product in late-nineteenth-century America. The California architects Newsom and Newsom, in the introduction to their book *Picturesque Californian Homes* of 1885, noted as a matter of course:

> The query "what shall we do with our walls?" has long since been answered. . . . White walls unrelieved by any color are relics of barbarism, and are almost a thing of the past. House-papering is now incorporated in building contracts, and a house is considered incomplete without these adornments.[1]

These prolific architects and their colleagues of greater stature regarded wallpaper as an important and expressive element within their designs. The public followed the lead of architects—the elite of the design hierarchy—both by using quantities of wallpaper and by regarding it as a worthy subject for theoretical discussion. During the last quarter of the nineteenth century, published analysis of wallpaper focused not only on its visual qualities but also on the symbolic, moral, and even religious implications of various kinds of patterning. Popular journalists, critics writing in art and architectural magazines, and novelists, as well as the authors of books and articles on interior decorating and domestic economy and those who wrote for the building and decorating trades, devoted an unprecedented and unsurpassed quantity of space and mental effort to the subject of wallpaper.

Serious discussion of wallpaper design appeared prominently in an astonishing variety of American publications well into the 1880s. However, only a generation later, avant-garde critics and architectural designers had totally rejected wallpaper as a subject worthy of analysis. They had also rejected it as the answer to the question posed by Clarence Cook in 1880. By 1915, white walls were no longer "relics of barbarism," but instead signaled incipient modernism in architectural circles. At the same time, in homes of the rich and fashionable that conformed to more traditional architectural styles, wallpaper was beginning to appear outdated, in one of the inevitable turns of fashion away from whatever has immediately preceded it and toward whatever seems fresh and chic.

Since its late-nineteenth-century heyday, wallpaper has never again been accorded the serious attention of students and practitioners of contemporary architecture and design. Until a very few years ago, this group tended to see all wall patterns as amorphous masses of ornamentation. Well into the 1960s ornament was not recognized as a legitimate concern by most architects, who subscribed to the theories that took hold in the 1920s and produced the International Style. The antipathy toward all decoration within buildings only recently began to lose its grip on contemporary architectural practice. Now, in the era of Post Modernism, avant-garde designers are appreciating and understanding anew the forms and richly ornamented surfaces of more traditional architectural monuments, and recognizing that to follow function is but part of the architect's challenge.

Had this book on wallpaper been published ten years ago, it would have reflected, in a vaguely apologetic tone, the still-prevailing climate of Modernist scorn for the subject. In such a climate, architectural preservationists, museum curators, collectors of antique furniture, and professionals in the decorating trades would have been the only audiences expected to make practical use of such a book.

At that time, many connoisseurs of eighteenth- and nineteenth-century decorative arts tended to blur distinctions between wallpaper styles almost as completely as did the Modern architects. Although the antique collectors may have been quick to distinguish furniture of one period from that of a few years later, they were often slow to recognize that differences within the two-dimensional format of the paper's surface are often as great as the differences in furniture designs.

The antique collectors' indifference to wallpaper is consistent with the fact that wallpapers can rarely serve their purpose, since old papers seldom survive in condition good enough to endow them with the status of antiques bringing high prices in auction sales rooms. Wallpapers were manufactured in multiples, then glued to walls, and it was seldom intended that they should be treasured for a long time, or ever be moved to other walls. Even in the earliest days of its production wallpaper was removed from the hand of the designing craftsman by a process of replication. In contrast, each piece of American furniture of the eighteenth and early nineteenth centuries was, and has been valued as, a product of the hand craftsmanship of an individual or group of closely allied individuals. In an industrialized world, where most things are machine-made, the relatively few products of hand craftsmanship have been endowed with ever-increasing status. The cabinet maker's wares have commanded the interest and the dollars of collectors, and have generated a body of literature to serve the collecting audience. Coincidentally, that literature is also of use to students of design and to the growing number of cultural historians who look to the object as another kind of historical document that can be "read" in an endeavor to learn more about the culture that produced it.

There is no comparable market for old wallpaper or literature about the wallpaper used in this country. Yet it was an important feature of interior architecture, sometimes dominating a furnishing scheme, but more often forming the background against which other decorative arts were shown to best advantage according to the lights of then-current fashion. It is a feature too often neglected or replaced by inappropriate patterns in museum "period rooms" and in restored historic houses. It is hoped that this book will serve as a tool for curators and restorers, helping them to correct and avoid mistakes where authenticity is important.

The fundamental intention of this book is to present a survey of the most distinctive styles in wallpapers and the ways they were used in America from the seventeenth century, when their presence here can first be documented, until the First World War. World War I has been chosen as the cutoff date for this book because with it the major movements in nineteenth-century design that affected wallpaper came to an end. But even before the war new themes, concerns, and motifs had been introduced that were not to be fully exploited by wallpaper designers until the 1920s. Well before the war, the seeds of incipient Art Moderne and Art Deco styles are to be found in some patterns, and the beginnings of the phase of reproducing historical patterns are evident. These themes dominated the fashionable wallpaper trade from the 1920s through the middle third of the twentieth century. They constitute a whole new chapter that awaits assessment and evaluation.

Through the period covered here, wallpaper's stylistic development closely follows and reflects that of the other decorative arts, especially textiles, decorative wall paintings, and architectural ornaments carved or made of plaster. Indeed, these more costly decorations frequently furnished the models for the cheaper wallpaper imitations. The styles of celebrated examples of sculpture and paintings were also reflected in wallpapers.

No attempt will be made to include all the painted, paneled, stenciled, and fabric-hung wall treatments that were popular in America during the eighteenth and nineteenth centuries. However, the focus on wallpaper should not be misread as an argument for its exclusive importance as a wall finish during the periods covered here. Rather, that focus simply reflects the fact that wallpaper in America is a large —and largely undocumented—subject, appropriate for and deserving of book-length treatment. Even within the scope of this book, there is no way to represent the full range of wallpaper types and the thousands of individual patterns that were available to the decorating homemaker in any given year.

Nevertheless, it will be possible to characterize distinctive styles and important trends in wallpaper manufacture, marketing, and use in America in chronological order. Within such a framework, English, French, and Oriental papers occupy a position of importance, since imported papers were probably the only ones available in America until the 1760s, and the imports continued to dictate the styles for American-made papers throughout the periods considered in this book. Paper hangings from England, France, and the Orient acted as style carriers, bringing to colonial American houses superficial coatings of the latest European fashions, rendered in full color.

In that important role as a transmitter of style, a role that continued into the later years of the nineteenth century, wallpaper merits the wider attention of students of the decorative arts and of folk art. Because it was relatively inexpensive and so easily transported—a few rolls could be tucked into almost any conveyance—wallpaper served as a rapid and effective disseminator of decorative styles to the very frontiers of America. Wallpaper, therefore, had a significant influence on other arts, whether "fine" or decorative, that were produced in relatively isolated spots. Those who value the traditional qualities of folk art will perhaps view this influence as a negative one. Cheap, easily available wallpaper was one commodity that helped make fashionable, store-bought newness more desirable than traditional, homemade qualities and therefore caused folk art to languish.

For the social historian, the nineteenth-century use of wallpaper in frontier regions gives evidence of a more comfortable domestic life, more quickly reflective of eastern fashions, than tales of cowboys and Indians might have led us to expect. And its use among the poorer classes in cities perhaps bears witness to their aspiration toward middle-class respectability. Physical evidence of the use of wallpaper has survived in surprising quantity, considering the ephemeral nature of paper itself and the probably short expectation of its duration when it was hung. But once on the walls, wallpaper often outlasted all manner of neglect and abuse for the simple reason that it was easier to cover it over than to strip it off a wall. Unlike furniture, wallpaper was seldom removed from the original context in which it was used. Though battered and discolored, many fragments of wallpaper remain in place, where they were hung two hundred years ago, telling us something about the tastes of ordinary people who seldom wrote about their aesthetic preferences. By looking at such evidence of its distribution, as well as by perusing wallpaper pattern books, one can trace the paths of wallpaper styles right across this country. It is possible to follow the high styles born in European cities as they were commercialized—diluted, distorted, and cheapened—and to assess, at least in part, how widely they were accepted through the eighteenth and nineteenth centuries, becoming part of a common popular culture.

The original intention in doing research for this book was relatively narrow—it focused on answering the practical questions of architectural restorationists and curators by gathering and ordering documentation on the use of wallpaper in America, and by describing and illustrating the styles of successive eras. However, my studies over the past nine years or so have revealed that the subject of wallpaper holds unexpected interest not only for this primary audience, and for a more general public, but also for specialists in other fields.

The art historian can find much of interest in the nineteenth-century theories about wallpaper design. As early as the 1840s, English critics began to formulate standards for wallpaper patterns, emphasizing that a wallpaper should look like what in fact it is: a two-dimensional arrangement of color, line, or patterning on a thin sheet of material covering a flat surface. This high evaluation of thing-like-ness and of abstract qualities in wallpaper design preceded by half a century the acceptance of similar values by "fine" artists for their easel paintings.[2]

The social historian interested in women's history can discover something about the position of women in the nineteenth century from the writings addressed to them about choosing wallpaper, among other decorations. These decorations were seen by contemporary theorists as special vehicles for the exertion of the moral influence peculiar to women.

The cultural critic can find in nineteenth-century decorating and design manuals evidence of the ways in which consumerism, especially the consumption of wallpaper, was justified and popularized in moral, aesthetic, patriotic, and even religious terms.

"What Shall We Do With Our Walls?" hardly seems a question likely to challenge the student of intellectual history. Yet it was a question to which the answers had surprising implications during the period covered in this book. In an era when many people were persuaded that material things revealed character, and could be used to form character, in times when it was widely believed that aesthetic sensibility had not only moral but also religious significance, what one might do with one's walls was a question of unexpected import. In this attempt to discover how Americans of the seventeenth, eighteenth, and nineteenth centuries answered Clarence Cook's question, we find that wallpaper tells us a good deal not only about their visual sensibilities but also about their values that we did not expect to learn.

A great collection of wallpapers like that at the Cooper-Hewitt Museum serves as a colorful index of decorative motifs popular during the past four centuries. But even more, it delights us with unexpected improvisations on a seemingly infinite number of themes, beguiles us with a whimsical variety of visual conceits, and simply exists as wonderful evidence of human creativity. Wallpaper is well worth looking at with new eyes, purely for itself.

Figure 1-1 (overleaf) This piece of embossed and gilded leather from the Stephen Gifford House, Bean Hill, Norwich, Connecticut, was probably made prior to 1700, somewhere in Europe. Its polychrome pattern relies on baroque scrollwork to give visual cohesion to the floral elements. It is colored with a bronze gold, and with dark green, burgundy red, and off-white over a ground of dull bluish green. The sample, 20 inches by 26½ inches, is preserved in the collections of the Society of the Founders of Norwich, Connecticut, and is published through their courtesy.

1 THE EARLIEST AMERICAN WALL HANGINGS

ENGLISH PRECEDENTS

The English colonizers of the East Coast of America drew upon the architectural and decorative styles of their homeland for the forms and ornamental details of nearly all their buildings and furnishings. Virtually every major study of American architecture and decorative art has demonstrated how consistently these transplanted Englishmen worked to make a "new" England on the edge of the wilderness.[1]

Although a great deal is known about the colonists' chairs, tables, and chests, there is less evidence about what they used to finish interior walls. Paper and cloth wall coverings were perishable commodities that have not survived in quantities equaling those of antique tables and chairs. Records of their presence in houses are also much rarer because wall finishes, once applied or installed, lost their value as "movables" that could later be sold. They were therefore not listed on inventories of estates—documents that have provided historians with a great deal of information about other objects commonly found in colonial households. However, scattered bits of evidence in colonial trading records and advertisements, as well as in eighteenth-century correspondence and journals, support the conclusion that the colonists closely imitated English styles in wall coverings, as they did in everything else.

In England during the seventeenth century, walls were finished with plaster, paint, or wooden paneling, called wainscoting. Exposed stone was also an accepted interior finish for a wall. All of these materials could provide highly decorative surfaces, and even when not elaborately ornamented, they were considered adequate for most rooms. In fact, decorating interior walls with "hangings," a term used to describe textiles, leather, or paper tacked, glued, or otherwise secured to cover a wall, was the exception during this era.[2]

The most luxurious way to decorate seventeenth-century walls was to hang them with woven tapestries of the kind that had been favored since the Middle Ages for the halls of the British nobility. Some paper hangings evolved as inexpensive imitations of tapestries and of other materials—including woolens, damask, velvet-woven silks, leather, and canvas—that were also occasionally hung on English walls during the seventeenth century.

None of the wall treatments that involved hangings was cheap. Very few people below the ranks of the gentry had the wealth to purchase wall hangings of any sort. Since the gentry constituted only about five percent of the population alive in Britain during the early seventeenth century, of whom only about two hundred families made up the nobility,[3] the "fashions" in wall hangings mentioned here must be understood to constitute decorations used by a very limited number of people.

In England as well as in Europe, those of the lesser ranks sometimes pasted sheets of paper decorated with pictures on their walls. These poor men's substitutes for the framed paintings of the rich were printed by the same craftsmen who made book illustrations. When they began to decorate the single sheets with patterning that formed a complete motif, or repeat only when juxtaposed with other sheets, they set in motion one of the processes from which modern wallpapers evolved.

The earliest document fixing a date for the English use of a proto-wallpaper was found in 1911 in the Master's Lodge of Christ's College, Cambridge. Scraps of a sixteenth-century paper were removed from the beams of the entrance hall and are now carefully preserved at the Victoria and Albert Museum in London. They bear the rebus of their maker—an H and a Goose—incorporated in the patterning. Hugh Goes, who devised this mark for his work, is known to have been active in York around 1509. He used the reverse side of a text, which can be approximately dated to that same year, 1509, as a surface on which to print a stylized pomegranate imitating textile patterns typical of the sixteenth century.[4] Such indications of date, maker, and place of manufacture are rare indeed for wallpapers of so early a period. Although there are other examples of wallpaper surviving from the seventeenth century, it is difficult to date them with precision. The only evidence on which such a judgment can usually be based is the ornamental style of the patterning, and that rarely gives a precise indication of the age of a paper.

Hugh Goes was a maker of woodcuts and a printer of books. His paper was block-printed in black ink, just as many book illustrations were. It is one of those seventeenth-century papers that were the products of printers rather than craftsmen specializing in making decorations for walls. The patterned sheets made by printers were used as endpapers for their volumes and as lining papers for boxes and trunks, where they survive more often than they do on walls *(figure 1-4).*

A quite different group of English craftsmen, more accustomed to decorating textiles than to working with printing presses, evolved another kind of wall covering that was used during the seventeenth century. This group had learned how to imitate cut velvets by applying flocking to canvas and paper. By the seventeenth century the fame of London-made flocked wall coverings had spread to the continent, and some of these were reaching America *(figure 1-2).* Nevertheless, although both printed and flocked papers had been introduced in England during the early sixteenth century, their use was still limited throughout the seventeenth century. Wallpaper was not to achieve real popularity until the middle of the eighteenth century.

Ascertaining what was hung on the walls of rich seventeenth-century colonists is difficult, because the evidence is scanty. However, scattered among written records are a number of clues to American decorative practices. These are often vague and elusive. A Boston reference to the sale in 1641 of a "dwelinge Howse . . . wth the appurteinances and the hangings of the parler"[5] provides a rare mention of hangings but imparts no information about them. In this context, "hangings" could have referred to textile wall hangings, or even to hangings of paper or leather, but it could just as well have meant window hangings, and what they looked like we are left to guess.

Fortunately, there are other, more informative references. A letter of William Fitzhugh of Stafford County, Virginia, documents the use of tapestries in America at an early date. In 1682, Fitzhugh wrote to an English correspondent:

> Please to procure me a Suit of Tapestry hangings for a Room twenty foot long, sixteen foot wide, and nine foot high and half a dozen chairs suitable.

It is significant that he specified that the chairs suit the wall hangings; this provides unusually early evidence for the practice in America, in good English fashion, of matching wall coverings with the coverings of seating furniture, a practice that was to become a fixed feature of eighteenth-century decoration in the colonies.

Four years later, Fitzhugh described in a letter

> my own Dwelling house, furnished with all accommodations for a comfortable and gentile living, as a very good dwelling house with 13 Rooms in it, four of the best of them hung.[6]

No doubt the tapestries from his earlier order were among these hangings.

As Fitzhugh's comments suggest, covering the walls with imported tapestries was a luxury only the rich could afford. Walls that were "hung" were a matter of pride, something for only the best of rooms, even in the house of this very prosperous gentleman.

Other American documents indicate that tapestries and woven textiles were used to cover at least some American walls through the colonial period. In 1745 Benjamin Church of Boston offered "Beautiful Arras-Hangings for a Room."[7] The word "Arras," taken from the name of a town in northern France that was a great fifteenth-century center of tapestry production, had become synonymous with the word "tapestry." A New York upholsterer, Stephen Callow, in 1749 advertised that he would hang rooms "with Paper or Stuff [textiles] in the newest Fashion."[8] Later, in 1767, a Philadelphia upholsterer from London, John Webster, mentioned among his services "rooms hung with paper, chintz, damask, or tapestry, & c."[9]

Some prosperous early Americans imported from England not only tapestries, but also imitations of them. Though "imitation" or "mock" anything has assumed a largely negative connotation in our twentieth-century world of plastic and polyester, the same sense of derogation was not implied in the eighteenth century. In 1737, Sir William Pepperrell of Portsmouth, New Hampshire, was unusually explicit in describing what he wanted when he ordered "mock tapestory" for a chamber (bedroom). In an order he sent to his London agent, he mentioned one method for making some of these imitations of tapestry. Pepperrell wrote:

> I desire you to geet mock tapestory or pant[d] canvis lay[d] in oyle for hangings for y[e] same [chamber] and send me.[10]

What Sir William wanted may have looked something like the English flocked canvas wall covering shown in *figure 1-2,* or the German canvas in *figure 1-3.*

Very few textiles hung as coverings for American walls in the seventeenth and eighteenth centuries have survived. One of the rare survivals, probably dating from about 1700, was found in the Thaxter House in Hingham, Massachusetts. It has been described as "pure, coarse linen . . . the design quite simply painted with some block-printed details."[11]

Another luxurious wall hanging used in America was leather, which was sometimes tacked, sometimes glued, to the walls *(color plate 1).* A proposal for hanging leather in the Governor's Palace in Williamsburg survives in the Journal of the Council for 1710: "That the great Room in the second story be furnished with gilt

Figure 1-2 An elaborate early-eighteenth-century flocked canvas hanging (opposite) combines imitations of velvet hangings and of paneled wainscoting. It was probably made in England and represents some of the early wall coverings available to wealthy Americans around the turn of the eighteenth century. The horizontal line that marks off the bottom third of the illustration imitates a chair rail. On the wall, it occurs at a height of 26½ inches. The dado (the covering of the lower part of the wall) follows a standard configuration found in fielded wooden panels. Here the panels are embellished with medallions and scrolls. The whole wall hanging is rendered in dark-green flocking on off-white canvas, brightened with painted accents of orange and green. It now hangs at the Winterthur Museum in a room of 1725–30, with furniture in the Queen Anne style. Its monumental architectural scale and the power of its luxuriant pattern attest to the grandeur and confidence of the baroque decorative tradition. Courtesy of the Henry Francis du Pont Winterthur Museum

1-2

leather hangings: 16 chairs of the same."[12] While leather for walls was not widely advertised, there are references to its availability. In New York in 1762, one Roper Dawson was selling "gilt Leather for Hangings" along with paper hangings and a variety of other goods.[13] A rare survival of leather hung in an American house during the colonial period is illustrated in *figure 1-1.*

THE FIRST WALLPAPERS IN AMERICA

By 1700, hangings made of paper were available in the stock of a Boston merchant. Although a precise date fixing the year of the first imported wallpaper to this country cannot be cited, a document of 1700 provides firm evidence that wallpapers had been available in this country during the last years of the seventeenth century, if not earlier. This evidence is included in an inventory made following the death of a Boston bookseller, Michael Perry. Entered in the list of his stock on hand in 1700 were: "7 quires of painted paper and three reams of painted paper."[14] ("Painted paper" is an Anglicization of the French term *papier peint,* which is still translated "wallpaper." In eighteenth-century documents "painted paper" consistently means what we now call wallpaper.) The inclusion of painted paper on this inventory strongly suggests that there was a market for such goods. By 1700 individuals in Boston no longer needed to send to London for paper hangings, but could purchase them from a local merchant.

However, it is difficult to determine just how poor an American might have been and still have ranked as a potential customer for wallpapers sold by Perry or one of his rivals. When wallpaper is mentioned in an early American document, it is inevitably a document relating to a merchant, a rich or famous man, or a public building. This is not surprising, since the letters, journals, and business accounts of powerful and important persons, and records of their building projects, enjoyed the best survival rate as treasured heirlooms or official records of state. The only available document that gives any kind of quantified indication of the economic circumstances of an ordinary householder who owned paper hangings during the colonial period in America dates from after 1750. It is the inventory of 1766 listing the possessions of a currier of Roxbury, Massachusetts, and will be discussed later in the text (see page 158). For the period prior to 1750, such information is sparse. While studies of seventeenth- and early-eighteenth-century inventories preserved in court records have given a good sense of the kind and quantity of other furnishings in the homes of ordinary people, they have yielded but little information about the wallpapers

Figure 1-3 J. B. Nothnagel of Frankfurt, Germany, printed the black outlines of this floral pattern and painted or stenciled in the orange, pink, blue, and white flowers with green foliage over a light-blue ground. It was executed around 1760 in oil paints on a coarse linen base. The portion shown here is approximately 34 inches wide. Americans of the seventeenth and eighteenth centuries might have described similar wall hangings as "Painted canvis lay[d] in oyle" or "printed canvas." Although there is very little evidence of the importation of German wall coverings to America during the colonial period, this example is included to suggest the whole category of early painted and printed canvas hangings because it is stylistically appropriate and is in unusually good condition. Cooper-Hewitt Museum, 1955-50-1; gift of Deutsches Tapetenmuseum

Figure 1-4 The two patterns lining an English Bible box were printed in ordinary black inks of the kind used for printing books. The patterns, incorporating heraldic, allegorical, and grotesque devices with strap- and scrollwork, represent types that were probably available in America before 1700. Sheets of patterning used to line such boxes were also pasted on walls. The Metropolitan Museum of Art; gift of J. Pierpont Morgan, 1927

1-3

1-4

used. Such inventories were often ordered by the courts following the demise of a property owner. Unfortunately, since the hanging of paper made it part of the fabric of a house, it was, as indicated above, not itemized.

Since there is but little documentary evidence about the use of wallpaper before the mid-eighteenth century, and few samples of papers actually hung on American walls at earlier dates, one can only surmise that during this period wallpaper was an unusual, but not unknown, wall decoration in America. Fragile bits of evidence such as gleanings from the documents already cited, and the survival, usually as scraps, of a variety of early wall coverings, attest to the fact that American walls before 1750 were whitewashed, painted, wainscoted, and hung with a variety of materials. But wallpaper, though less expensive than leather or textile hangings, was still beyond the budget of most householders.

The mixture of wall treatments in colonial America during this early period was well summarized by an English visitor, James Birket. He described the situation to which the available evidence gives substance: Wallpaper figured as but one among a number of possible options for wall finishes when it was first used here. In 1750 Birket published his impression of houses in Newport, Rhode Island:

> The houses in general make a good Appearance and also as well furnished as in Most places you will meet with, many of the rooms being hung with printed Canvas and paper &c which looks very neat. Others are well wainscoted and painted as in other places.[15]

WALLPAPER STYLES OF THE SEVENTEENTH AND EARLY EIGHTEENTH CENTURIES

Before the 1760s, the history of wallpaper styles in America is a history of imports. Most papers came from England, some were imported from the Orient (these were called "India Papers," even when they came from China), and a few from Europe, especially from France.[16] Not until the 1760s did manufacturers begin to make wallpaper in America.

Pieces of wallpaper that can be documented both to have been hung in an American house and to date from before 1750 are extremely rare. For this early period, therefore, it is exceedingly difficult to create a neat scheme tracing coherent stylistic development in the kinds of papers Americans were importing. The problem of pinning down dates for specific styles in wallpaper patterning, either prior to 1750 or at a later time, is complicated by factors common to studies of style in almost any category of decorative objects in almost any period.

In the first place, unlike many works of sculpture or painting, furnishings are rarely signed and dated. Second, economic factors complicate research: wallpapers that appear to be particularly early because of a simple, even crude, pattern and quality of finish may in fact have been made at a surprisingly late date. There is a temptation to look for progress from the simple and crude toward the complex and highly finished. But, in some cases, wallpaper patterns were simplified so they could be made cheaply, even though more sophisticated styles and techniques for printing were current. Or again, crudeness, simplicity, or old-fashioned patterning may reflect nothing more perplexing than the fact that the skill of a craftsman was limited, along with his knowledge of and concern for contemporary fashion. The taste of the customer can also add difficulties to the researcher's attempt to use style as a tool in establishing the date of a wallpaper. Old-fashioned patterns enjoyed long-lived popularity on levels below the trend-setting. A Philadelphia advertisement of 1770 offering "Paper-Hangings (not lately imported)"[17] suggests another related complication: wallpapers may have remained on merchants' shelves for many years before they were sold, or may have been well seasoned in household closets or attics after purchase and before they were actually hung on the walls.

Even though these problems pose difficulties, the life spans of wallpaper styles can still be roughly delimited. It is possible to find enough examples to make a few generalized statements about the earliest pattern types that would have been available for importation and were probably used in America.

English styles, as mentioned earlier, dominated the pre-revolutionary wallpaper market in America, a domination well protected by the Acts of Trade and Navigation. But even without this protective colonial policy, English papers would probably have been favored, since they were far better made than those produced elsewhere in Europe. Well into the eighteenth century, English papers enjoyed a reputation as the best available. They were particularly popular in France, even in court circles. Not until late in the eighteenth century, with heavy borrowing of techniques from the English, were the French to produce papers of equal quality.

Describing houses in Portsmouth, New Hampshire, in 1750, James Birket, the English traveler whose comments on wall decorations in Newport were quoted earlier, remarked on the use of English wallpapers. In houses "of Modern Architecture," he said, "the rooms are well plastered and many Wainscoted or hung with painted paper from England."[18]

1-5

Since good secondary sources do exist that illustrate early examples of English papers, particularly samples preserved at the Victoria and Albert Museum in London and at the Whitworth Art Gallery in Manchester, no attempt at a comprehensive survey of these papers will be attempted here.[19] *Figures 1-4, 1-5, and 1-6* suggest the character of major early pattern types. English prints, book illustrations, and paintings serve as the best visual records of the look of the earliest papers probably hung on American walls, yet even for a period as late as the first half of the eighteenth century, such primary visual documents are rare. No sample books of patterns are known to survive from the seventeenth or early eighteenth century, either in England or in this country. Not until relatively late in the history of wallpaper production, and then only very rarely, were designs for wallpapers published in pattern books. Makers of paper hangings doubtless drew on the same engraved ornament books that provided models for nearly all craftsmen. Study of sixteenth- and seventeenth-century French and German ornament books does convey a sense of the general vocabulary of ornament available to printers of papers to be hung on walls. However, very few illustrations of patterns specifically intended for wallpapers were ever published, whether in this earliest period of their limited use, or even well into the nineteenth century.

Figure 1-5 (left) Orange, yellow, and gray-green washes were used to brighten this pictorial pattern, printed in black ink. An English example, it represents another pattern type available in America before 1700 for lining trunks and boxes and for wall decorations. The fragment shown is 19¼ inches wide and 12½ inches high. Courtesy of the Colonial Williamsburg Foundation

Figure 1-6 This portion of a wallpaper border survives at the Morris–Jumel mansion in New York City, General Washington's headquarters during the Battle of Harlem Heights in 1776. It borders the burlap lining of a leather-covered trunk said to have belonged to Aaron Burr. The early- to mid-eighteenth-century paper resembles many of the early examples preserved in collections of French wallpaper. It was crudely but boldly made by block-printing the outlines in black printer's ink and brushing in splotchy watercolor —pink for the blossoms, dense green for leaves—probably using a stencil. Courtesy of the New York City Department of Parks

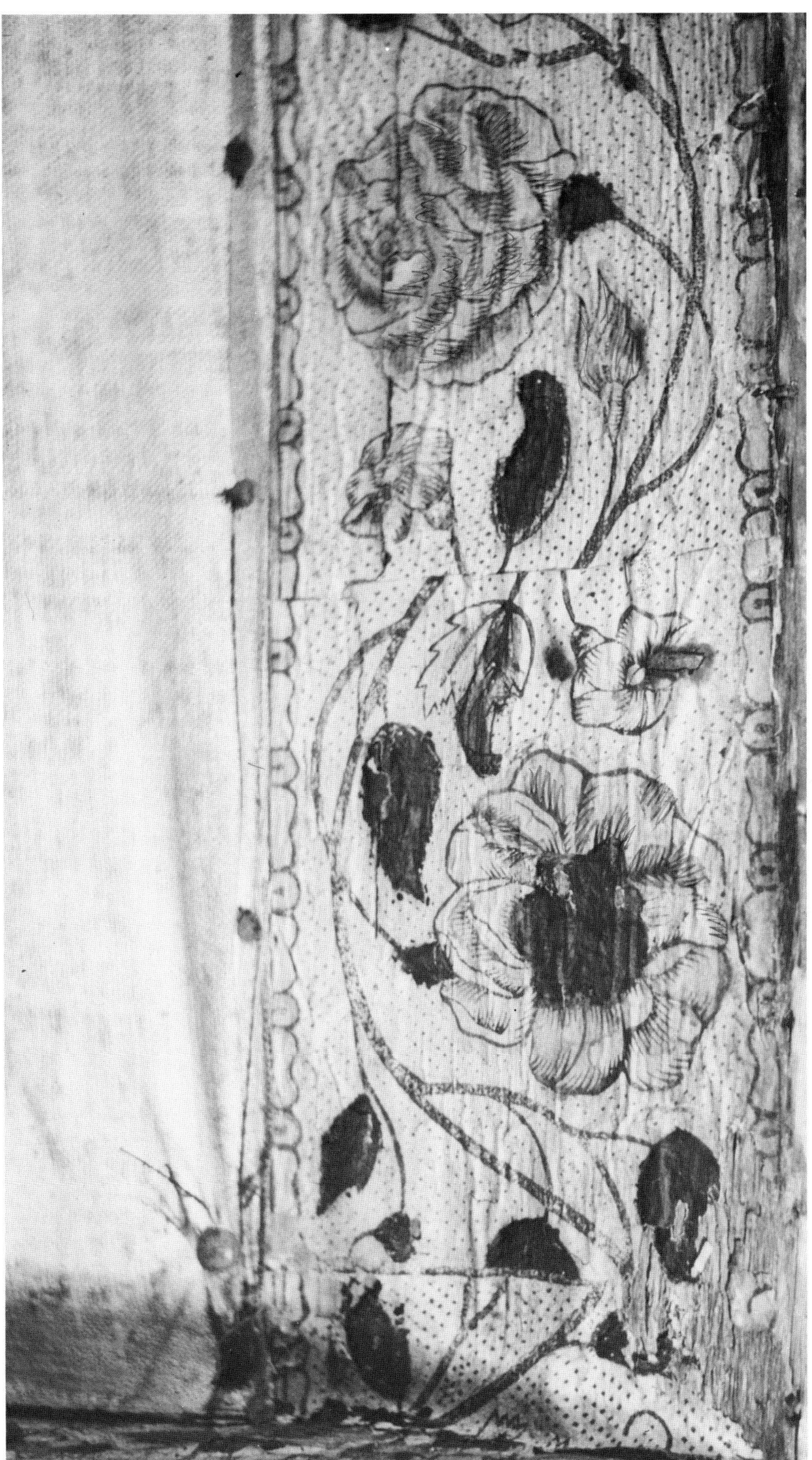

1-6

Although there is no equivalent in wallpaper history for Thomas Chippendale's *Gentleman and Cabinet-Maker's Director* (1754), which serves furniture historians so well, it is significant that in 1762, just at the end of this period being broadly considered the early prelude to wallpaper history in America, Chippendale himself included two pages showing wallpaper border designs *(figure 7-2)* in a new issue of the *Director.*

Among the very earliest paper patterns available for decorating American walls were the English ones described above as printed in the ordinary black ink used by printers of books. Such papers have most frequently survived as box linings. However, many of the patterns were not completed within the confines of one sheet of paper, but formed a full repeat only when placed with several other sheets. Therefore it is presumed that the larger repeats were designed for covering walls.

In many examples from this early period, it appears that ornamentation for the paper's surface was produced by the simple process of juxtaposing a conglomeration of pictures—pictures closely akin to the illustrations for cheap, popular books of their era. Such patterns, one of which is shown in *figure 1-5,* exhibit little evidence that serious attention was paid to the basic problem of effectively and harmoniously covering the large expanse of a wall's surface with a repeating pattern. One could hardly expect such concerns to disturb makers of books, who, as a sideline, printed patterns on small sheets, most often to be used as endpapers or lining papers. When these sheets were used to decorate walls, they had to be pasted up one by one.

In *figure 1-4,* the strong geometric structuring of the two patterns, which imposes order on the combinations of strapwork, scrollwork, and heraldic motifs, relies on borders and emphasizes segmentation of the design. The fact that it features some elements while subordinating others suggests a greater concern for overall composition and cohesion of design than the example of *figure 1-5.* The same could be said for others of the more sophisticated lining papers/wallpapers of the sixteenth and seventeenth centuries that feature strap- and scrollwork and grotesqueries derived from Mannerist, Elizabethan, or Tudor vocabularies of architectural ornament, and

Figure 1-7 This elaborate English handbill of about 1820 illustrates some wallpaper patterns available to American importers in the early eighteenth century. Heavy borrowing from textile designs is apparent in the illustrated patterns and is itemized in the advertisement. Imitations of expensive architectural detailing are mentioned as well. These two sources—textile patterning and architectural ornament—proved to be the most important for designs of wallpaper throughout its history. Courtesy of the Bodleian Library, Oxford

AT THE BLUE-PAPER
Warehouſe in Alder-
manbury *LONDON.*
BEING,

Sturt Sculp.

The Original Manufacture Warehouſe:
ABRAHAM PRICE, makes and ſells the true ſorts of FIGUR'D
Paper Hangings Wholeſale and Retale,
IN PIECES OF TWELVE YARDS LONG; IN IMITATION OF
Iriſh Stitch, Flower'd Sprigs and Branches: Others Yard-wide in
Imitation of Marble, and other colour'd Wainſcots, fit for the
Hangings of Parlours, Dining-Rooms, and Staircaſes:
And a Curious Sort of Imboſs'd Work Reſembling Caffaws,
and Bed Damasks; with other Things of Curious
FIGURES AND COLOURS; CLOTH TAPESTRY HANGINGS ETC.
All *which* are diſtinguiſh'd *from any* Pretenders *by theſe* Words, *at ye end of*
each Piece The Blue Paper Manufacture.

1-7

from contemporary German and French ornament books. But these boldly printed black-and-white patterns, marked either by the abstraction and geometry of heraldry or by the peculiarities of Mannerist ornament, seem more appropriate for the confines of small areas like the interiors of boxes and trunks than for walls. Indeed, many of the sixteenth- and seventeenth-century patterned papers have an overpowering optical effect when pasted in repeat across large surfaces.

In most of the earliest eighteenth-century wallpapers that have come to light in this country, the boldness of geometric patterns in stark black on white has by now been abandoned for the relative softness and subtlety of more naturalistic floral patterning derived from textile designs. This reliance on textile prototypes had been apparent in the abstracted and flattened floral motifs on flocked papers of the sixteenth and seventeenth centuries. These papers, on which the early fame of English paper stainers rested, were derived from a baroque vocabulary of ornament, as interpreted by Italian weavers of cut velvets and damask-woven silks. Later versions of related patterns on wallpapers appear in the 1720 handbill of a London wallpaper dealer, illustrated in *figure 1-7.* The patterns running vertically along the two edges of this illustration were derived from Italian textiles. Indeed, the very earliest English wallpaper pattern that can be dated, the Christ's College pomegranate pattern printed by Hugh Goes of York around 1509, was based on Italian textile designs.

By the early eighteenth century, specialists in block-printing, many of whom had learned their craft decorating textiles, took over wallpaper production from book printers, and textile patterning came to dominate wallpaper design even more. In fact, the same blocks could be used to print on papers as well as on woven fabrics. The similarities to contemporary textiles are particularly striking in the two patterns flanking the door in *figure 1-7.* These are based on printed cottons or perhaps even on delicate floral-sprigged brocades. In addition, a panel of paper imitating flame-stitched embroidery is displayed between the windows to the left of the door, and the text of the handbill mentions wallpaper imitations of yet another kind of embroidery—"Irish Stitch"—as well as "Flowered Sprigs and Branches."

Since textile designers had long since worked out subtle techniques for repeating the elements in their patterns, the relatively new wallpaper trade was quick to make use of this experience in solving not only its need for skilled craftsmen but also its design problems.

2 HOW WALLPAPER WAS MADE

Since technology can greatly affect style, by setting limits as well as opening new possibilities, it is important to understand how wallpaper was made. Early craft techniques and terms of the trade are the subject of this chapter.

A man who produced wallpaper in the eighteenth century was known as a "paper stainer." In most cases, he did not make the paper stock on which to print patterns. Instead, it was his job to transform into "paper hangings" the special class of blank paper received from a mill or warehouse, paper known in its undecorated state as "hanging paper." In eighteenth and nineteenth-century sources, the order in which these words "hanging" and "paper" are coupled is usually significant. Hanging paper was generally a product sold at wholesale to the craftsmen who added color and patterning. A householder could buy at retail paper hangings for his rooms.

THE PAPER STOCK

Robert Dossie, in *The Handmaid to the Arts,* first published in London in 1758, gives us the most informative mid-eighteenth-century account of English paper-staining techniques. He described "the kind of paper employed for making the paper hangings" as "a sort of coarse cartoon manufactured for this purpose."[1] "Cartoon" as a term denoting a grade of paper suggests cheap stock similar to that on which an artist might produce working drawings or cartoons. Hanging paper was of an inferior quality by eighteenth-century standards, since it was made of unbleached rags, giving it a murky gray, tan, or blue shading. This discoloring was of little consequence because it would be covered over with the paper stainer's colors. In comparison with modern wallpaper stock, it was strong, high-quality stuff.

The paper maker started with rags of cotton, linen, and other fibers and reduced them to a pulp, achieved by soaking, beating, and processing the rags with chemicals in vats filled with water *(figure 2-l).* Then he took a rectangular mold made up of a wooden frame, or "deckle," and a bottom of wire cloth, and plunged it into the vat. With this frame, he scooped up a small quantity of pulp, manipulating the deckle to spread the mushy mixture over the wire surface. This surface often incorporated a decorative element or lettering that left a faint impression—a "water mark"—in the paper itself that served as the maker's signature or trademark. When the water drained through the porous bottom, a thin layer of pulp remained, to be further dried, pressed, flattened, and smoothed. The deckle for making paper could be no larger than an individual paper maker could handle, and the size of a deckle determined the size of a sheet of paper.

Although accounts of the eighteenth-century craft published as early as 1836 and as late as 1924 give "standard" sizes for sheets of early wallpaper, actual measurements of sheets making up papers preserved in museum collections reveal considerable variation from any prescribed norms. Among the secondary sources, one cites a French paper of 1700 measuring 16 1/2 inches by 12 1/2 inches as indicative of the usual size for its period. Another gives 22 inches by 32 inches as the norm for "elephant size" paper—that used for individual sheets of painted papers sold in America early in the eighteenth century. In 1836, the writer of an English engineers' and mechanics' encyclopedia stated: "Elephant size handmade paper (28 inches by 23 inches) is used almost exclusively for the manufacture of paper hangings, being joined together and printed on."[2]

During the early eighteenth century in America as well as England, stained paper was only occasionally sold by the sheet. A 1712 advertisement published by Thomas Hancock of Boston offered painted paper by the "quire"—a quire was made up of about twenty-five sheets of paper. Most wallpapers, however, were sold in rolls. Between 1712 and 1736, accounts kept by Daniel Henchman of Boston record not only "quires" and "reams" of painted paper but also "rowls" of painted paper.[3] Through the 1730s, Henchman's paper hangings were most frequently advertised for sale in rolls, as "Stampt paper in rolls, for to paper rooms" or "roll paper for rooms." The individual sheets of paper were glued together to make up these rolls.

Many paper stainers had close ties with paper mills and some in fact owned their own mills. These mills supplied stainers with raw hanging paper in sheets, and by the eighteenth century, the normal first step of the stainer's job was to glue together those sheets.[4]

The standard length of a "piece" or roll of joined sheets was firmly established by English excise officials during the eighteenth century at 12 yards. *The Compleat Appraiser,* an English manual, in its fourth edition of 1770, recorded: "A Piece of *Wall-Paper* is (in SIGHT) generally, 21 *Inches* or 1 *Foot* 9 *Inches* WIDE, and 12 *Yards* in LENGTH."[5] During the eighteenth century, Americans apparently accepted that 12-yard length as their standard for a "piece" of wallpaper, as indicated by advertisements of the period. A Bostonian offering his wallpapers during the 1780s emphasized "Each piece warranted to contain 12 yards," and another in 1791 described his papers as "English measure." In yet another Boston advertisement, dating from 1789, local manufacturers warned "The Public are cautioned against being imposed on by those who make their rolls only 8, instead of 12 yards in length."[6]

When examining examples of old wallpapers, a knowledge of paper-making tech-

Color plate 1 A duplicate image of a single Flemish leather panel has been reversed, to give a complete repeat of the pattern, which is symmetrical on either side of a vertical axis. Embossed, painted, and gilded leather work of this kind was a specialty of both Spanish and Flemish craftsmen during the sixteenth and seventeenth centuries. Italian designs provided the sources for the High Renaissance vocabulary of ornament displayed here. The panel is 59 inches high. Cooper-Hewitt Museum, 1966-64 -1; gift of Harvey Smith

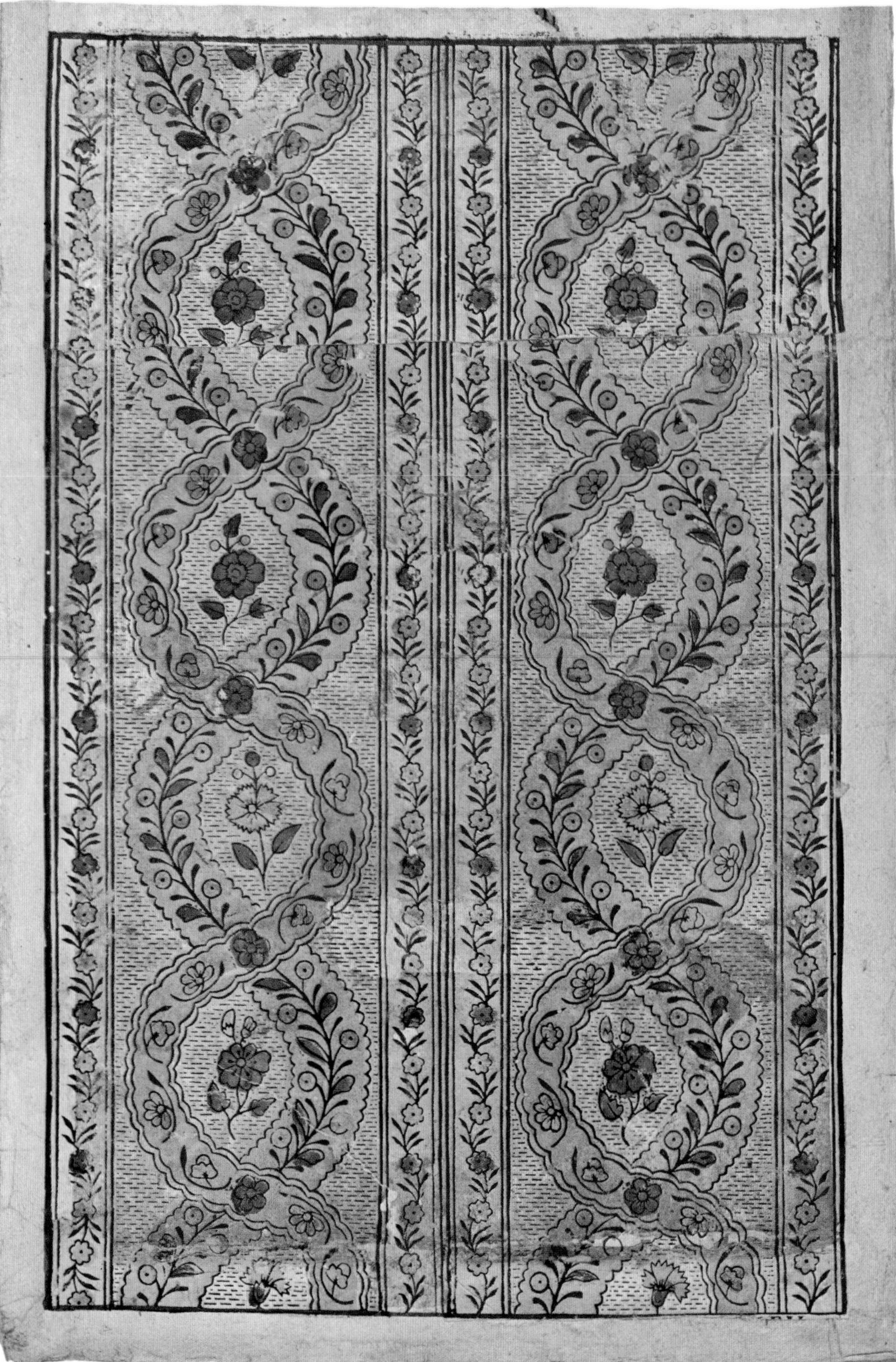

2

3

Color plates 2 and 3 Printed in black ink and colored with washes, this "domino paper" (left) was made in France during the mid-eighteenth century. It is 14 inches wide and 20¾ inches high. An enlarged detail (above) centering on a sprig, 2 inches tall, shows that thin washes, stenciled, color the elements outlined in ink. The thin-bodied colors give blurred edges. Cooper-Hewitt Museum 1928-2-75; gift of Eleanor and Sarah Hewitt

4

5

Color plates 4 and 5 A French wallpaper made about 1800 is block-printed over a layer of coloring. The paper is 23½ inches wide. A 2-inch-wide detail (above) shows that the ground as well as the printed colors have been applied over a horizontal joint, seen as a line across the center between two handmade pieces of paper. The detail also shows the layering of colored shapes, produced by carved wooden blocks with raised surfaces. The thickness and chalkiness of the distemper colors are characteristic. The opaque hues cover one another and the brown background. Holes pitting the surfaces of printed areas (see right-hand leaf) are the result of tiny air bubbles, which popped as pressure was applied to the block. When thinner colors were applied by block-prints a sunburst pattern of streaking is sometimes apparent: the impact of the block has pushed color outward from the center toward the edges of a shape, where it gathers in thicker outline. When thin colors were applied by machines using rollers, the streaking is vertical *(plate 68)*. Sharp-edged impressions are characteristic of wood block printing, in contrast to the blurred edges produced by stenciling *(color plate 3)*. Cooper-Hewitt Museum, 1972-42-189; gift of Josephine Howell

6

Color plate 6 (left) A pattern derived from damask-woven textiles has been executed in flocking on a varnish ground. An apparently original price is inscribed on the back in English pence. Though found in France, the paper represents a mid-eighteenth-century flocked type for which the English enjoyed a good market in France as well as in America. When the words "Caffy," "Caffoy," "Caffaw,-s," or "Cassaw" were used to describe wallpapers in eighteenth-century documents, they were probably referring to papers similar to this one. Cooper-Hewitt Museum, 1972-42-187; gift of Josephine Howell

Color plates 7 and 8 Two variants of the mid-eighteenth-century English pattern formula in which sinuous, leafy vines putting forth exotic blossoms were combined with diaper work—diamond grid patterning—are shown here. Other closely related patterns appear in *figures 3-1 through 3-5.* The diaper designs were derived from woven patterns in textiles, most probably from Italian examples. The exotic blossoms relate closely to those typical of (East) Indian chintzes, the painted and printed cottons so fashionable in Britain during the seventeenth and eighteenth centuries. These two samples survived in the Jeremiah Lee mansion in Marblehead, Massachusetts, remnants of decorations hung about 1765. Part of the green piece *(color plate 7)* has been folded to show an excise mark ("GR" crowned, over "55") stamped on the reverse. Each wallpaper is about 21 inches wide. Cooper-Hewitt Museum, 1938-62-24, -25

8

7

Color plate 9 (See overleaf) Oliver Phelps chose the finest French wallpapers by Jean-Baptiste Réveillon for the wing he added to his Suffield, Connecticut, house in 1795. As is apparent on the stair landing shown here, he had them skillfully hung. His workmen centered the pattern elements with care and finished the decorative scheme with combinations of borders—wide (10½-inch) horizontal bandings at baseboard and cornice levels, narrow (3¾-inch) edgings for tracing the intricate outlines of the windows and pilasters. Courtesy of the Antiquarian and Landmarks Society, Inc., of Connecticut

9

niques can prove useful in making the most basic distinction between the earliest and those produced after about 1835. Horizontal seams spaced at regularly recurring intervals usually indicate that the wallpaper was made before machine-made paper was generally adopted by wallpaper manufacturers. The trade began to use "seamless" or "continuous" paper between 1835 and 1840.[7]

Seams in the early handmade papers are sometimes visible even if the papers are still on walls *(figure 3-4).* They often show through the applied layers of ground coloring and patterning. Strong side lighting, or simply holding a flashlight very close to the surface of the wall and shining it across an expanse of paper, sometimes makes the seams between individual pieces more apparent. Other indications that the paper stock is handmade and probably early include "deckled," or irregular, nearly ragged, edges, as well as water marks and the imprint of the paper maker's woven screen.

COLORING MATERIALS

For the earliest known English wallpapers, printing inks, most often black but sometimes in colors, were used. In many of the early examples, outlines are printed in black ink, with color added by brushing on washes, applied freehand, or with the aid of stencils. By the mid-eighteenth century, distemper colors—pigments carried in a water-based medium of glue or sizing—were coming into use in the trade. Color printing from blocks was replacing the practice of adding color with a brush.

In *The Handmaid to the Arts,* Dossie specified: "By *distemper* is meant all paintings on scenes, hangings of rooms, or other parts of buildings, where SIZE is used." The word "distemper" is derived from the Latin verb *temperare,* to mingle. Andrew Ure described the process of making distemper colors for wallpaper in his *Dictionary of Arts, Manufactures, and Mines,* first published in 1839:

> All the colors are rendered adhesive and consistent, by being worked up with gelatinous size or a weak solution of glue, liquefied in a kettle. Many of the colors are previously thickened, however, with starch.[8]

Distempers, which produced matte, thick-bodied, opaque effects and a chalky appearance (frequently they contained an admixture of whiting, which was indeed chalk), far outnumber non-water-based colors in eighteenth- and early-nineteenth-century wallpapers. They were less expensive than oil and varnish-based colors and dried more quickly. However, what Dossie described as "varnish colors" were also used. He advised that they "must be formed of oil of turpentine, and the resins and gums which will dissolve in that menstrum."

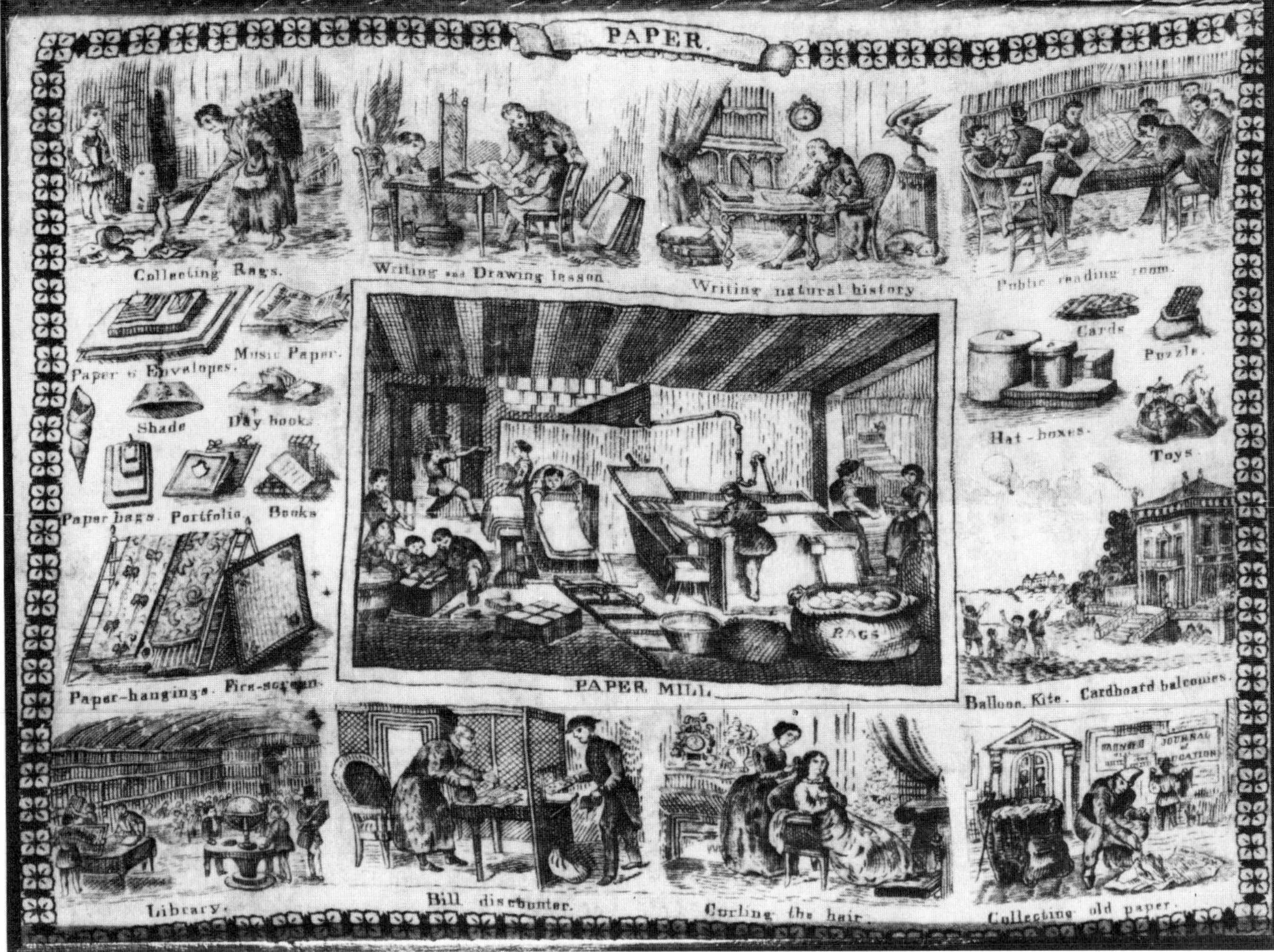

2-1

Figure 2-1 On a cotton handkerchief (8¾″ × 11¼″) of the early nineteenth century, illustrations intended for youthful instruction about the craft of paper making have been copperplate-printed. A paper mill is shown at the center, where stages in the conversion of the bag of rags to the many useful products shown in the borders, including paper hangings and fire screens, are represented. A workman leans over the vat of pulp, holding his wooden frame, or "deckle," with its screen for collecting pulp. To his left, another workman strains over the alternate stacking of felt and damp pulpy sheets to be flattened while water is forced out of them in the press shown on the far left. Courtesy of the New-York Historical Society, New York City

The non-water-soluble varnish colors were used as ground colors for flock patterns, which will be treated more fully below. In addition, some colors, such as the green made from verdigris, could be rendered more brilliant in the varnish medium. In early American-made wallpapers, details like leaves are frequently found in a glossy varnish green that contrasts with the matte finish of other colors in a pattern.

In his *Handmaid,* Dossie published a list of "Colours proper to be used for Paper Hangings," which is reproduced as Appendix A. Similar lists published during the nineteenth century make clear that newly discovered, and in many cases cheaper, pigments like chrome yellow, Scheele's green, Schweinfurth green, and artificial ultramarine were adopted by paper stainers. Selections from these later lists are found in Appendix A. When chemical analysis reveals the presence of these "datable" colors, we can establish that the wallpaper under examination could not have been made any earlier than the time when that new pigment became available to the wallpaper trade.

GROUNDS

Brushing on a layer of color, called the "ground," was the first step in decorating "hanging paper" and transforming it to "paper hangings." Dossie recommended that a single layer of whiting carried in a medium of glue or size be laid on the paper with "a proper brush in the most even manner. This is all that is required, where the ground is to be left white." To achieve a white background, it was often necessary to cover over the unbleached paper stock.

A single coating of colored grounding might also suffice for paler colors "such as straw colours or pinks . . . by mixing some strong colour with the whiting. But where a greater force of colour is wanted," Dossie instructed that after the first had dried, a second layer of ground, a layer more vividly colored, must be applied over the white layer, which then became an under ground.

Varnish grounds were needed to carry the adhesives for flocking (see page 50), but they might also be used for non-flocked patterns, though they were more expensive. Dossie recommended a varnish base to produce grounds brighter than were possible using distemper, particularly for a green, a yellow, and a pink. After laying the ground color, craftsmen had to allow it to dry thoroughly before further decorating the paper.

APPLYING PATTERN BY PAINTING

The use of the artist's brush to paint, or in period terms "pencil," designs on

2-2

Figure 2-2 An illustration printed at the top of a bill of sale dated July 5, 1804, shows steps in paper staining. On the far right, two craftsmen prepare and mix colors in barrels. The workman wielding a large brush in either hand is laying ground color on a piece of paper that was made up of a series of small handmade sheets "joined," or pasted together. Immediately to the left of the tablet describing Clough's manufactory, a printer with his left hand under the handle of a "print," or carved wood block, raises a mallet with his right hand to strike a firm impression. The "tere" boy standing to his left (see page 110) prepares the color block between each impression, distributing color to be picked up by the wooden "print." The man at the far left, with the assistance of yet another boy, is probably rolling and trimming paper in standardized lengths for sale. An earlier version of the same scene appeared on Clough's bill head of 1795, although the exhortation encouraging American manufactures is an addition to the bill head dated 1800. Courtesy of the Massachusetts Historical Society

Figure 2-3 Three wooden printing blocks, "prints" in eighteenth-century terms, are shown (right). Each is about 8 inches high, 24 inches long, and 1½ inches thick, and was used in France during the mid-nineteenth century to print different colors in a swag border pattern. "Made in such a manner that the figure to be expressed is made to project from the surface by cutting away all other part" (Dossie), the one at the bottom was used first, to cover larger areas of patterning with color. After that first color had dried, the second block was used to print more detailed areas of the pattern in a second color. Not only were the wooden surfaces of the second block carved in relief to stamp the impression, but bits of metal were also driven into the block, whose protrusions formed the scalloplike rows of dots at the bottom. The third block, shown at the top, added the fine details, which were applied last, over the first two colors. This block's printing surface is formed entirely of bits of metal. In the top corners of each block are points used to guide the printer in matching or registering his impressions. Cooper-Hewitt Museum 1956-1641-2, -3; purchased in memory of Mrs. Lloyd S. Bryce

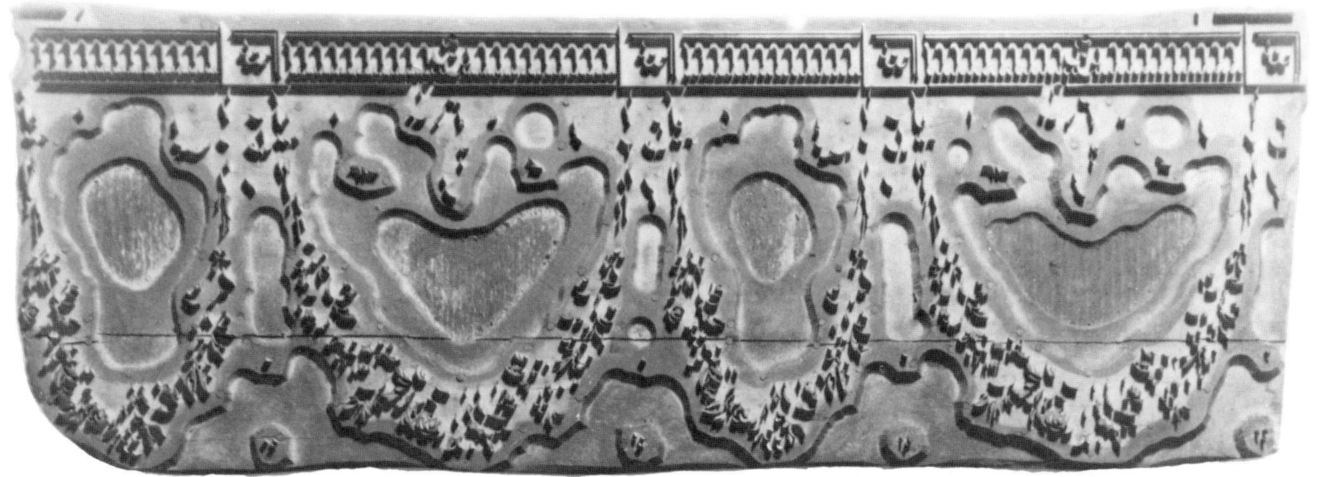

2-3

wallpaper was an expensive and relatively unusual method for patterning eighteenth-century papers in the West. Painting as a wallpaper-decorating technique had been brought to greatest perfection by Chinese craftsmen who produced designs especially for export to the West during the period. Using traditional techniques that permitted no mixing of colors with opaque whites, they applied pure, transparent colors. By varying the pressure and angle of their strokes, the liquidity and quantity of pigment carried on their brushes, and other skillful manipulations akin to modern watercolor techniques, they produced variations of hues and shading. Color once applied to the absorbent mulberry paper used in Chinese paper hangings could not be reworked. Western imitators of the Chinese designs were unable to master these demanding techniques, and their productions can usually be distinguished from their Chinese models by opacity and murkiness of coloring, reliance on overpainting, and by the quality of brush-stroking, usually more hesitant, tending to go back over itself.[9]

The Western wallpaper trade used painting as a technique not only for imitating Chinese papers but also for producing non-repeating scenes—mural paintings rendered in distemper colors on paper. Examples are illustrated in *figures 3-9, 3-10, and 3-12.* These one-of-a-kind paintings were usually produced to specifications for a particular room, and were expensive.

For more ordinary wallpapers with repeating patterns, paper stainers might paint in details and highlights on printed patterns. However, Dossie commented:

> Pencilling is only used in the case of nicer work, such as the better imitations of the India [Chinese] paper. It is performed in the same manner as other paintings in water or varnish.

APPLYING PATTERN WITH STENCILS

Paper stainers sometimes used stencils made of leather, oilcloth, or pasteboard to decorate wallpapers.[10] The craftsmen used large brushes to spread their distemper or varnish colors over the cutout figures of the stencils. Robert Dossie explained that though stenciling was cheap, it was too difficult to control to get effects equal to those achieved in printing:

> Stencilling is indeed a cheaper method of ridding [getting through, getting rid of] coarse work than printing, but without such extraordinary attention and trouble as to render it equally difficult with printing, it is far less Beautiful and exact in the effect.

Surviving wallpapers give evidence that for "coarser" work it was used, sometimes in combination with printing. Solid shapes of coloring could be quickly stenciled on a paper; then more precise outlines of shapes, and continuous elements, could be printed around and over the stenciled colors. Frequently this order was reversed: the outlines first printed in black were filled with stenciled color. Dossie explained that because of the difficulty printers had in registering or exactly fitting each separate color to be printed in a pattern, "in common paper of low price it is usual . . . to print only the outlines, and lay on the rest of the colours by stencilling." This saved the expense of making blocks and could be "practiced by common workmen, not requiring the great care and dexterity necessary" in printing *(color plates 2,3).*

APPLYING PATTERN BY BLOCK-PRINTING

Most paper staining in the eighteenth and early nineteenth centuries was done by printing from carved blocks of wood "cut in such manner that the figure to be expressed is made to project from the surface, by cutting away all the other part" as Dossie explained *(figure 2-3).* The paper stainer's method for using these blocks can most simplistically be compared to using a rubber stamp with raised letters to mark "First Class" on an envelope. Dossie called wood blocks carved with raised printing surfaces "wooden prints" or simply "prints." Each color to be printed in a single pattern required the carving of a separate block.[11]

To make the printing blocks, craftsmen glued together three or more separate boards, forming layered blocks about two inches thick. The boards were stacked so that the grains of their woods were successively at cross angles to prevent warping. The bottom board on which the pattern elements were carved was most often pear wood; early sources also mention that sycamore was frequently used as well. Andrew Ure's *Dictionary* of 1839 specified that poplar should be used for the two supporting boards that made up the printing block. A strap or handle was fixed to the back or top of the block so that it could be lifted *(figure 2-4).*

Robert Dossie described the process of using the printing blocks, or "prints." It may help in visualizing the process to keep the rubber stamp comparison in mind. Dossie reported that undecorated paper was laid on a surface—a "block"—covered with leather, presumably to absorb some of the force of the blow of the descending printing block. Another piece of leather held the distemper colors, just as an ink pad carries the color for a rubber stamp. As detailed in the eighteenth-century craftsman's handbook:

THE PAPER STAINER AND PAPER HANGER.

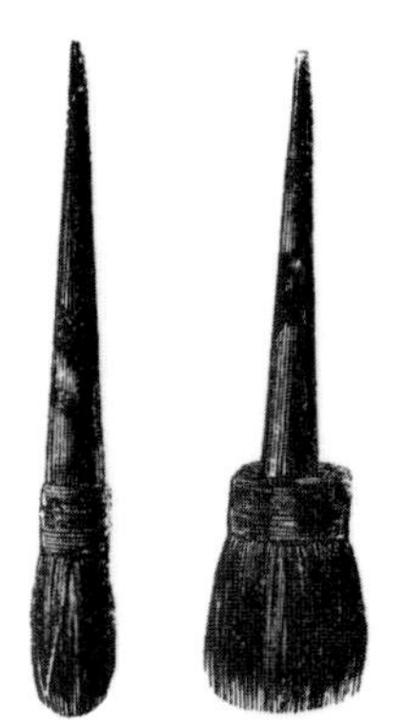

302, 303. PAINT BRUSHES.

304. DRUM FOR LAYING ON FLOCK.

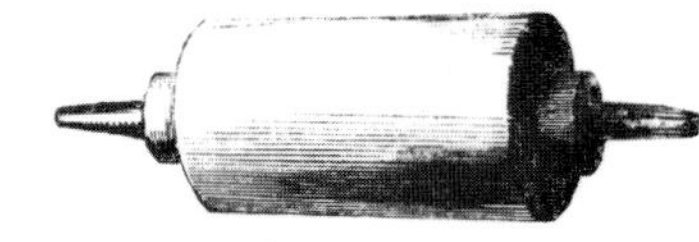

305. COLOUR DRUM.

306. PAINT POT.

307. SIZE CAN.

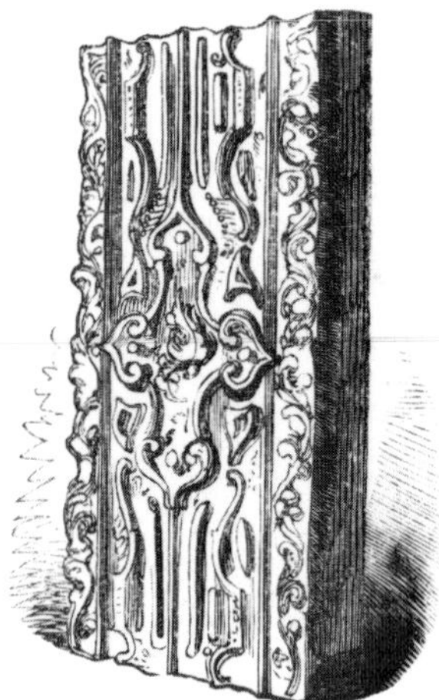
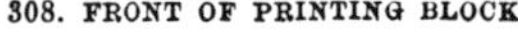

308. FRONT OF PRINTING BLOCK.

309. PRINTING PRESS.

310. BACK OF WOODEN BLOCK FOR PRINTING.

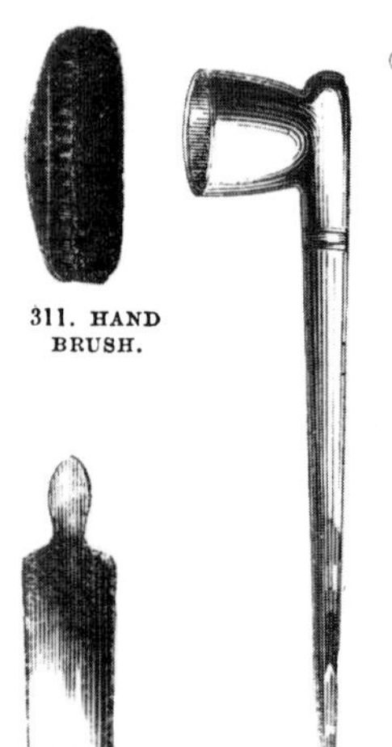

311. HAND BRUSH.

312. SPAT.

313. LADLE.

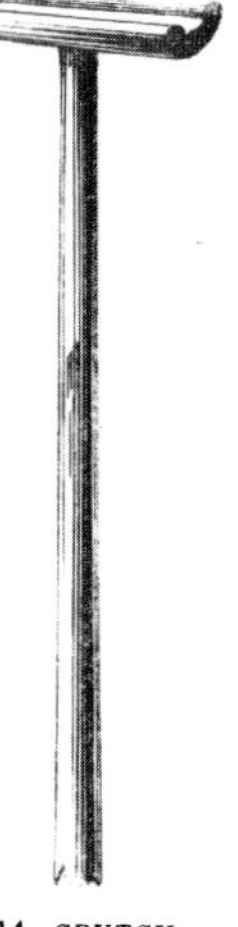

314. CRUTCH.

315. COLOUR-GRINDING MACHINE.

316. COLOUR SIEVE.

2-5

Figure 2-4 A page from an 1860 edition of Charles Tomlinson's *Illustrations of Trades* (left) indicates that block-printing continued relatively unchanged in principle from Dossie's day well into the nineteenth century. The use of carved wooden blocks with raised printing surfaces has been improved with mechanical devices designed to lighten the workman's task. A wooden structure with levers carries the weight of the printing block in the central vignette. The workman has only to guide it. At the bottom, color grinding and mixing machines have been modified and mechanized. The drum for laying on flocks, shown center top, is very like that described by Andrew Ure, writing in the 1830s (see page 51). Photograph courtesy of the Science and Technology Research Center, New York Public Library, Astor, Lenox, and Tilden Foundations

Figure 2-5 Scientific American for November 26, 1881 (above), illustrated a flocking trough based on eighteenth-century models. The man on the left block-prints a pattern in highly adhesive varnish on paper that is then fed into the flocking trough with its flexible, drumlike bottom. Boys beating on the bottom of the trough raise clouds of finely powdered wool shavings to spread them for even adherence over the surfaces bearing the printed varnish pattern. Photograph courtesy of the Science and Technology Research Center, New York Public Library, Astor, Lenox, and Tilden Foundations

> The paper . . . is laid on a proper block, on which a piece of leather is strained. The colour . . . is spread on another piece of leather, or oil cloth, laid on a flat block[12] somewhat larger than the print; which is done by a boy or man, who attends for that purpose, and having the colour by him in a pot, spreads it with a brush on the block betwixt every stroke and impression the printer makes. . . . The printer then takes the print either in his right hand, or, when too heavy to be so managed, in both, and drops it gently on the block, just charged with colour; from whence he again immediately raises it in the most perpendicular direction, and lets it fall in the strongest, though most even manner, he can on the paper, increasing the force by all the additional velocity he can give the print *(figure 2-2, color plates 4,5).*

A separate "print," or block, was used for each color in a pattern. Therefore this procedure had to be repeated as many times as there were colors in a given paper. After each color was applied, the paper had to be hung up, usually on poles suspended from the ceiling, to dry.

To ensure the proper registering, or fitting into place, of all the separate colors and elements in a pattern, Ure explained:

> Each block carries small pin points fixed at its corners to guide the workman in the insertion of the figure exactly in its place. An expert hand places these guide pins so that their marks are covered and concealed by the impression of the next block; and the finished piece shows merely those belonging to the first and last blocks.[13]

APPLYING FLOCKED PATTERNING

Dossie described flocking as "raising a kind of coloured embossment by chopt cloth." He instructed that papers to be flocked must be prepared with a varnish ground

> . . . as the flock itself requires to be laid on with varnish, the other kind of ground would prevent it from taking on the paper, and render the cohesion so imperfect that the flock would peel off with the least violence.

Flock was usually made of chopped woolen rags, either cut up by hand or in a mill. Other cloth might occasionally be used; for example, silk flock was put on paper hangings during the mid-eighteenth century in Lyons, France.[14]

Manufacturers added flock to their hangings by printing or stenciling a pattern in varnish to which the chopped cloth particles could adhere. Dossie's 1758 publication simply instructed the craftsman to use varnish to block print a pattern on the prepared paper, then to remove the paper to another table "to be strewed over with flock, which is afterwards to be gently compressed by a board or some other flat body, to make the varnish take the better hold of it."

Some eighty years later, Ure described a slightly more specialized method for spreading flock on paper hangings using a device that seems to have been closely related to the one of the 1880s pictured in *figure 2-5:*

> Upon the workman's left hand, and in a line with his printing table, a large chest is placed for receiving the flock powders . . . It has a hinged lid. Its bottom is made of tense calf-skin. This chest is called the *drum* . . . [an assistant draws a length of the paper printed with varnish] into the great chest, sprinkling the flock powder over it with his hands; and when a length of 7 feet is printed, he covers it up within the drum, and beats upon the calf-skin bottom with a couple of rods to raise a cloud of flock inside, and to make it cover the prepared portion of the paper uniformly.[15]

Ure added that to shade the flocking after it had dried, additional colors were printed on the flock; in effect, the wool was dyed in place *(color plate 38).*

In addition to proper flocking, Robert Dossie had mentioned "a kind of counterfeit flock paper" made during the mid-eighteenth century. He described the product as having "some pigment, or dry colour . . . being well powdered . . . strewed on the printed varnish."[16]

3 EIGHTEENTH-CENTURY ENGLISH WALLPAPER STYLES

By the middle of the eighteenth century, products of England's fledgling paper-staining craft were reaching this country in some quantity. In addition to larger numbers of American wallpaper advertisements dating from the 1760s, there is a dramatic increase in the number of period papers that survive from use in American houses. Examples of the flocking and color printing techniques described by Dossie are included among the best-preserved samples from this era, many of which bear English tax stamps *(color plate 7).*

FLORAL PATTERNS

Whether printed in distemper or in varnish colors, or whether flocked, floral motifs derived from textile prototypes form the largest category of repeating patterns in this first relatively large group of wallpapers known to have been used in America. Among the floral patterns, a distinctive sub-group is characterized by thick-stemmed vines growing sinuously up a length of paper. The vines put out large leaves and flowers. Between the elements displayed on the vine, infills of diaper patterning (patterning based on a diamond grid) appear. In some examples, portions of the infill diaper are boldly outlined to emphasize strong shapes that become major pattern elements themselves *(figure 3-5).* However, in most examples from this "sinuous-flowering-vine-with-diaper" group *(color plates 7,8; figures 3-1 through 3-4),* the diamond grid remains a subordinate design element, filling the background space. In the textiles from which these wallpaper patterns were derived, the woven versions of the diaper were not emphasized as the dominant elements into which they were sometimes transformed when imitated on paper.

What seems to have been a fairly standard size scale for such patterns is seen in the Dutch watercolor illustrated in *figure 3-2.* An enlargement of that already impressively large scale appears in the flocked example of the pattern type shown in *figures 3-3 and 3-4,* where the repeat length is just over 6 feet. *Figure 3-5* shows an unused sample of a similar eighteenth-century floral paper, where symmetry on either side of a vertical axis has been imposed on the pattern formula, and the infill patterning between the sinuous vines has become a very important element in the scheme. In contrast, the diapered infill of *figure 3-1* is a weak element. The diamonds are softly rendered spots merely forming a background for the flowering vine.

A closely related group of floral patterns of the mid-eighteenth century features flowering vines unencumbered by diaper patterns or vertical symmetry. In *figure 3-6,* one of these is illustrated. It represents a large group of patterns closely related to India chintz patterns.[1]

Another group of eighteenth-century English floral wallpaper patterns derived from textile models can be characterized as flower and ribbon patterns, many of which took the form of floral stripes. In these, sprigs of flowers or vines were interspersed among vertical bandings that imitated the bandings woven in textiles—the vertical stripes formed by horizontal ribbing that sometimes looks like grosgrain ribbon. In addition to the stripes, sometimes the "ribbons" or bits of lace were arranged in a more capricious, sinuous, or spotted manner. An example of the type is shown in *figure 3-7.*

Damask woven silks provided the models for other more formal, symmetrical floral patterns in wallpapers. The flowers in these are usually arranged in bouquets. The example shown in *color plate 6* imitates an effect characteristic of damask weave—the contrast between a shiny and a dull surface. In textiles, this contrast was created by varying the weaving technique. In wallpaper, the matte flocking contrasted with the smoother area of varnish ground coloring, and also formed a raised surface in low relief above the ground. This paper could be of the type described in an English advertisement of 1702 as "a curious sort of Flock work in imitation of Caffaws."[2]

The word "Caffaws," appearing in scattered eighteenth-century documents as a word descriptive of wallpaper, needs definition. It occurs in an unusually detailed and informative invoice of papers purchased from a London paper stainer in 1761 for a house in Albany. While in England during 1761 and 1762, Philip Van Schuyler bought wallpapers from a London paper stainer, William Squire. Squire had specifically listed "Caffaws" on his trade card, illustrated as *figure 1-7.*

The list of Van Schuyler's purchases from the London shop begins:

> Bought of Wm Squire:
> 8 Pieces Tulip Green flock
> 8 Dover D^{o} [Meaning of "Dover" unknown]
> 8 Royal Caffy Blue
> 8 Tulip D^{o}
> 8 D^{o} Yellow
> 8 Royal Caffy Crimson
> 8 New Flock D^{o}
> 56 Pieces flock paper a 7d £ 19-12-0[3]

"Caffy" in this listing is probably a variant of "Caffaw" or "Caffoy," meaning silk damask. When these words were used to describe paper hangings during the eighteenth century, they seem to have been associated not only with flock, as indicated in the advertisement quoted above, but also with damask patterns. One eighteenth-century English letter writer described a room hung with "a pearl-colored caffoy—the paper is like a damask; pictures look extremely well on that paper."[4] The paper illustrated in *color plate 6* is perhaps of a type that would have been variously called caffaw, caffy, or caffoy during the eighteenth century.

LANDSCAPE PAPERS

In addition to floral patterns, more grandiose English wallpaper decorations, especially non-repeating views surrounded by elaborate wallpaper or *papier-mâché* "frames," were stylish among those who built mansions in America during the 1760s. One of the most interesting documents indicating that eighteenth-century mural-like wallpapers were ordered from England is the Van Schuyler listing of purchases from William Squire, part of which was cited earlier. The 1761 document continues with entries that include wallpaper views, panels, and borders. "Stoco" refers here to wallpaper or possibly *papier-mâché* imitating plaster work. Two words on this list elude precise definition: "Nickolls" and "Tripoly's"

–80 Doz Borders
– 8 Pieces Feston Gothic Stoco
– 8 Nicholls Do
–24 Dozn Stoco Borders
–10 Paintings of ruins of Rome
1 Room – 9 Ornaments of Pannells
– 6 Tripoly's
– 1 Picture of a Philosopher for door piece
–48 Sheets Top & bottom festoons
a Neat Mache ceiling to plan for a Room 25 ft by 20[5]

These papers sent from London to Van Schuyler have been destroyed, but this document supplements the evidence of the only two surviving sets of such papers that relatively elaborate wall decorations were imported. It also indicates that these decorations were produced to order—"to plan"—for specified rooms. The notation to the left of one portion of the list—"1 Room"—makes this clear. A particularly intriguing aspect of the document is the fact that the entry "Paintings of ruins of

Rome" could be used to describe the elaborate wallpapers that were hung in the house of Van Schuyler's neighbor and kinsman, Stephen Van Rensselaer. The papers Van Rensselaer had in his Albany house, hung about 1768, are preserved in the collections of the Metropolitan Museum of Art in New York *(figure 3-9).* Papers incorporating many of the same elements found in the Van Rensselaer papers and described in the Van Schuyler invoice survive in place in the Jeremiah Lee mansion in Marblehead, Massachusetts *(figure 3-10).* The Lee papers were also installed during the 1760s. Both of these sets of wallpaper surviving from the 1760s include views of Roman ruins, painted in grisaille and surrounded by painted frames and trophies.

The word "paintings" in the Van Schuyler invoice is important. Both the Van Rensselaer and the Lee wallpaper views of Rome were painted, not printed. Many publications of the early twentieth century, however, erroneously attributed both papers to a London printer of scenes for use as wallpapers, Jackson of Battersea.[6] Although Squire seems a probable source for the papers at Marblehead and in the Metropolitan Museum, he is not the only possible maker. Other London paper stainers advertised "paintings of Landscapes" during the mid-eighteenth century.[7]

Apparently, these landscape papers of the 1760s did not go out of style immediately. In 1774, the colonial governor of New Jersey, William Franklin, prepared instructions for the interior decoration of the house he was to occupy, the Proprietary House in Perth Amboy. These instructions took the form of a plan of the house, drawn in his own hand, specifying wall coverings to match furniture he planned to bring to the house. In the large central space marked "Hall," the governor wrote:

> To be papered with Paper made on Purpose to suit the Pannels, Chimney, etc. And if it won't add much to the expence (which tis thought it will not) to have the Falls of Passaick & Cohoes represented in black and white in Imitation of Copperplate on a Buff coloured ground . . . the same kind for the Staircase on which the Falls of Niagara to be painted or stained.[8]

The man delegated to obtain the wallpapers, the New York importer John J. Roosevelt, did order all but the ambitious waterfall schemes, which proved too costly.

IMITATIONS OF ARCHITECTURAL ORNAMENT

The painted landscape wallpapers that survive from the 1760s feature illustrations of architecture, for the most part classical and ruinous. During this period architecture and its ornamentation served as sources for wallpaper designs almost as often as did textiles. *Papier-mâché* imitations of architectural ornament formed an important part of the wallpaper trade during the eighteenth century. The "Neat Mache ceiling to plan for a Room 25 ft by 20" that William Squire furnished to the specifications of Philip Van Schuyler would have been the kind of custom work Squire was prepared to provide as a regular service to his affluent London and colonial customers. Imitations of "stoco"—stucco or plaster work—were furnished by the paper stainers in both two-dimensional paper versions, either printed or painted, and in the three-dimensional relief of *papier-mâché.* Models for these paper imitations of plaster ornament were readily available in illustrations published in architectural pattern books. An example is provided in *figure 3-13,* which shows stucco work designs from an architectural book by John Crunden.

Documentary evidence for American use of *papier-mâché* ornaments is relatively plentiful. In 1757 Richard Washington sent from London for the use of George Washington several sets of *papier-mâché* ornaments for ceilings.[9] An advertisement in a Charleston paper in 1765 offered imported "Machee Ornaments for cielings, &c to imitate Stoco Work"; another in a Philadelphia newspaper in 1768 listed "Paper Hangings, with Paper Machee Borders."[10] The use of *papier-mâché* ornaments with landscape papers like the Van Rensselaer and Lee examples is suggested by a New York newspaper advertisement of 1764. The description there of a "Farm on Staten Island" includes details about the house in which the parlor was "19 × 26 hung with Landskip Paper framed with Papier Machee."[11]

PRINT ROOMS

The idea of decorating with wallpaper imitations of framed pictures was a popular one during the middle of the eighteenth century, not only for large-scaled wall treatments like those in the Lee and Van Rensselaer houses, but also for small-scale ornamentation. The English during this period delighted in pasting engraved prints directly on the walls and framing them with wallpaper borders. Rooms so decorated were called "print rooms." Horace Walpole described his print room at Strawberry Hill in 1753 as "a bed-chamber hung with yellow paper and prints framed in a manner invented by Lord Cadogan, with black and white borders printed."[12]

Print rooms offered the opportunity for exercising ingenuity, originality, and personal taste in the combining of prints of any description with wallpaper decorations and hanging them to best advantage on one's own walls. In one of the most

11

Color plate 10 In the master bedroom of the Phelps–Hatheway house, another handsome French arabesque pattern (page 57) is finished with an intricate combination of border patterns. The vertical repeat measures 47 inches. Courtesy of the Antiquarian and Landmarks Society, Inc., of Connecticut

Color plate 11 In the sitting room on the first floor of the Phelps–Hatheway house wing, another stylish French wallpaper of the 1790s (previous page) was hung with a combination of borders. The pattern is of relatively small scale—the repeat length is 21½ inches. Courtesy of the Antiquarian and Landmarks Society, Inc., of Connecticut

Color plate 12 This elegant arabesque pattern (right) bears the mark of Barabé & Cie., Paris. The wallpaper was probably purchased in Paris in 1792 for use in the parlor of a mansion John Brown had built between 1786 and 1788 in Providence, Rhode Island. Courtesy of Mr. Norman Herreshoff; photograph courtesy of *Antiques* Magazine

12

13

14

Color plate 13 This French paper (above, left) represents a style derived from the more elaborate arabesque papers that became popular in America during the 1790s. This example was found in Lorenzo, a house in Cazenovia, New York, built in 1807–8. The paper was mounted on linen and hung on the walls with tacks. It duplicates one of the patterns used in the John Brown house in Providence, Rhode Island. The block-printed colors in this sample from Cazenovia have darkened, dirtied, and faded on the walls. Traces of the original bright green show along margins that had been protected from light. It is 20¼ inches wide, 41¾ inches high. Cooper-Hewitt Museum, 1905-5-3; gift of Mrs. Charles S. Fairchild

Color plate 14 *La Chasse au Fauçon* is depicted on this handsome product of the Parisian wallpaper factory of Jean-Baptiste Réveillon. The multiplicity of glowing colors in a palette representative of his factory (the blue ground is typical, as are the bright touches of orange) suggests Réveillon's accomplishment in color printing using wood blocks. The sample shown is 21¼ inches wide. Cooper-Hewitt Museum; gift of Eleanor and Sarah Hewitt

15

16

17

18

Color plate 15 The flowering tree, displaying fantastic blossoms, birds, and insects, and rising from contorted rockery (previous page, top left), was a popular subject of the painted Chinese wallpapers purchased by a few rich Americans. The panels are 96 inches long, covering the walls from chair rail to ceiling. Cooper-Hewitt Museum, 1971-85-1; gift of Mrs. Harry Payne Bingham

Color plate 16 Made in the late eighteenth or early nineteenth century, this Chinese example (previous page, top right) is another in the large category of export wallpapers that combined flowering trees with birds and insects. Painted on joined sheets, the portion shown is 43 inches wide. Cooper-Hewitt Museum, 1954-168-1; gift of John Judkyn in memory of Florence Judkins

Color plate 17 A detail from an unused portion of an Oriental wallpaper hung in 1865 at Castlerock in Nahant, Massachusetts (previous page, bottom left), reveals a carelessness in brush-stroking and a nearly mechanical technique in the rendering of highlights when compared with earlier examples like *color plate 15.* This probably results from haste of production for an expanded Western market. Inferior painting technique, in combination with the harshness of coloring, is thought to distinguish later papers from the eighteenth-century examples. Cooper-Hewitt Museum, 1953-189-1; gift of Mrs. Samuel Hammond

19

Color plates 18 and 19 The spectacular landscape wallpapers hung in the "Chinese Parlor" at Winterthur are representative of another major category of eighteenth-century Chinese export wallpapers. The wealth of exotic details—people in strange costumes going about their work or enjoying their leisure, bits of architecture, animal species unknown to Europeans, and unfamiliar plants—proved to be of endless fascination to Western eyes. Courtesy of the Henry Francis du Pont Winterthur Museum

Color plate 20 Some of the wallpapers advertised as "India figures" and "chintz figures" in America during the mid-eighteenth century probably resembled this French pattern of the 1760s (top left). This wallpaper sample is 22¼ inches wide. Cooper-Hewitt Museum, 1931-45-10; gift of Eleanor and Sarah Hewitt

Color plate 21 "Mock India picture" or "China figure" might have described this French block-printed wallpaper of the 1770s (below) when it was new. Using the Pillement engraving shown in *figure 5-4* (below), the designer incorporated the Chinoiserie vignette in a very Western pattern with classical urns. It is is 20½ inches wide. Cooper-Hewitt Museum, 1931-45-13; gift of Eleanor and Sarah Hewitt

Figure 5-4 Jean Baptiste Pillement's publications of the 1760s and 1770s provided pattern makers with models for Chinoiserie motifs and Indian florals. This engraving (bottom left), plate 26 in Pillement's *Nouveau Cahier de six feuilles . . . Sujets Chinois Histoires,* was published in 1773. Cooper-Hewitt Museum, 1921-6-D198

20

5-4

21

splendid surviving English print rooms of the eighteenth century, that at Woodhall Park, Hertfordshire, the wallpaper-framed prints were artfully arranged above chair rail level, over the entire expanse and height of the walls; the walls were also bordered and decorated at intervals with paper swags and urns on tall pedestals.[13]

There are scattered bits of evidence that rooms were hung in this way by American decorators. In an advertisement of 1784, Joseph Dickinson seems to have been describing his readiness to create print rooms in Philadelphia. He offered to "superintend or do the business of hanging rooms, colouring ditto, plain or any device of prints, pictures and ornaments to suit the taste of his employers."[14]

Surviving wallpapers from an upper chamber in the Moffatt–Ladd house in Portsmouth, New Hampshire, present the most interesting evidence about a "print room" in America. Unfortunately, only parts of the scheme survive, and they have been removed from the walls of the room in which they are known to have hung. A sequence of hunting scenes engraved on individual sheets of paper *(figure 3-14)* is thought to have formed a decorative frieze around the top of the walls. Apparently, the prints were used in combination with the repeating pattern illustrated in *figure 3-15.* Since the expression of originality and individual taste were important aspects of the appeal of print rooms, it has been all but impossible to know precisely how to reconstruct this one. No standard models exist.

GOTHIC PATTERNS

The repeating pattern on the yellow wallpaper *(figure 3-15)* used with the hunting prints in the Moffatt–Ladd house exemplifies an eighteenth-century interpretation of Gothic motifs in the manner of Batty Langley (1696–1751). This indefatigable popularizer of architectural features climaxed his career in 1742 with a pattern book in which Gothic architecture was "improved" by being classified into "orders" like those of classical architecture. The relationship to Gothic forms may seem obscure to twentieth-century eyes, but Batty Langley with his brother Thomas popularized such designs and the bottom border, *color plate 23.*[15] Such a pattern as that illustrated in *figure 3-15,* with its bizarre mixture of floral naturalism, scrolls and swirls borrowed from the rococo style, and elements from classical architectural ornament like the beading along one edge, passed for Gothic during the eighteenth century. The Langleys specialized in taking the shape suggested by a Gothic arch and subjecting it to their interpretations of classical proportioning and symmetry.

During the 1760s, the Gothic style was frequently mentioned in advertisements for wallpaper. In 1761, Charles Digges of Annapolis advertised "stained paper for rooms, in the Gothic and Chinese Taste"—the same combination favored in Chippendale's *Director.* Advertisements Thomas Lee placed in Boston newspapers during 1764 and 1765 included, from London, "a fine Assortment of Gothic Paper Hangings."[16] The Van Schuyler invoice of 1761 also included borders in the Gothic style.

PILLAR AND ARCH PATTERNS

One distinctly English wallpaper style that probably originated during the 1760s continued in vogue in America through the turn of the nineteenth century. During the 1790s the pattern formula was copied by several American manufacturers. Called "pillar and arch patterns" by the eighteenth-century trade, these patterns featured a round-headed arch carried on two columns or pillars. These architectural elements framed a central motif, usually a single figure, a group of figures, an urn, or a vase holding flowers. One or two motifs were repeated one over another, and side by side, or were drop repeated.

During the 1790s, Zecheriah Mills, a Hartford paper stainer, advertised: "Large and elegant Pillar and Arch figures for spaceways, halls, &c."[17] Fragments of an English paper found in the Samuel Buckingham house built in 1768 in Old Saybrook, Connecticut, fit the same description *(figure 3-16).* The English wallpaper from the Buckingham house was printed in somber shades of gray, black, and white. Mills apparently used the same pattern as the basis for a paper that bears his mark *(figures 3-18, 3-19),* although Mills enlivened the design with a blue ground color.

Other "large pillar and arch figures" have been found in American houses. One pattern from the Paul Revere house in Boston may well have been among the imported paper hangings Revere advertised in 1784 as for sale at his silversmithing shop.[18] Examples of imposingly large pillar and arch patterns have been preserved in several New England houses *(figures 3-19 through 3-22).*

The use of pillar and arch patterns was not entirely restricted to halls and entryways. In a ballroom in Bristol, Connecticut, fragments survive that make up one of these patterns illustrating Judith kneeling before Josiah. In *figure 3-19* another pillar and arch pattern is seen as it was hung in a parlor of the Johnson house in North Andover, Massachusetts.

The appeal of eighteenth-century pillar and arch patterns seems to have waned after the first decade of the nineteenth century. *Figure 3-21* illustrates a pattern

3-1

Figure 3-1 In these fragments, diaper patterning like that in *color plates 7 and 8* has been rendered as mere dots. The wallpaper is block-printed in black, white, and blue on lighter blue. The width of the larger fragment is 22 inches. Cooper-Hewitt Museum 1938-62-26; gift of Grace Lincoln Temple

3-2

3-3

Figure 3-2 A floral pattern similar to wallpapers from the Jeremiah Lee mansion *(color plates 7 and 8, figure 3-1)* appears among a wealth of details documenting an interior and its occupants in a Dutch watercolor of about 1765–70. In the watercolor, the pattern on the walls appears as a faded tan. In scale, it is more typical of surviving examples of such patterns than is the pattern illustrated in *figures 3-3 and 3-4.* Courtesy of the Colonial Williamsburg Foundation

Figure 3-3 A flocked version of a flowering vine with diaper pattern is shown in a bedroom of the Webb house, built in 1752 in Wethersfield, Connecticut. Originally a bright red, the flocking has now darkened. The wallpaper is supposed to have been hung in 1781, a relatively late date for the style, in preparation for a visit by George Washington. Courtesy of the National Society of the Colonial Dames of America in the State of Connecticut

Figure 3-4 Here, the enormous size of the repeat, 72½ inches long and 38½ inches wide, contrasts with the narrow border, just under 2 inches wide. Seen on another wall of the Webb house bedroom, the scale of this pattern is unexpected in so small and low-ceilinged a room. Many of the horizontal as well as vertical seams between individual sheets of paper, each about 21 inches wide and 24 inches long, are apparent in this photograph. Courtesy of the Colonial Dames of America in the State of Connecticut

3-4

probably similar to the one that inspired James Fenimore Cooper's (1789–1851) none-too-flattering description of a wallpaper in his novel *The Pioneers,* published in 1823. The house in that novel is thought to have been modeled on the house of Cooper's father in Cooperstown, New York. Cooper observed:

> The walls were hung with a dark, lead-colored English paper that represented Britannia weeping over the tomb of Wolfe. The hero himself stood at a little distance from the mourning goddess, and at the edge of the paper. Each width contained the figure, with the slight exception of one arm of the General, which ran over on the next piece, . . . some difficulties occurred that prevented a nice conjunction; and Britannia had reason to lament, in addition to the loss of her favorite's life, numberless cruel amputations of his right arm.[19]

OTHER STYLES

During the eighteenth century, relatively styleless, often small-scaled patterns made in England were being imported along with the more distinctive floral, landscape, and architectural papers. The simplest patterns had long been printed on decorative papers. The elements in these patterns included little diaper configurations *(figure 7-6)* and florals made up of tiny sprigs bearing an abstracted leaf or two or a blossom over a background of dots *(figure 7-8)* or dashes; basic geometric forms; and stripes, plaids, and dots. Earlier, papers with these basic patterns had been used as endpapers in books and as lining papers for boxes and trunks. By the eighteenth century, their usefulness as wall decorations was well established. In France, a group of craftsmen known as *dominotiers* specialized in the production of single sheets of such patterning, called "domino papers" *(color plate 2).*

Because the simplest patterns, whether French or English (or by the late eighteenth century, American), look so very much alike, it is difficult to specify just where a given example of one of these common, cheap, almost timeless papers might have been made. They were relatively inexpensive because they usually were printed in only one color, or very few colors, requiring limited time and skill in handling the printing blocks. The fragments pictured in *figure 7-6* would not be recognizable as English if they did not bear English tax stamps.

The marks stamped by British officials to indicate payment of excise duties can be helpful in cataloguing wallpapers found in America *(color plate 7)*. Still, most of the marks provide little information more positive than proof of English origin and of a date falling within the reigns of one of the kings George—that is, between 1714, when George I ascended to the throne, and 1830, when George IV died. The monograms "GR" of the four successive monarchs cannot be distinguished from one another in the marks that law required be stamped on every sheet making up a "piece" of hanging paper. After 1786 some of the tax stamps did include dates. With the accession of William IV in 1830 his monogram came into use, but tax stamps including his "W" are found only on papers made during the next six years. The duty was removed in 1836.

The burden of taxation on the manufacturers retarded the development of the paper-staining craft in England. It encouraged the production of cheaper papers rather than improvements in quality. The complex system of enforcement of the English duties required the presence of a customs official in the factories. Lest anyone doubt that there was a serious attempt to enforce the complicated, but apparently often abused, legislation, it should be noted that in 1806 the falsification of wallpaper stamps was added to the list of offenses punishable by death.[20]

Among the distinctly English styles introduced to America during the pre-revolutionary period, colors were generally more somber, and palettes more limited, than those that came to America from France later in the century. The English clung, or perhaps were limited by their technical abilities, to palettes that were often dominated by shades of gray; this seeming preference is apparent in the painted grisaille landscapes, in the pillar and arch patterns, and even in floral patterns. In surviving examples, yellow has frequently been introduced as a foil to the grays. Even when blues, greens, reds, and yellows are found in these samples, they lack the clarity, brilliance, and inventive color combinations of later eighteenth-century French examples.

The Revolution by no means marked the end of American importation of wallpapers from England. Florence Montgomery, in her book *Printed Textiles: English and American Cottons and Linens 1700–1850* (1970), has explained how circumstances favored continued importation of English goods after the Revolution. Americans were accustomed to dealing with British merchants in whom they had built up trust; credit terms were favorable; English language and laws were familiar to American merchants; and Englishmen had learned to provide goods that appealed to specific American markets.[21] While English paper stainers failed to retain their hold on the American market as effectively as did the English cloth manufacturers, Americans continued to import some English wallpapers after the Revolution.

In addition, a recently arrived British citizen taking up residence in America during the late eighteenth century might naturally enough turn to familiar sources in Britain for furnishings. When Lady Jean Skipwith, who came as a bride from Scotland, set about decorating Prestwould, a house near Clarksville in Virginia, she sent to James Maury, her agent in Britain, for wallpapers. In a letter of 1795 she described the "sundry articles" she wished "to import to finish a House Sir Peyton [Skipwith] is building, all of which articles we would have of a very sufficient good quality . . ." She asked for samples of paper hangings, cautioning "We do not mean to go to the length of India paper, only plain English and Irish," and commented, "I am very partial to papers of only one color, or two at most—velvet paper I think looks too warm for this country."[22] By "India paper" she almost undoubtedly meant hand-painted Chinese papers and by "velvet paper," flocked paper.

Special orders to Britain were not the only way an American could acquire English wallpapers after the Revolution. American merchants continued to stock them, as their advertisements attest. One post-revolutionary advertisement from Richmond, Virginia, indicates that stylish European goods reached America more promptly than is sometimes suggested. A list of English products offered for sale in Richmond in 1788 included "Elegant Paper Hangings of Woodmason's pattent."[23] Unfortunately, the patent itself cannot be found among the British patent records, but a reference to Woodmason's wallpapers in an exactly contemporary English letter suggests that they were fully up to date and causing a stir of interest in England during the same year they were being offered in Richmond. In a letter of 1788, an English lady recounted her arrival at a country house, Oswestry, where "They showed us various patterns they had received from London, of Woodmason's new invented paper. Never more disappointed. Dingy. Wholly deficient in colour, lustre, and effect."[24] Such papers must also have made a poor showing in America, where the newly developed styles of the late-eighteenth-century French paper stainers' art were setting trends.

Figure 3-5 (opposite page, left) This tax-stamped English wallpaper of about 1765 (left) was found in a German castle, Schloss Weikersheim, in Württemberg. Similar patterns were exported from England to America. In the design, a degree of symmetry has been imposed on the flowering vine and diaper formula. The diapered areas have been boldly outlined and colored, so they take on distinctive shapes, arresting attention rather than receding into the background. This piece, 22¾ inches wide, has never been hung, so the pattern in black and gray on a varnish green ground survives in good condition. Cooper-Hewitt Museum, 1952-52-10

3-5

Figure 3-6 Another English floral pattern of about 1765 (right), this unused wallpaper sample was found at Schloss Weikersheim in Württemberg. The meander is block-printed in multi-colors with accents of powdered mica over a glazed gray ground. The paper is 23¼ inches wide. Cooper-Hewitt Museum, 1952-52-11

3-6

Figure 3-7 Like many wallpapers, this floral-striped pattern imitates textile prototypes—silk brocades and others. It is block-printed in white, pink, and black on a leaf-green ground. An English excise mark and a Dutch water mark indicate that the raw hanging paper was imported from Holland and transformed by English paper stainers into this paper hanging. Where an overlapping joined sheet has been peeled away at the top, bearing with it the ground color and patterning applied after the sheets were joined, a light strip of raw paper is visible along the edge. The portion shown here is 22½ inches wide. Courtesy of the Colonial Williamsburg Foundation

3-7

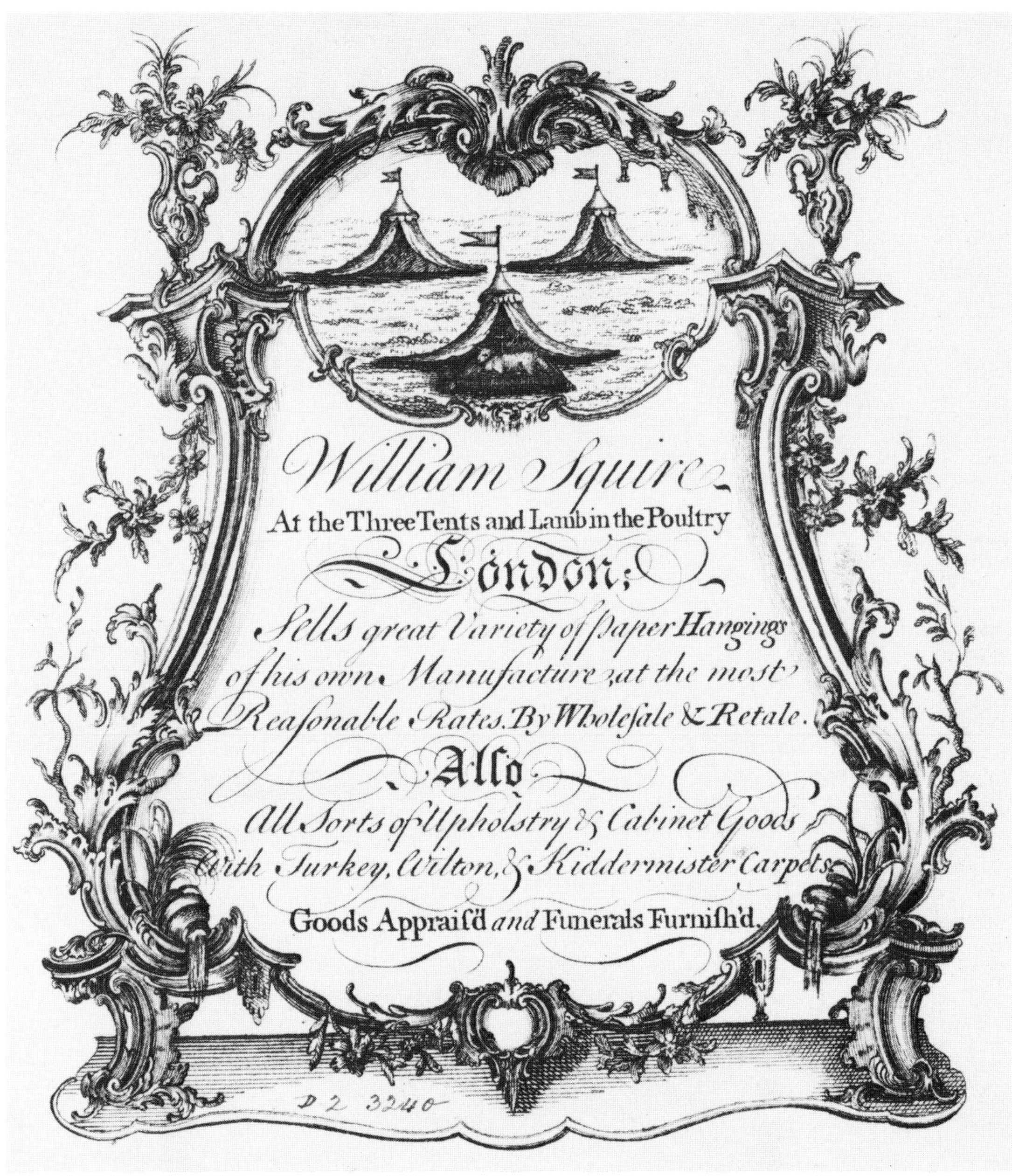

3-8

Figure 3-8 William Squire, the London paper stainer whose trade card is illustrated here, sold flocked "caffy" papers and "10 Paintings of ruins of Rome" to Philip Van Schuyler of Albany in 1761. The rococo frame around Squire's advertisement is an exercise in the same vocabulary of ornament seen in the wallpaper "frames" around landscape papers in the Lee and Van Rensselaer mansions pictured in *figures 3-9 through 3-12.* By permission of the trustees of the British Museum

3-9

Figure 3-9 Stephen Van Rensselaer ordered painted wallpaper (above) for a mansion he built in Albany, New York, between 1765 and 1769. Plans for the room were sent to London, and wallpaper was executed especially for it. It arrived from England in 1768. Surrounded by a mustard yellow ground, in the smaller panels, views of fall and winter, respectively, are rendered in grisaille. In the large central panels, Roman ruins are illustrated, again in shades of gray. They were based on an engraving of a painting by Giovanni Paolo Pannini (c. 1691–1765), an artist who was active in Piacenza and Rome during the first half of the eighteenth century. The same scene once was a part of the wallpaper decorations of the Lee mansion hall. The Metropolitan Museum of Art

Figure 3-10 Elaborate wallpapers (bottom right) are the original decorations in place where they were installed during the eighteenth century on the second-floor landing of the stair hall in the Jeremiah Lee mansion, built between 1767 and 1769 in Marblehead, Massachusetts. Rendered in shades of gray within painted wallpaper "frames," the ruins of Rome, based on engravings after Pannini, are painted in tempera. Narrow printed border papers were used around the doorway, above the chair rail, and under the wooden dentiled cornice appearing in this illustration. Courtesy of the Marblehead Historical Society

Figure 3-11 (top right) For the imitations of trophies in *figure 3-10* (the trophies flank the central panel with its Roman ruins), an English craftsman relied on "*Attributs Pastorals*" by Jean Charles Delafosse (1734–89). The engraving appeared as plate 92 in *Deuxième Recueil de l'Oeuvre de Jean Charles Delafosse* (1770). The Lee mansion papers include other motifs derived from plates in the same volume, as well as from designs by Pannini and by Joseph Vernet (1714–89). Cooper-Hewitt Museum, 1921-6-D89; Ex Coll. Decloux

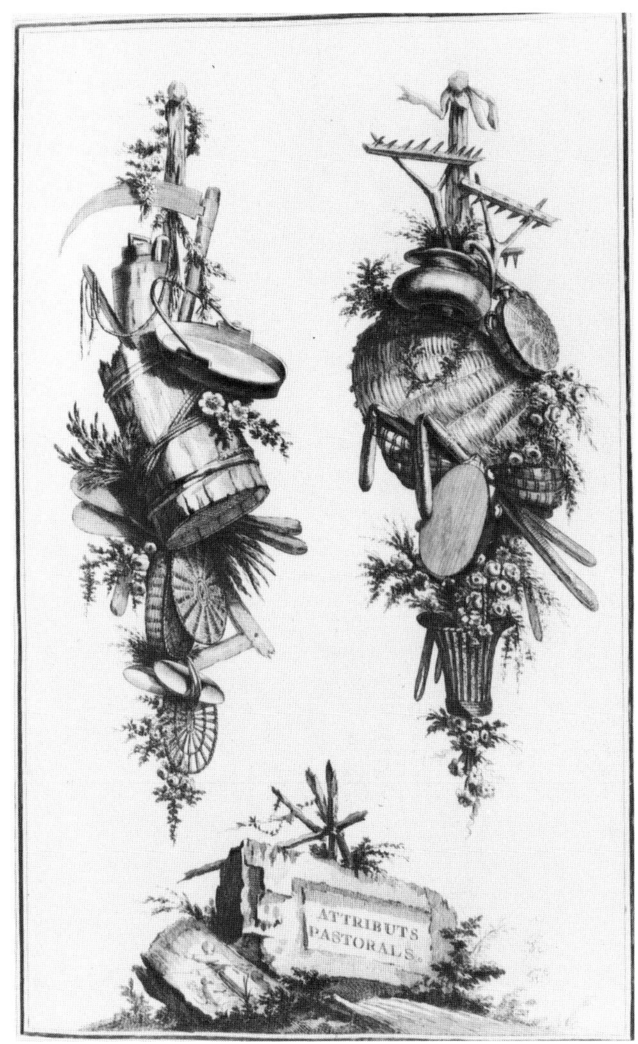

3-11

3-10

3-12

Figure 3-12 In the second-floor drawing room of the Lee mansion in Marblehead (above), additional wallpapers painted in grisaille survive. The scenes are based on engravings by Pierre Antoine de Machy (1723–1807). Yet another room in the Lee mansion retains similar wallpapers installed during the eighteenth century, with scenes after engravings by Vernet. The rococo frames painted around the ruins are similar to those in the Van Rensselaer wallpapers shown in *figure 3-9.* Courtesy of the Marblehead Historical Society

Figure 3-13 "The section of the four sides of a room ornamented with stucco work, proper for an eating parlor" is the caption on this illustration in John Crunden's *Convenient and Ornamental Architecture,* published in London in 1767 (right). Appearing as a double-page spread, plates 56 and 57 in the book, it illustrates designs for elaborate plaster-work ornaments—called "stucco work"—of the kind often imitated in wallpaper. "Stucco papers" and "stucco borders" are terms that appear frequently in the bills and advertisements of mid-eighteenth-century American paper stainers and wallpaper dealers. The Metropolitan Museum of Art, Harris Brisbane Dick Fund, 1928

3-13

3-14

Figure 3-14 This engraving, "In Full Chace," is one of a series of four hunting scenes thought to have been incorporated in a frieze around the tops of the walls in an American "print room" of about 1763. They survived in the "yellow chamber" of the Moffatt–Ladd house in Portsmouth, New Hampshire. The engraving shown here is after the work of an English artist, James Seymour (1702–52). It is 18 inches wide and 11½ inches high, printed on paper bearing English excise marks. Courtesy of the Society for the Preservation of New England Antiquities

3-15

Figure 3-15 A sample of wallpaper from the Moffatt–Ladd "yellow chamber" imitates stucco work in an eighteenth-century reordering and "improvement" of Gothic motifs. Batty and Thomas Langley included ornamentation like this in their *Gothic Architecture Improved by Rules and Proportions,* published in 1747. It includes similar quatrefoils formed of curves derived from Gothic arches, but filled with leafy ornaments and bordered by beadings and bandings derived from classical sources. The pattern shown here was engraved on plain paper; then a coating of yellow was applied. It was used in combination with prints, including the one shown in *figure 3-14.* This piece is 23 inches wide and 26 inches high. Courtesy of the National Society of the Colonial Dames of America in the State of New Hampshire

3-16

3-17

Figure 3-16 Block-printed in black and white on a gray ground, the English paper shown above was probably the original wallcovering used in the Samuel Buckingham house, built in 1768 in Old Saybrook, Connecticut. The remnant illustrated is about 24 inches wide. Cooper-Hewitt Museum, 1970-26-2; gift of Jones and Erwin, Inc.

3-18

3-19

Figures 3-17 and 3-18 A reverse of the Buckingham house pattern, the pattern shown to its right was block-printed in black and white on a blue ground. Part of the mark of "Mills & Co./Hartford" was found on the reverse when the paper was removed from a house in Haddam, Connecticut. In 1796 Zecheriah Mills advertised what could only have been similar papers—"large and elegant Pillar and Arch Figures for spaceways, halls, &c." This is about 22 inches wide. Cooper-Hewitt Museum, 1970-26-1; gift of Jones and Erwin, Inc.

Figure 3-19 An English "pillar and arch" pattern was photographed where it was hung about 1780 in the "best room" of an early-eighteenth-century house built by Timothy Johnson in North Andover, Massachusetts. Because of the bust surmounting each arch, it was known as the "Marie Antoinette Pattern." Although the house still stands, the paper was removed in 1926, but samples of the gray, black, and white pattern are preserved by the North Andover Historical Society, which owns the house. The repeat length is 47¾ inches and the border is 1¾ inches wide. Photograph, Cooper-Hewitt picture collection

3-20

Figure 3-20 This pillar and arch pattern is shown as it was photographed in 1914 on the walls of the Hancock–Clarke house, built in 1698 in Lexington, Massachusetts. It was found in the "west room" in a part of the house built during the early eighteenth century. So closely does this pattern correspond to one depicted on a billhead of Appleton Prentiss (*figure 6-2,* third pattern from the right) that it seems probable Prentiss made it. Fragments of a closely related English pattern that perhaps served as the model for the paper shown here survive in the Cooper-Hewitt collection (1938-5-1, gift of Paul F. Franco). The Cooper-Hewitt fragment was used as a book cover, and bears the handwritten dates "1777" and "1818." Cooper-Hewitt picture collection

Figure 3-21 The pillar and arch pattern formula (right) frames a patriotic vignette in a modern reproduction printed by the Thomas Strahan Company. It is based on samples of the late-eighteenth- or early-nineteenth-century original, which survived in a tavern in Lexington, Massachusetts, and in a house in Salem, New Jersey. Beside an Indian maiden, representing America, a patriot tramples British laws underfoot and extends the declarations of July 4, 1776, to Britannia, who weeps over a tomb. Original fragments in the Cooper-Hewitt collections are block-printed in black, white, and gray on a light colorless ground. The scale in the original is large—the maiden stands 11 inches high. Cooper-Hewitt Museum, 1953-198-15; gift of *Wallpaper* Magazine. Original fragments, Cooper-Hewitt Museum, 1960-25-1; gift of Mrs. A. Ingham Evans

3-21

3-22

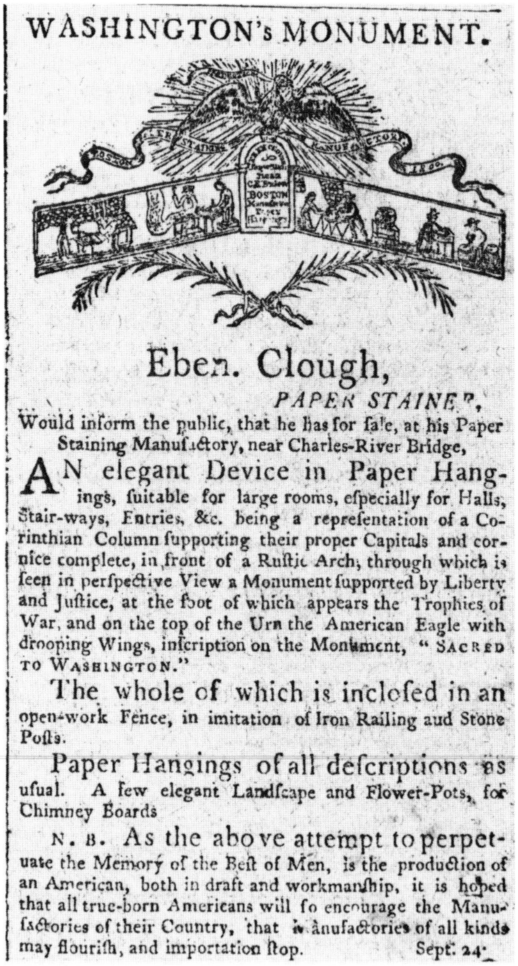

WASHINGTON's MONUMENT.

Eben. Clough,

PAPER STAINER,

Would inform the public, that he has for sale, at his Paper Staining Manufactory, near Charles-River Bridge,

AN elegant Device in Paper Hangings, suitable for large rooms, especially for Halls, Stair-ways, Entries, &c. being a representation of a Corinthian Column supporting their proper Capitals and cornice complete, in front of a Rustic Arch, through which is seen in perspective View a Monument supported by Liberty and Justice, at the foot of which appears the Trophies of War, and on the top of the Urn the American Eagle with drooping Wings, inscription on the Monument, "SACRED TO WASHINGTON."

The whole of which is inclosed in an open-work Fence, in imitation of Iron Railing and Stone Posts.

Paper Hangings of all descriptions as usual. A few elegant Landscape and Flower-Pots, for Chimney Boards

N. B. As the above attempt to perpetuate the Memory of the Best of Men, is the production of an American, both in draft and workmanship, it is hoped that all true-born Americans will so encourage the Manufactories of their Country, that manufactories of all kinds may flourish, and importation stop. Sept. 24.

3-23

Figure 3-22 (far left) Ebenezer Clough, a Boston paper stainer, combined the English pillar and arch formula with a memorial to George Washington. Designed in 1800, the wallpaper was block-printed in black and white on a ground that was originally blue. The same image, perhaps copied from Clough's wallpaper, is found in a memorial picture worked in silk embroidery (see Chapter 3, footnote 25). The sample shown is 21 inches wide. Cooper-Hewitt Museum, 1960-103-1; gift of the children of Edith Parsons Morgan

Figure 3-23 (left) An advertisement for the paper shown here appeared in the *Columbian Centinel,* a Boston newspaper, in September 1800. The image Clough used here repeats the one illustrated in *figure 2-2.* Courtesy of the New-York Historical Society, New York City

Figure 3-24 (right) The English pillar and arch pattern formula, after a drastic process of dilution and adaptation, is still recognizable in this early-nineteenth-century American wallpaper fragment. It is block-printed in blue, pink, green, and black on tan and is about 11 inches wide. The Metropolitan Museum of Art; gift of Mrs. Thomas W. Fletcher, 1932

3-24

Manufacture Royale de Papiers pour Tentures et Décorations, des Srs Arthur et Robert
Rue de Louis le Grand, au coin du Boulevart.

Le 29. Juillet 1790.

Fourni à Monsieur De Jefferson par ordre de Mr. Short Grille de chaillot

22.	Rouleaux cendre bleu unie	à 8#.	176	"
4.	Rouleaux draperie sur fond bleu fin	à 18#.	72	"
72.	cantonnieres coloriées	à 4#.	288	"
12.	Rouleaux No. 848. tors de Roses de 4°.	à 12#.	144	"
22.	Rouleaux No. 614. treillage Blanc mat à	6#	132	"
4.	Rouleaux draperie Imprimée sur Treillage	à 15#.	60	"
5.	Rouleaux No. 815. bordure couleur d'or	à 5#.	25	"
22.	Rouleaux cramoisi uni	à 4#.	88	"
4.	Rouleaux de draperie sur fond cramoisi	à 15#.	60	"
5.	Rouleaux No. 951. bordure grise	à 4#.	20	"
21.	Rouleaux No. 844. brique unie	à 2#. 10s	52	10
			1117	10

Nous avons reçu le Montant ci dessus par les mains de Monsieur Short dont quittance a Paris ce 19 Aoust 1790./

Arthur et Robert

4-1

4 EIGHTEENTH-CENTURY FRENCH WALLPAPER STYLES

Independence from England ended British colonial trading restrictions and cleared the way for a dramatic increase in the importation of French wallpapers by Americans. However, during a transitional period lasting through the 1780s, English papers maintained a strong position in the American wallpaper trade alongside the newly introduced French styles, if we may judge by the frequency with which both English and French imports were offered for sale. In 1787 the French removed export duties on wallpapers, thereby lowering the prices that Americans had to pay.[1]

The post-revolutionary popularity of French *papiers peints* was not simply the result of removal of British restrictions on non-British goods, and of the price drop on French papers when export duties were abolished. Nor was it just a byproduct of American gratitude for French assistance during the Revolution. These factors were doubtless important, but the quality of the French papers themselves was probably their most important selling point in America during the late eighteenth century. By the 1790s, the beauty of French wallpapers had captivated American taste makers, and they were gaining popularity among wider circles of consumers.

By 1780 in France the *papiers peints* of Jean-Baptiste Réveillon (active 1765–89) were emerging as unrivaled masterpieces of wallpaper design and the art of block-printing in color.[2] Through the first half of the eighteenth century, French workmen had made advances that transformed the craft of the *dominotiers*—makers of marble papers who also used inks to print sheets of patterning that were often colored by hand—into the art of creating large repeating patterns and panels in colors printed from wood blocks.

Jean Papillon (1661–1723) and his son Jean-Baptiste-Michel Papillon (1698–1776) made major refinements in the craft. They produced patterns that repeated beyond

Figure 4-1 (left) Thomas Jefferson, an innovator and trend-setter in American architectural and decorative styles, visited the showrooms of paper stainers while he was in Paris. Here Jefferson's purchases from Arthur et Robert, one of the leading manufacturers, are itemized. Plain blue and plain scarlet paper, with drapery ornaments printed on grounds matching the plain papers, a lattice pattern, a drapery design printed over a lattice, wallpaper valances, a brick pattern, and one gold and one gray border are listed. Reproduced from the collection of the Library of Congress

the confines of a single sheet of paper, and they improved printing techniques. Building on their advances, imitating English patterns, and adopting English techniques like those described by Robert Dossie in his *Handmaid to the Arts* of 1758, Jean-Baptiste Réveillon perfected methods for using wood blocks to print in distemper colors. The clarity of his colors and his inventive combinations of brilliant and pastel colors distinguish wallpapers by Réveillon from all that preceded his. He achieved great subtleties by using many blocks to print a multiplicity of hues and shadings in a given pattern. Designs printed on pastel blue grounds in soft multi-colors, heightened by brilliant but subtle accents of strong colors—particularly a bright orange—are characteristic of Réveillon's production *(color plates 9, 14).*

He based his papers not only on adaptations from textile patterns, but also on more ambitious, larger-scaled designs by important artists.[3] He employed more than three hundred workers in what had become not just a craftsman's workshop but a factory. This factory produced papers for three distinct markets. For the most extravagant purchasers, Réveillon's workers might use as many as eighty blocks to print each pattern classified as "*grand lux.*" For the bourgeois, they used seven or eight blocks to produce "*communs*" papers. For the broadest market, they might use only one color in printing papers described as "*ordinaires.*"

The most celebrated styles of Réveillon's factory were those based on arabesque designs. Raphael's decorations of the Vatican loggia provide the most familiar examples of this classical style, which Réveillon translated into wallpaper designs. The arabesques featured long vertical panels of foliage, flowers, and scrollwork, branching out symmetrically from a central stem, and adorned with figures, grotesques, animals, trophies, and architectural fragments *(figure 4-2).* In many of Réveillon's arabesque patterns, and in related wallpapers that repeated similar motifs on a smaller scale, the impact of the painted wall decorations that had recently been discovered at Pompeii and Herculaneum is apparent. The neoclassical elements in these designs appealed strongly to Americans, who associated them with the virtues of Greek democracy and the Roman republic. This generation found such wallpapers appropriate embellishments for the homes they were building for themselves in the same classical architectural styles they approved in the public buildings for which they were responsible.

The best examples of the period's use of Réveillon-style wallpapers in America survive in the Phelps–Hatheway House in Suffield, Connecticut. By 1795, Oliver Phelps, a prosperous merchant and speculator, had added a north wing to his house, which had been built about 1760. For that wing, Phelps purchased the finest of

French wallpapers and borders and had them hung in the most fashionable way *(color plates 9 through 11; figures 4-3 through 4-5).*[4] Borders were used to outline every architectural feature that interrupted the patterned walls. In their astonishingly fine state of preservation, these papers are remarkable not only as elegant examples of late-eighteenth-century French design and craftsmanship but also as masterpieces of refinement in stylish hanging of patterning with borders to complement an architectural setting. The careful centering of panels on walls and the infinite attention devoted to the use of borders to outline every bend in the framework of a window or a pilaster contribute enormously to the wallpaper's success as wall decoration. Wider borders run horizontally along the chair rail and at cornice level in each of the five wallpapered rooms, while narrower borders run vertically to outline pilasters and window and door frames, then complete the enframement of these openings in the wall by outlining their horizontal elements as well with the narrower strips.

Five variations on styles popularized by Réveillon were used in the Phelps–Hatheway house. The most formal and purely neoclassical of the patterns is an example in the hallway, which can be identified as a Réveillon paper *(figure 4-3; color plate 9).* It features an architectural framework surrounding a classical figure. The central figure within each repeat of the architectural surround is one of four that are used for the paper—a piper, a tambourine player, and a male as well as a female dancer. Birds, fantastic creatures, flowers, and foliage in delicate and orderly configurations embellish the larger elements in the pattern. The architectural structure is borne by rows of winged maidens. Arranged in vertical panels and rendered in an imposingly large scale, the pattern is hung with bold border patterns. The wider border shows Greek pots alternating with figures derived from paintings on such pots.

In the pattern originally used in the dining room, but now exhibited at the Winterthur Museum *(figure 4-4),* an arched surround for classical figures is composed of stylized floral motifs rather than architectural elements, as in the hall paper. Each repeat incorporates a flower-filled urn flanked by birds. The birds perch on delicate scrolls embellishing the space above the figure group. The emphasis on flowers in this panel scheme has made the pattern seem only slightly less formal than that used in the hall. In the Winterthur room, border patterns derived from architectural ornament are narrower and do not assume such visual importance in the decorative scheme as do the borders in the hall.

In the large-scaled pattern used in the master bedroom at the Phelps–Hatheway house *(color plate 10),* flowers, a figure, and a rich vocabulary of ornaments branching out from a central vertical axis in each paper width typify the arabesque formula. Floral borders with black grounds finish and set off the repeating pattern with a bold flair. This combination of a strong-figured, often darker, border with a more delicately colored repeating pattern represents a taste for which we have quantities of additional evidence from the eighteenth and nineteenth centuries.

In the wallpaper hung in the sitting room on the ground floor of the Suffield house *(color plate 11),* a simpler arabesque-style floral pattern incorporating only two major elements—a basket of flowers alternating with an urn—creates a less formal tone on a reduced scale. The roses in the baskets are picked up in a floral border, one of two borders used in this room.

The same floral border is used with a still simpler repeating pattern in the small bedroom over the sitting room *(figure 4-5).* In this pattern, the motifs are widely spaced, giving a light, informal effect. A more balanced, nearly horizontal arrangement of design elements has been substituted for the vertical emphasis of the arabesque patterns in the preceding illustrations. Floral garlands and delicately scrolling vegetation link the principal motifs: whimsical depictions of an arch surmounted by fat cupids supporting an urn amidst blossoms that hint of Chinoiserie, and a medallion showing a frieze of putti.

Some of these papers at the Phelps–Hatheway house can be documented as products of Réveillon's factory. But by the time Phelps purchased his wallpaper in the mid-1790s, Réveillon's business had been taken over by Jacquemart et Bénard, who carried on using Réveillon's printing blocks and his maker's mark. Other Parisian manufacturers were producing wallpapers in the styles originated at the Réveillon factory. In 1790, Thomas Jefferson wrote to Paris requesting that his correspondent William Short have the "old man" at Arthur's send "specimens of his good Arabesques, noting the price of each."[5] The spectacular pattern in this style which was used in the John Brown house built in Providence, Rhode Island, between 1784 and 1788 bears the mark of yet another Parisian manufacturer, Sr. Barabé & Cie. *(color plate 12).*[6] From the same house, another high-style French arabesque pattern of the period duplicates a pattern from a house in Cazenovia, New York, a sample of which is now in the Cooper–Hewitt collections *(color plate 13).*

Arabesques on wallpapers caught the attention of an English visitor to Philadelphia in 1794. Henry Wansley remarked on the magnificent house of a Mr. Bingham and noted that the dining room was "papered in the French taste, after the style of the Vatican in Rome."[7] During the 1790s, these high-style French wallpapers traveled at least as far west as Kentucky, where Jefferson's friend, George Rogers Clark, built a house called Locust Grove near Louisville. Restorationists working on the house during the 1960s found in the ballroom fragments of an arabesque wallpaper designed about 1786 by François Cietti (active by 1758), and printed by the Réveillon factory.[8]

In addition to such elaborate and expensive papers by craftsmen like Réveillon, more ordinary wallpapers also arrived here from France during the years around 1800. In the 1790s, wallpaper advertisements from all the major cities along the East Coast, from Charleston to Boston, featured the French imports. A Baltimore advertisement of 1798 is typical: "An assortment of elegant French Paper Hangings of a superior quality, comprising a great variety of new and tasty figures."[9]

The comments of another English tourist who was in America during the mid-1790s provide additional evidence of the newly stylish status of French wallpapers. William Strickland's published travel diary includes a description of Clermont, the home of Chancellor Livingston, on the Hudson River near Hudson, New York. He visited the house on October 12, 1794, observing:

> The principal rooms which are of good dimensions are hung with french papers, which are chiefly used in this country, the patterns being much more beautiful and elegant, and lively, than what are manufactured in and exported from England and much cheaper in proportion to their merit . . .[10]

Although their coloring may well have been livelier, some of the more ordinary patterns imported from France were closely related to standard patterns of English and American manufacture. Therefore, they will be described in Chapter 7, which deals with internationally common styles of wallpaper.

4-2

Figure 4-2 A design by Jean-Baptiste Fay (see Chapter 4, footnote 3) was the source for this wallpaper panel (left), block-printed at the Parisian manufactory of Jean-Baptiste Réveillon about 1788. The arabesque pattern shown here was rendered in multi-colors, including many shades of rose and green, on a cream ground. The panel is 8 feet 2 inches high and 32¼ inches wide. Cooper-Hewitt Museum, 1925-2-332; gift of Eleanor and Sarah Hewitt

Figure 4-3 (right) In the stair hall of the wing added in 1795 to the Phelps–Hatheway house in Suffield, Connecticut, the principal pattern was made in the Réveillon factory, probably under the direction of Réveillon's successors Pierre Jacquemart and Eugène Bénard. The incorporation of architectural elements and its large scale (each vertical repeat is 46¾ inches) make this pattern comparable to English pillar and arch designs, also popular for American halls *(figures 3-16 through 3-21).* Courtesy of the Antiquarian and Landmarks Society, Inc., of Connecticut

4-4

4-5

4-6

Figure 4-4 (top left) The original woodwork and wallpaper from the dining room of the Phelps–Hatheway house has been removed and installed as the "Federal Parlor" at the Henry Francis du Pont Winterthur Museum. A reproduction of the original dining room replaces it in its original site. The wallpaper, probably made by his successors Pierre Jacquemart and Eugène Bénard, bears the stamp of Jean-Baptiste Réveillon. A duplicate of the pattern hung within a few years of the installation of the wallpaper in Suffield survives in the parlor of the Quincy homestead in Quincy, Massachusetts. Courtesy of the Henry Francis du Pont Winterthur Museum

Figure 4-5 The smaller bedroom on the second floor of the Phelps–Hatheway house wing of 1795 preserves additional French wallpapers as originally installed (top right). The pattern was printed in a multiplicity of soft greens, pinks, beiges, and blues on a ground now off-white, which may originally have been pale blue. The roses featured in the border, which duplicates that used in the sitting room, are pink. The repeat length is 21½ inches. Courtesy of the Antiquarian and Landmarks Society, Inc., of Connecticut

4-7

Figure 4-6 (opposite page, bottom) A charming example of a small-scale repeating pattern derived from French arabesque-style wallpapers is perhaps American-made. It was found in the Lazarus LeBaron house, built in 1794 in Sutton, Massachusetts. Block-printed in orange and black on a light blue ground, a color scheme inspired by Réveillon papers, it was hung with a matching swag border 6¾ inches wide and with a narrow floral edging 3½ inches wide. Old Sturbridge Village, Sturbridge, Massachusetts

Figure 4-7 (above) Free-hand painted imitations of French wallpapers, like those illustrated in *figures 4-4 and 4-5* and in *color plates 10 and 11,* embellish the parlor of the Sherwood–Jayne house, built in 1767 in East Setauket, Long Island, New York. The artist carefully followed the French precedent in giving his pink and green flowers a blue background, and showed swag borders as well as narrow edgings like the paper ones standardly hung with French patterns during the 1790s. Now restored, the original painted decorations date from about 1800. Courtesy of the Society for the Preservation of Long Island Antiquities

4-8

Figure 4-8 This French wallpaper would seem to have been designed to appeal only to French patriots. However, American advertisements of the 1790s attest to the fact that wallpapers incorporating such specific references to French patriotic themes, and to political events in France, were used in the United States. During 1790, Francis Delorme advertised for sale in Philadelphia "handsome Paper-Hangings from Paris, in the latest Taste, some emblematic of the late Revolution." (See Chapter 4, footnote 11.) Three years later, Philadelphia wallpaper manufacturers Burrill and Edward Carnes announced "The New and beautiful figure of the destruction of the Bastille lately received from Paris, is now finished." (See Chapter 4, footnote 12.) The sample of wallpaper shown here is 21¾ inches wide and 14¾ inches high. It is attributed to Jacquemart et Bénard, Parisian manufacturers. The pattern is block-printed in multi-colors on a dark blue ground. Cooper-Hewitt Museum, 1925-1-370

5 ORIENTAL WALLPAPERS

The beautiful painted papers made in China for export to the West played an important part in stimulating the popularity of wallpapers in England and America from the late seventeenth century through the nineteenth. Because each Chinese paper was a unique painted work, not a print made in multiple, it was expensive. Such papers were hung only on the walls of houses belonging to rich Americans. However, they set a high standard for design and craftsmanship, evoked enthusiasm for wallpapers in general, and probably whetted the appetites of many American consumers for the non-repeating French block-printed scenic papers that became more widely available early in the nineteenth century.

The Chinese designs were usually made in sets of twenty or twenty-five non-repeating panels each 4 feet wide, 12 feet long. The panels were made up of joined sheets of mulberry paper. Chinese paper makers could produce larger individual sheets of paper than did the mills in the West; the average size of sheets used was 24 by 60 inches.[1] The panels were sometimes mounted on silk and individually rolled on wooden rollers like scroll paintings. Many were shipped to the West in this form. Some were pasted on heavy paper mounted on coarse linen or canvas.

Many of the Chinese designs were painted on ungrounded papers. The paper itself might have been colored in the pulp, but after the second quarter of the eighteenth century, when laid-in grounds enjoyed a vogue, ungrounded papers were preferred, since they incorporated less paint on the surface, which might crack when rolled for shipment. The Chinese used simple carbon inks to outline the many elements included in these designs, and colored them with washes, often reserving areas of uncolored paper to add highlights.[2] Through the eighteenth century, a tendency developed toward thinning of paints, giving greater luminosity.

Although standardization of patterning was avoided in these individually rendered paintings, the Chinese wallpapers do fall into several major categories. The largest category of Chinese panels that survive from eighteenth- and nineteenth-century installations in American houses features blossoming trees, or bamboo, often growing from fantastic rock formations *(color plates 15, 16).* Branches and backgrounds were bedecked with birds, insects, and animals rendered in intricate detail.

Where an additional bird or butterfly was wanted after the paper was hung on the walls, whether for visual effect or to hide a stain or a hole in the paper, an extra from those sold with the sets of paper was cut out and pasted on. In one well-known set of Chinese papers hung during the first quarter of the nineteenth century, in England, at Temple Newsam house, Leeds, birds added to the painted Chinese specimens have

5-1

Figure 5-1 A standard variation on the flowering tree theme in eighteenth- and nineteenth-century Chinese papers is represented in this illustration—the incorporation of a procession of underscaled human figures at the base of the panels. In this example, probably nineteenth century, bamboo is substituted for the more usual flowering tree, but blossoms still appear among its shoots. The paper was once in the stock of a New York antique dealer, Josephine Howell, who gave this photograph to the Cooper-Hewitt Museum.

5-2

Figure 5-2 Hand-painted in China for export during the early nineteenth century, the pattern shown below is of a type frequently copied in repeating patterns printed in the West. The flowers are multi-colored, although red and white predominate, and they are painted over a green ground on the 21-inch-wide fragment. Cooper-Hewitt Museum, 1939-71-1, purchased in memory of McDougall Hawkes

recently been identified as ones that were carefully cut from the double elephant folio edition of John James Audubon's *The Birds of America* (1827–38).[3]

Another group of papers, which features flowering trees or bamboo rising through the height of a panel, is distinguished as a recognizable type by the addition of underscaled figures. The figures in their Chinese costumes seem doll-like when they are shown between the bases of the trees or the bamboo *(figure 5-1).*

During the early nineteenth century, another variant within the flowering-tree category gained a popularity it was to retain for many years: Chinese papers in this group have the common feature of a balustrade at the base of the panels. This feature helped to "finish" a room as did Western wallpaper imitations of wainscoting.

"Factory papers," as they are called by the modern antique and wallpaper trades, were another major category made by the Chinese for export. In these, the Chinese catered to Western fascination with goods coming from the Orient. Narrative and anecdotal in tone, they depicted intriguing details in the several stages of production of pottery, silk, tea, or rice. Like the flowering-tree and bamboo papers, these were non-repeating paintings, but they differed in the arrangement of motifs in space. Most of the flowering trees and bamboo rise from the base of a panel of paper right up its length, existing in a single, non-confusing plane of flattened space. The factory papers, however, show individual scenes in distinct spaces stacked one over another. All of those scenes are flattened and brought right up to the surface of the picture plane.

This highly stylized, anti-perspective handling of space characterizes another major group of Chinese wallpapers, those detailing daily life amidst Chinese landscapes and buildings *(figure 5-3).* In the Chinese landscape papers now exhibited at the Winterthur Museum *(color plate 18),* such handling of space is modified by a slight diminution in scale in the vignettes as they ascend up the paper's height, and by a suggestion of atmospheric perspective.

In addition to these major categories, a few painted Chinese imitations of repeating textile patterning were produced for export and probably reached this country. In some of these patterns, the influence of India chintzes is apparent, reflecting the fact that trade between China and India carried styles between these countries.

Eighteenth-century American documents reveal that Chinese papers were called "India papers." The loose usage of the word "India" to refer to anything from the Orient is not so much a reflection of recognition of stylistic interchanges in the Far

East as of the fact that many Oriental goods first came to this country in ships of the British East India Company. A New York advertisement of 1790 demonstrates the confusing way the word "India" was used, even though the advertiser clearly was offering goods brought directly from China: "Landed this day from Canton, India Paper Hangings of exquisite beauty, fit to adorn the most superb saloon in America."[4]

A sense of the high esteem accorded Chinese papers is conveyed in other advertisements of American upholsterers and paper hangers. One from a Philadelphia newspaper of 1786 suggests that a hierarchy was recognized, classifying the Oriental, European, and American paper hangings in that order: Thomas Hurley offered to "hang any Paper from the most elegant imported from the East Indies or Europe, to the most indifferent manufactured in this country."[5] Lady Skipwith's request for papers in 1795, cited earlier, confirms the suggestion that Chinese papers were deemed superior to the European imports. Lady Skipwith had explained: "We do not mean to go the length of India Paper, only plain English and Irish."[6]

Consumers admired the Chinese papers for their beauty, and certainly Western craftsmen admired the technical virtuosity displayed in their execution. The Englishman Robert Dossie had recognized the superiority of the painted Chinese papers when he observed in his *Handmaid to the Arts,* "Pencilling [painting by hand] is only used in the case of nicer work, such as the better imitations of the India paper."[7] One Philadelphian of 1785 described his wallpapers as "made equal to India." Published works on English wallpaper history note that not only did French and English craftsmen strive to imitate the Chinese papers exactly, but they also went so far as to bring Chinese craftsmen to France and England to paint papers on the spot.

More often than they attempted to paint imitations of Chinese papers, Europeans resorted to printing blocks and plates to provide outlines for motifs and then filled in the colors. In 1832, however, the Zuber factory in Alsace produced its *"Décor Chinois"* *(color plate 35)* in the thick, block-printed distemper colors typical of their other work , making no attempt to imitate the Chinese technique of outlining the design. In *"Décor Chinois"* the flowering trees with exotic birds are derived directly from Chinese models like that illustrated in *color plate 16,* and from the stylized rockery of papers like that shown in *color plate 15.*

In eighteenth-century wallpaper references, the confusion resulting from use of the words "India" and "China" is compounded by the fact that wallpaper patterns based on textiles made in India also enjoyed great popularity in America. "India figures," a phrase found in American documents from the 1750s and continuing through the century, could have referred to printed, repeating floral patterns made in the West, derived from "India chintzes" *(color plate 19).* Chintzes were cotton fabrics, painted and printed with patterns featuring fantastic blossoms and foliage. Wallpapers imitating these Indian textiles were also called "chintz figures" in eighteenth-century documents. For instance, an invoice of wallpaper that was sent from London to George Washington in 1757 included "96 yards Chintz paper"; 8 dozen borders to ditto."[8]

In addition, phrases like "India figures," "mock India pictures," "Chinese pieces," and "Chinese papering" could also have referred to patterns that would be described today as "Chinoiserie." Chinoiserie designs—Western interpretations of Chinese motifs—were derived from goods and illustrations brought from China, including painted wallpapers. In designing their "Chinoiserie" motifs, Europeans often made significant changes in the character of their Oriental subjects. No student of Chinese art would ever mistake the Chinoiserie motifs published by Jean Baptiste Pillement (1728–1808) for the real thing *(figure 5-4, page 64).* But by providing makers of textiles and wallpapers with a series of pattern books, no matter what liberties he took with the Chinese prototypes, Pillement played a role similar to the one Chippendale served for furniture makers in popularizing a taste for things Chinese. In his ornaments, useful to craftsmen who made many kinds of decorative objects, Pillement adapted Oriental motifs to Western tastes. Evidence for the use of these books by wallpaper manufacturers appears in the wallpaper illustrated in *color plate 21,* where the motif of the bell ringer is clearly derived from the Pillement engraving of 1773 shown in *figure 5-4.* In addition to borrowing from Chinese arts, Pillement also made use of Indian motifs. The pattern in *color plate 20* has floral elements very much like those Pillement derived from India chintzes.

The influence of Pillement is apparent in patterns surviving from eighteenth-century use in America. Some of the survivals are French, and others bear English tax stamps. An English example bearing an excise stamp was preserved as the cover for a 1780 "Tryal Docket for the Court of Equity, Hillsborough District" in the North Carolina state archives. The pattern is a simple linear one showing fragments of garden structures made up of latticework, an out-scaled dragonfly, a smaller peacocklike bird, and floral bits and pieces.

Chinoiserie papers were not only imported; advertisements indicate that Americans also manufactured wallpapers in the Oriental styles. One advertisement published in Baltimore in 1775 offers wallpapers described as "Mock India Pictures . . . all entire the manufacture of this country."[9] An inventory made in a Boston manufactory of the 1780s and 1790s mentioning "China fig." provides further indication that Americans printed their own Oriental-style wallpaper patterns.[10]

In America, fascination with Chinese and Chinoiserie wallpapers was long-lived. As early as 1737 Thomas Hancock of Boston wrote one of the most vivid descriptions of wallpapers used in this country that might well have been in the Chinese style. In a letter to a London stationer, he included an order for a "Shaded Hanging" to be done after a pattern he sent from Boston. The pattern had "Lately Come over here, and it takes much in ye Town and will be the only paper-hanging for Sale wh.am of opinion may Answer well." Hancock wanted his own paper made "more beautiful by adding more Birds flying here and there, with Some Landskips at the Bottom."

Hancock's further comments make clear that papers made in England, rather than in China, were wanted. He described the wallpapers of a Boston friend, "Done in the Same manner but much handsomer . . . made by one Dunbar, in Aldermanbury." Hancock continued his description: "In other part of these Hangings are Great Variety of Different Sorts of Birds, Peacocks, Macoys, Squirril, Monkys, Fruit, and Flowers, &c."[11]

Although few of the actual papers have survived in America, it is possible to document the constancy of the appeal of Chinese papers, and imitations of them, from this 1737 Boston reference right through the eighteenth and nineteenth centuries. This documentation comes in the form of advertisements from north and south, from every area that was beginning to boast grand houses built by the rich in imitation of current English fashion. In wallpaper advertisements, whether published in Charleston, South Carolina, newspapers of the 1750s and 1780s, in the press of New York, Boston, or Annapolis in the 1760s, or in Salem, Massachusetts, in the 1780s, Oriental papers and imitations of Oriental papers are distinguished from the usual offerings by superlatives and by the fact that the number available is often small and is given quite specifically. For instance, John Arthur offered New Yorkers in 1773 "an assortment of paper hangings, in which are two elegant India patterns."[12]

Figure 5-3 A detail from another painted Chinese landscape paper shows scenes of rural life in a continuous panorama in which little attempt is made to create a three-dimensional illusion of recession into distant space. Rather, figures in the background are vignetted and rendered at large scale so they can be seen in detail. During the middle of the twentieth century this paper was sold by the antique dealer Josephine Howell, who gave this picture to the Cooper-Hewitt Museum for its study files.

5-3

Over a quarter century later, advertisers apparently still counted on offerings of India papers to arouse interest. In 1804 Henry W. and Lewis Phillips of New York advertised: "Chinese Papering—A few sets very elegant for drawing rooms."[13] The following appeared in a Baltimore paper of 1806:

> For sale, TWO SETS of elegant PAPER HANGINGS, just arrived from the East Indies—This paper is thought by judges to be superior to any in the United States. Each set contains about 130 square yards. Enquire of R. & S. WINCHESTER, at Fredericksburg, or of Mr. GHEQUIERE of Baltimore.[14]

The rich and powerful in America could secure the Chinese papers from such local advertisers, or through friends or agents abroad. A bill of lading of May 8, 1768, records the shipment from England to Nathaniel Lytl. Savage at "Pleasant Prospect" in Northampton County, Virginia, of "6 or 8 pieces of good and handsome paper hangings, a suitable pattern for a large passage @ 3/pr—brown India."[15] Shortly after the Revolution, in 1784, Americans entered the China trade, and wallpapers were imported in increased quantities by firms like A. A. Low of New York and by ship captains in the China trade. George Washington's interest in such wallpapers is an indication of their status. In 1787, Washington wrote to Robert Morris:

> It is possible I may avail myself of your kind offer of sending for India Paper for my new Room, but presuming there is no opportunity to do it soon; I shall not, at this time, give you the dimensions of it.[16]

By the middle of the nineteenth century, the popularity of the Chinese papers was waning. Arbiters of taste in London were perhaps reacting to overexposure to the Oriental styles when an article appeared in *The Art Union* of 1844 reporting, "the Indian or Chinese papers are now chiefly in use for screens, or for very dark rooms, and are not likely to be again very popular."[17] Since English styles still set the tone for interior decorating in America, such opinions doubtless affected American taste. The Civil War with its blockades made international trade difficult during the 1860s, and direct imports from China were curtailed. Although they had lost some of the glamor of novelty and of exotic mystery, painted Chinese wallpapers continued to be hung through the nineteenth century. Fragments of Chinese papers installed in an American house during 1865 are illustrated in *color plate 17.* Nor have such papers ever entirely disappeared from the American wallpaper trade. At this writing, one firm in New York not only deals in old Chinese papers, restoring and installing them, but also supports an atelier of Chinese craftsmen who continue to paint new wallpapers in the style of the flowering tree patterns of the early nineteenth century.

6 EIGHTEENTH-CENTURY AMERICAN WALLPAPERS

THE BEGINNINGS OF THE CRAFT

The 1756 advertisement of a "silk-dyer and scourer, lately from Dublin" provides the earliest documentation yet located for the printing of wallpaper in America. The advertisement is that of one John Hickey, who as a "dyer and scourer" would have been primarily concerned with the dyeing, cleansing, and printing of textiles. But in the *New York Mercury* for December 13, 1756, Hickey specified that paper hangings were among his products. He informed New York's mid-eighteenth-century public that he "stamps or prints paper in the English manner and hangs it so as to harbour no worms."

The paper-staining business apparently made only tenuous progress in this country in the years immediately following Mr. Hickey's first efforts. Nine years after his advertisement for the stamping and printing of paper, the production of wallpaper in America was still considered a noteworthy and novel enterprise. On June 6, 1765, *The New York Gazette* . . . reported as news from London that "a new manufactory for Paper Hangings is set up at New York, which is an article in great demand from the Spanish West Indies." On December 12 of that same year, the *Boston Newsletter* reported that "John Rugar produced several Patterns of Paper Hangings" made in New York "at a numerous meeting of the Society for promoting Arts &c.," also in New York. The *Annual Register,* London, 1766, carried a similar report of the presentation of "Paper Hangings of domestic manufacture . . . in 1765, to the Society of Arts Manufactures, and Commerce instituted in New York . . . which were highly approved and, when offered for sale were rapidly bought up." These reports, which all probably refer to John Rugar's activities, seem to document a larger-scale, more specialized attempt to produce paper hangings than John Hickey's sideline of stamping papers, which had only been an adjunct to his more important activity of decorating textiles. In 1769 Plunket Fleeson, a Philadelphia upholsterer who had been in business since 1739, first announced that he had for sale "American Paper Hangings . . . manufactured in Philadelphia . . . not inferior to those generally imported."[1]

By 1788 there were enough paper stainers in New York to muster a contingent to march in the Federal Procession of July 23 in honor of the Constitution of the United States. Their showing included:

> A flag displayed, representing a piece of paper of a verditer blue ground, printed with a figure of Gen. Washington, with the words "New York Manu-

6-1

Figure 6-1 Some of the early products made by the Bumstead family of Boston, wallpaper manufacturers from the 1790s through 1898, are pictured on this bill head. In a diary of 1842, the founder's son recounted the beginnings of the unusually long-lived enterprise. He stated that Josiah Bumstead recalled taking a store on Marlboro street in 1791, when he was twenty-one years old, and "had only one or two hundred dollars worth of goods to commence with." Although upholstery goods were featured in advertisements of 1793, in 1796 the emphasis of the business was changed: Josiah announced the establishment of his paper-staining manufactory at West Boston. Courtesy of the Manuscript Division, Baker Business Library, Harvard University

> facture" in blue letters, on a gold ground, borne by Mr. John Colles, attended by an apprentice in a coat and cap of paper laced with bordering, and others, carrying decorated tools. . . . On the borders of the flag "Under this Constitution we hope to flourish."[2]

The paper stainers of Boston were also numerous enough to participate as a group in a parade of the 1780s that greeted George Washington when he came to Boston in October of 1789. Led by Joseph Hovey, they bore a flag with "three panthers heads and a pelican" and with the motto "May the fair daughters of Columbia deck themselves and their walls with our own manufactures."[3] By 1790 Boston is reported to have produced 27,000 rolls of paper hangings, "which sufficed not only to supply the state, but to furnish considerable quantities for other states."[4]

"Factories of paper-hangings are carried on with great spirit in Boston, New Jersey, and Philadelphia," recorded Lord Sheffield in his account of commerce of the United States, published in 1791.[5] Although overlappings of partnerships among individual paper stainers complicate the task of counting the numbers of manufactories in each city, directory listings and advertisements from New York, Philadelphia, and Boston document the existence of at least thirty paper stainers' manufactories before 1800. Each city supported approximately ten paper-staining establishments, which enjoyed varying degrees of success and longevity. During the eighteenth century, there were also two paper stainers in Albany, one manufactory in Hartford, probably one in Baltimore (the local advertisements fail to specify whether the products of a Baltimore or of a Philadelphia manufacturer are being offered), and one in Springfield, East (New) Jersey (See Appendix B.)

The inclusion of paper hangings in tariff legislation protecting American manufacturers attests to the fact that the domestic craft was early established with enough strength to gain political recognition. Legislators apparently considered wallpaper among domestic manufactures to be encouraged by taxing the foreign competition. In 1786 the legislators of Massachusetts prohibited the importation of paper hangings into that young state.[6] A few years later, following the strengthening of the federal government, the first American tariff law of 1789 placed a duty of seven and a half percent on imported paper hangings. This rate of duty was a percentage point below the average rate of all duties imposed by that law, a fact suggesting that while protection was deemed a good thing, it was not felt that strong protection of paper hangings was essential or that fine foreign goods could be kept off the American market.

Subsequent tariff legislation probably contributed to the steady growth of American paper-staining manufactories. The duty on paper hangings was raised to ten percent in 1790. While the rates on other types of paper remained at that level, a fifteen percent duty was imposed on imported paper hangings in 1792. An additional "temporary" two and a half percent duty, which proved to be permanent, was imposed on all paper in 1804.[7]

Not only did European wallpapers come to this country and set styles; so did European paper stainers. Several eighteenth-century American paper stainers advertised that they had trained in Europe. Training in Dublin was boasted not only by that earliest of the American craftsmen in New York, John Hickey, but also by one Edward Ryves, a Philadelphia paper stainer of the 1780s. Another Philadelphian, Joseph Dickinson, advertised during the 1780s that he was a manufacturer from London with experience in Paris. When Messrs. Colay (Le Collay), Chardon, and Orinard arrived in Baltimore in 1789, the *Daily Advertiser* of New York City noted: "They have been regularly bred and employed in that business [the printing of cotton and linen cloth and paper hangings] in the city of Nantz, in France, where the art of callico and paper hanging printing is improved beyond any part of Europe." Le Collay and Chardon regularly mentioned their French training in advertisements that ran in the Philadelphia papers of the late eighteenth century.[8]

Notices that called for employees to work in paper-staining establishments sometimes specified the need for a person who could speak French as well as English, as in one of 1789 from a Philadelphia paper. In other cases, the advertisements were printed in both French and English, as was a New York example of 1794. Cornelius and John Crygier described their business to New Yorkers in an advertisement of 1799, proudly noting "they have some of the best workmen from Europe." Anthony Chardon listed his paper-staining business in Philadelphia directories of the 1790s as "French paper mfr."[9] European craftsmen not only founded and were employed by the American manufactories, but also were prized by them.

However, a system of apprenticeship under which boys eleven to sixteen learned the craft as assistants to the journeymen paper stainers quickly produced native-born craftsmen. "Tere boys" or "tearers" were the English names for these assistants, who spread color on the color block between each imprint of the printing block, helped spread flocking, hung up the paper to dry, mixed colors, and generally helped around the manufactory. These names were apparently English corruptions of the French word "*tireur*"—drawer, puller.[10] Advertisements for these apprentices punctuated eighteenth-century paper stainers' notices. As many as three or four at a time were needed by one manufacturer who required "active, sprightly, and honest boys." By the 1790s, Boston-trained paper stainers like Zecheriah Mills and Thomas Webb were setting up manufactories in Hartford and Albany. In 1807, a New Bedford manufacturer advertised that he had placed his factory under the direction of "an experienced workman from Boston."[11]

In their advertisements for skilled workmen, manufactories sought craftsmen with two distinct groups of skills. First, they needed "engravers in wood," "print-cutters" with an ability to make wood blocks—to draw as well as carve the designs on the "prints." Moses Grant of Boston included on his billhead of 1811 the boast that he kept "constantly employed an artist at cutting new patterns."

Manufactories also needed a second group of workmen: "stampers," "printers" who could use those carved pieces of wood to actually impress the patterns on paper.[12] Although printing skills bring to mind associations with the production of books and engravings, in fact the skills of these early American paper stainers—carving the blocks and printing with those blocks—were more closely allied to the skills of the linen stampers, who block-printed designs on textile fabrics. In fact, the paper stainers probably would have been ill-trained to use the letter presses and copper plates of the book and newspaper printing offices.

There is scant information about the sizes of the earliest American paper-staining establishments. Occasionally a workshop would publicize the number of workmen employed; for instance, in 1792 Burrill and Edward Carnes advertised that they constantly employed thirty workmen in their Philadelphia establishment. A year later, they gave that figure as forty employees, a figure repeated the following year by Anthony Chardon, who had taken over the business.[13]

Some of the first American manufacturers started their factories to supply businesses that began as import and retail shops. William Poyntell of Philadelphia, proprietor of a book and stationery store during the 1780s, seems to have become a wallpaper manufacturer early in the 1790s. By this time a substantial businessman, he doubtless acted as entrepreneur, leaving to others the craftsman's job of actually staining paper. William Mooney, a New York upholsterer who also sold paper hangings from the 1780s into the nineteenth century, had an interest in their manufacture, in association with Mackay and Dixey of Springfield, East (New) Jersey. In 1790, he advertised:

> Calico Printers and Paper Stainers: If any there are who are perfect masters of the above branches, by applying to William Mooney, Upholsterer, . . . they will hear of immediate employment.

6-2

6-3

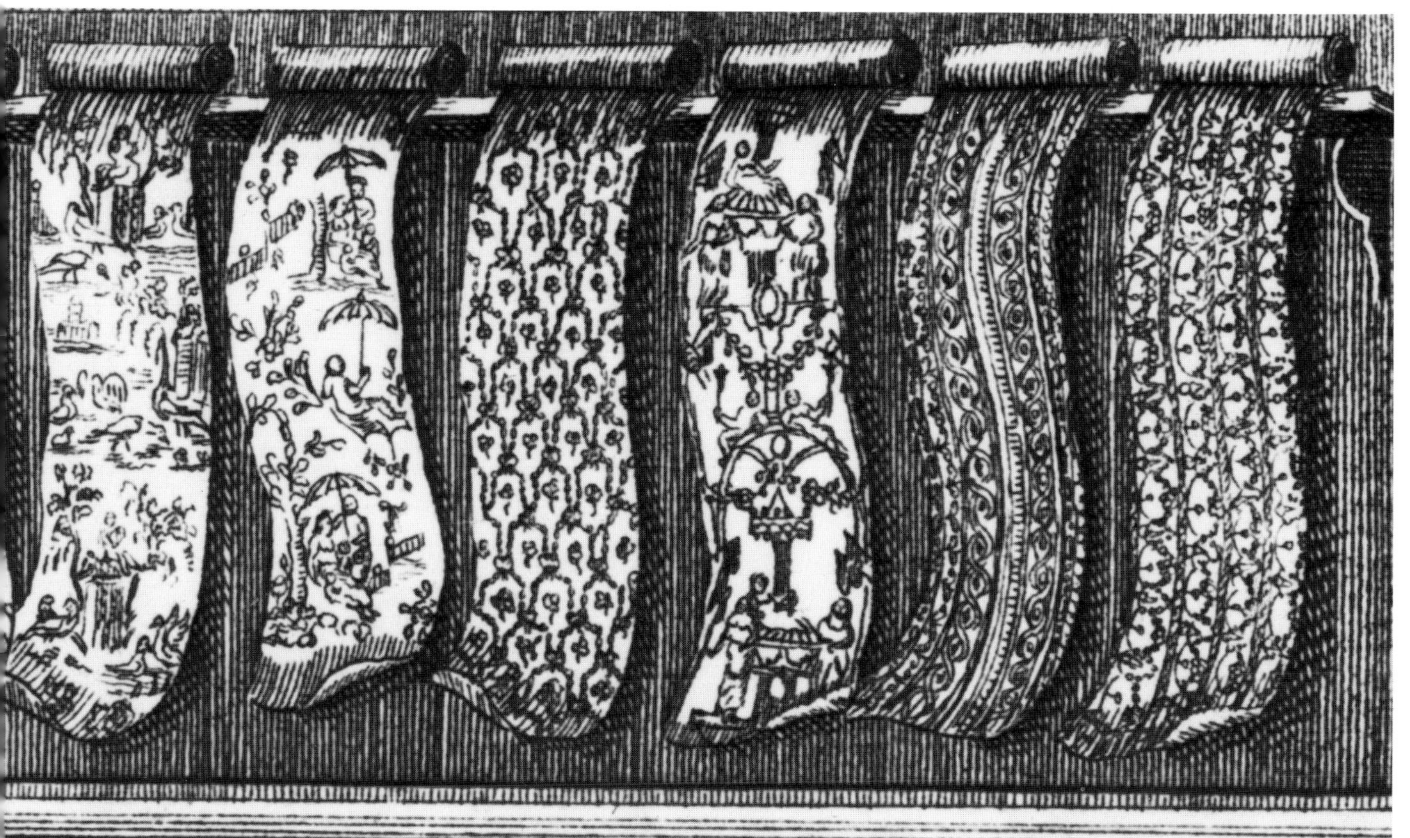

Figure 6-2 This is a detail from the billhead of Appleton Prentiss, Boston paper stainer. The transaction recorded on the bill is dated 1794. So carefully and accurately has the engraver detailed each of thirteen patterns by Prentiss that Abbott Lowell Cummings and Richard Nylander of the S.P.N.E.A. have been able to use them as the basis for attributing to his factory a number of wallpapers found in houses throughout New England. The variety shown ranges from a simple diamond diaper on the far left, through a stripe (second from the right) and through a pattern based on the English pillar and arch designs (third from right), as well as four swag borders, printed side by side (far right), to an arabesque pattern derived from papers by Réveillon (sixth from left). Perhaps the pattern shown as the fifth from the left is the "Bird in Ring" listed on an earlier inventory of Prentiss's factory (page 118). "Two Birds," on that same inventory, may be the one shown seventh from the left, and "China fig" the one fifth from the right. Collection of the Society for the Preservation of New England Antiquities

Figure 6-3 Fragments of a pattern block-printed in black and white on gray (left) preserve a variety of motifs in a loosely structured configuration. The paper was found, with the 1½-inch border shown here, in a house in Deerfield, New Hampshire. Based on English prototypes, it was perhaps made in Boston during the 1780s or 1790s by Appleton Prentiss, who included a pattern very much like it on his billhead (*figure 6-2,* sixth from the right). Cooper-Hewitt Museum, 1968-85-1, -2; gift of I. M. Wiese

6-4

Figure 6-4 This mark, stamped on the reverse of the pattern of *figure 6-5,* dates it between 1794 and 1813. Advertisements make it possible to bracket the paper-staining career of Zecheriah Mills in Hartford between 1793 and 1816. The omission of "& Co." in this mark may date it about 1801, when Mills announced the dissolution of a partnership and dropped "& Co." from his ads. The mark in *figure 3-18* is probably earlier. Courtesy of the Connecticut Historical Society

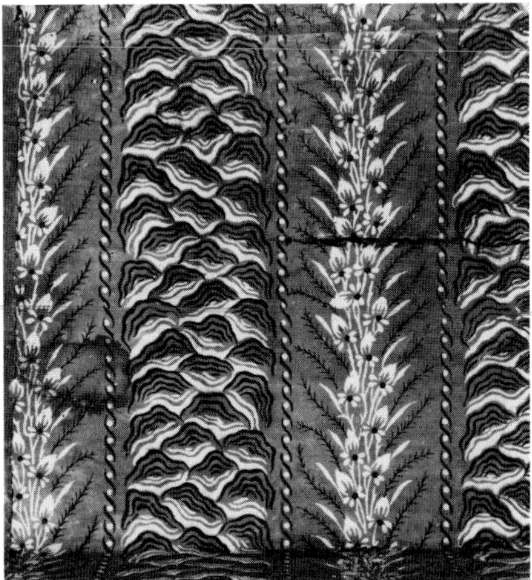

6-6

Figure 6-6 Another pattern by Zecheriah Mills of Hartford, active 1794–1813, this stripe is printed in black and white on gray. It duplicates an example at the Society for the Preservation of New England Antiquities with a blue ground. Still a third example, illustrated in *Antiques* Magazine, September 1952 (page 216), bears a mark like that of *figure 6-4,* with the numbers "264" substituted for "278." The example shown here lines a trunk owned by the author, which was found in New Haven, Connecticut.

6-5

Figure 6-5 (left) Made and marked about 1800 by Zecheriah Mills in Hartford, this paper was found in Lebanon, New Hampshire. A duplicate pattern is at a Deerfield, Massachusetts, house at Historic Deerfield, Inc. This sample is block-printed in black, white, and glossy varnish green on a ground now cream-colored in exposed areas; but where it had been protected by the border, it remains a gray-blue. The sample is 27½ inches wide; the border—orange, pink, white, and light blue—is 2⅛ inches wide. Courtesy of the Connecticut Historical Society

By the time this advertisement appeared, William Mooney was more to New York than a mere upholsterer; he was a founder of what was to become an important power in New York politics. In 1786 the St. Tammany's Society, then a patriotic and "revelrous" group, subsequently an infamous political machine, had chosen as its first Grand Sachem Mr. William Mooney.[14]

Prior to the Revolution, booksellers, stationers, and upholsterers sold most of the products of the earliest American paper stainers alongside imported wallpapers. Cabinet makers, house painters, and merchants or importers of almost any description also dealt in paper hangings. Wallpapers were occasionally included in the auction listings of imported and domestic goods that so frequently appeared in eighteenth-century newspapers.

Movement toward specialization among the growing numbers of shops of every variety in American cities produced "Paper Hanging Warehouses" by the 1780s. In 1789 Joseph Hovey had in Boston "a large room at his Manufactory, hung round with 60 of his best patterns, for the inspection of those Ladies and Gentlemen who will please to honour him with their custom." A pattern book, a familiar enough feature of modern wallpaper retailing, was ready for inspection at the shop of William Poyntell of Philadelphia in October of 1792.[15]

Imported as well as domestically produced paper hangings were widely distributed to every settled part of the country during the eighteenth century. A Philadelphia manufacturer appealed to wholesalers in 1786 with the promise that "merchants trading to any part of America may depend upon being supplied to advantage." American manufacturers sold their products to wholesale customers not only in the north, but also to "the Southern Market" as "4000 elegant Paper Hangings with rich borders to suit" were specifically earmarked in a 1799 New York advertisement. In the 1790s, advertisements for Boston paper hangings appeared not only in Providence and New York newspapers, but also in newspapers farther afield, in Baltimore and Savannah. During the same period, newspapers in Baltimore, Charleston, and New York carried advertisements for Philadelphia-made paper hangings. In Hartford during the first decade of the nineteenth century, Zecheriah Mills advertised the newest Philadelphia, New York, and Boston patterns, and also listed out-of-town agents who sold wallpapers of Mills's own manufacture in New Haven, Litchfield, Middletown, Norwich, New London, Windham, Springfield, and Northampton.[16]

From the earliest days of manufacture here, Americans probably exported paper hangings in limited quantities to a small foreign market. In 1765, a British report of a New York paper-staining manufactory, which we have already mentioned, had described paper hangings as "an article in great demand from the Spanish West Indies." Joseph Hovey, a Philadelphia manufacturer advertised in 1788 that his paper hangings "produce a good profit by exportation to the Southern States and the West Indies."[17]

STYLES AND MOTIFS

American paper stainers of the eighteenth century aspired to perfect the skills of the thrifty imitator. By underpricing imports, these craftsmen hoped to gain a foothold in the domestic wallpaper trade. Their early appeals to American customers offered cheaper goods that looked like and were made as well as English and French papers. A 1784 advertisement of Joseph Dickinson of Philadelphia summarized the almost universal claim of American paper stainers as they expressed it in the many notices they placed in newspapers: He bragged that he sold his own products "at lower prices than can be imported to America from any other country whatever." His advertisements published three years later continued in the same vein, declaring that he was "determined to undersell all imported paper twenty percent," which he explained was possible "on account of [not paying] duties, freight, &c." A Bostonian, Josiah Bumstead, was offering the products of his manufactory twelve years later "at half the price that they can be imported for."[18]

American paper stainers did not boast of inventing radically new styles, nor do their designs bear the earmarks of patriotism except in extremely rare cases. The eagles, George Washingtons, and other patriotic motifs that today intrigue lovers of Americana were unusual indeed, and for the most part these appeared on wallpapers of the nineteenth, rather than the eighteenth, century. Nor did American paper stainers claim the mastery of craft techniques surpassing or varying the quality of those used to make the imported papers. Their selling points were far more practical. A New York advertiser of 1787 argued persuasively:

> One very great advantage will attend the purchasing of Paper Hanging manufactured here, is, that it can always be matched again; many persons have been obliged to new paper their Rooms for the want of a few yards of Paper Hanging, that has been imported, being damaged on their walls; and another very great advantage is, they can be sold much cheaper than the imported Paper Hangings, and warranted to be equally as good.[19]

"Cheap" rather than "fine" is the word emphasized in almost all the American paper stainers' notices. They might have claimed to equal the quality of paper, color, and printing of the imports, but in the light of available evidence those claims may seem somewhat exaggerated. Among papers bearing the marks of American craftsmen that are available for study, the printing is often cruder, the colors thinner, the elements of a pattern more often printed out of register than in the imported patterns that served as models. In making such a statement, I have doubtless disturbed the shade of the long-buried Joseph Dickinson of Philadelphia, who advertised in 1786:

> . . . notwithstanding some fallacious reports have been propagated by foes to this country, that Paper cannot be made equal to European, I am determined to prove the contrary and willing to shew colour for colour, paper for paper, cheaper than can be imported from any part of Europe.[20]

Those imports exemplified a wide variety of styles, and American craftsmen printed their own versions of nearly all of them. Given the limited numbers of skilled artisans in America, the relatively small sizes of their workshops (when contrasted with the veritable factories of eighteenth-century France like that of Réveillon in Paris, which employed more than three hundred workmen), and the simplicity of their tools, American paper stainers printed an impressive variety of patterns. The range possible within an American workshop of the eighteenth century is suggested by an advertisement of Burrill and Edward Carnes of Philadelphia. In the spring of 1793, they announced that they had ready for sale at their manufactory "A most beautiful Assortment of 15,000 Pieces Paper, in 600 different patterns, from two to twenty-six colors."[21] One reason such variety was possible is that it was a relatively simple matter to combine and recombine printing blocks and colors.

For the most part, the generalization holds that American papers of the eighteenth century looked like the imports, though perhaps a bit cruder. Such a generalization is forced on us because the wording in most advertisements for wallpapers is vague, because there exist few other verbal descriptions of wallpapers, and because there are very few surviving samples of American-made wallpapers that can be precisely fixed to dates within the eighteenth century or to a given maker. However, there are a few exceptional documents that provide enticingly specific information about papers made here during the period. In rare instances, an advertiser would seek to attract customers by characterizing in relatively concrete terms just what he was making. One example of this is provided by William May's advertisement of 1791 offering his fellow Bostonians "An infinite number of variegated Papers on the subjects of War, Peace, Musick, Love, Rural Scenes &c."[22] Although this advertisement provides an unusually specific description of an American paper stainer's wares, if one could not refer to the French neoclassical patterns of the same period *(color plates 9 through 14)* it would be impossible to guess the forms such papers might have taken.

A document providing an unusually descriptive listing of the productions of one eighteenth-century Boston paper stainer, Appleton Prentiss, has been preserved at the Baker Business Library of Harvard University. A four-page invoice of "Sundry Paper-Hangings, Prints, and Utensils remaining on hand at the manufactory, No 43 Marlborough Street" suggests the kinds of patterns being made during the closing decades of the eighteenth century. Prentiss's billhead, *figure 6-2*, illustrates some.

The compilers of this list used brief phrases or pattern names to denote each lot of paper. These names were repeated on the list in irregular order, apparently as the papers were found. Some appeared only once or twice, while others, such as "rools [rolls] Peacocks paper," appeared as many as fourteen times. It is probable that duplications of entries for the same pattern implied different colorings, and in some cases colors were noted beside the pattern names. The appearance of "blue" beside a pattern listed elsewhere without color notation was frequently accompanied by an increased evaluation for the pattern.

More than twenty-five distinct names for individual designs—motifs or figures, abbreviated on the list as "fig"—were included. "Diamond fig" must have denoted a geometric pattern. "Curtain fig" probably referred to an imitation of drapery, "Velvet fig" to a flocked pattern, probably simulating designs familiar in textiles, like the "caffoy" patterns discussed on page 53 and illustrated in *color plate 6.* "China fig" was the name assigned to what was probably a pattern with Chinoiserie motifs.

Six bird patterns were included, designated "Bird in Ring," "Two Birds," "Four Birds," "Parroquicts," "Peacocks," and "Swans." Among other living things depicted on Prentiss's wallpapers were "Dalphin" and "monky." Many of these papers may have been loosely assembled configurations of motifs like the pattern illustrated in *figure 6-3* and perhaps attributable to Prentiss. "Tree fig" and "Vines" patterns found on the list possibly were constructed in a similar way, as might have been the "Basket," "Bird Cage," "Nott," "Star and Ribon," and "Farm House" designs. "Bunch" and "Purple bunch" perhaps stood for patterns depicting grapes.

"Flower Pots," "Pots," and "Small Flower Pots" probably were printed as fire board decorations, like that illustrated in *figure 8-1.* "Canopy," a word used elsewhere to describe border patterns, may have been used to designate borders in this listing also. Other borders included "Festoon" borders—spelled alternatively "Festune"—and "Festune blue borders." These may have duplicated the pattern illustrated in *figure 7-4.* "Narrow borders" also were listed. The entry "54 rools English paper" simply may have stood for the imported papers Prentiss had for sale as distinguished from papers of his own production. "Green" perhaps referred to a plain-colored paper. "Arches" and "Piller Paper" on the list were probably similar to *figures 3-16 through 3-22.*[23]

Although such unusual documents give a great deal more information about the kinds of wallpapers early American craftsmen were printing than do the more standard, formulaic advertisements for "Paper Hangings in the Latest Fashion," we must turn to illustrations of contemporary English and French papers to translate those rare descriptions into visual images. In addition, reference to European precedent is particularly necessary for grasping any real sense of the configurations of myriad motifs in the broad range of styles that inspired American imitation. Some surviving fragments of wallpaper bearing the marks of American makers do document what some of the papers looked like, but advertisements and manuscripts, including the Prentiss invoice, suggest that the range of patterns was broader than the limited number of American survivals can document.

Just as it was necessary to refer to European wall hangings of the seventeenth century to find examples of what was used in America at that time, so for the eighteenth century one must continue to look at the more plentiful and better-documented French and English wallpaper collections to get a sense of the wide range of styles made in eighteenth-century American wallpaper manufactories. A New York agent for papers from a Boston factory proudly noted in 1790: "The figures are chosen from the newest European patterns."[24] Almost undoubtedly, the patterns he referred to took the form of actual imported wallpapers rather than any patterns printed in books. Lacking published designs specifically adapted to wallpapers, American paper stainers, who also advertised imported papers, probably used the very papers they were also selling as sources for their own products. Tangible evidence of this practice is provided by the papers illustrated in *figures 3-17 and 4-6.*

Examples included in preceding chapters, which survey the most distinctive, high-style French and English patterns, can only begin to suggest the range of more fashionable papers available for copying by American paper stainers. The next chapter will present some of the kinds of wallpapers that were internationally produced and internationally popular, as well as some of the more standard papers that Americans copied. Most of the patterns printed by American manufactories were probably like the more common examples printed in a limited number of colors. They are described in the final section of the next chapter.

7 THE PAPER HANGER'S STOCK IN TRADE

The shelves of wallpaper warerooms, whether in London, Paris, or New York, were stocked with standard items made by English, French, and American paper stainers alike. Some decorative elements made of paper and some simple patterns were in such general currency in Europe and America that it is difficult to distinguish their national origin. Trying to do so seems less important than recognizing the elements as parts of schemes typical of papered rooms in all parts of the West, and the simplest patterns as internationally popular styles of patterning during the late eighteenth century.

"ARCHITECTURE PAPERS"

Paper decorations imitating architectural details were among the wallpapers most precisely described in American advertisements. The phrases "Architecture Papers for entry-ways,"[1] "Large Arch Papers for Halls,"[2] or simply "Arches,"[3] all found in wallpaper advertisements of the 1790s, probably refer to patterns of the kind illustrated in *figures 3-16 through 3-22.* These papers depicted large architectural objects such as columns and arches, reduced to small scale.

In contrast, imitation at full scale was exploited to fool the eye with cheap substitutes for elements of interior architectural finish, such as moldings, plaster work, and various other ornaments. Among this second, and much larger group, of "architecture papers," "Pannel papers" were particularly popular *(figure 7-1).* William May in a 1789 Boston advertisement described his "Pannel Papers" with the phrase "to imitate Wainscotting."[4] By the 1780s, lining the walls of a room with wooden panels defined by decorative moldings—that is, wainscoting the walls—was largely confined to the dado level, the lower area of the wall, between the chair rail and the baseboard. In 1792, Appleton Prentiss of Boston advertised "handsome wainscot or dado patterns."[5] Two years earlier, Joseph Hovey, also of Boston, had offered "elegant Pannel Papers suitable for wainscoting and stair cases."[6] Pannel papers for staircases may have been printed to rise along the diagonal, like the patterns of a much later period illustrated in *figures 16-18 and 16-26.* In Philadelphia, William Poyntell mentioned "Stucco and other colored PANNELL PAPERS" in an advertisement of 1793.[7] During 1796 and 1797 Zecheriah Mills of Hartford repeatedly referred in print to the "Panel Figures for Wainscoats" he was making.[8] Perhaps published engravings like those illustrated in *figures 3-11 and 3-13* inspired the ornamental figures placed in the centers of his mock panels.

William May of Boston spiced his newspaper notices with even more elaborate descriptions of the architectural ornaments he was making. His offerings for 1791 included "A rich variety of papers, ornamented with every different order of architecture . . . Tutanick, Tuscan, Dorick, Gothick, and Composite orders."[9] Bits of architectural fakery were advertised as imported wallpapers and were copied by American manufacturers. Householders of the 1790s could have pasted on their walls elaborate architectural details, which included "Cornices and pillars, with statues for halls and entries," advertised by William Mooney of New York,[10] "Pilasters" offered by Thomas and Caldeleugh of Baltimore,[11] and "fancy pieces . . . in imitation of cornice work,"[12] advertised by Moses Grant and Co. of Boston. (For illustrations of similar papers, printed in the nineteenth century, see *color plates 39 through 42.)*

The most common wallpapers depicting architectural ornaments were imitations of carved and plaster-work cornices and of simpler moldings designed to run along the tops of walls marking the juncture with the ceiling. As wallpaper borders, they perhaps carried the "look" of high-style architectural detailing to the broadest group of consumers. "Stucco" wallpaper borders, imitating ornamental plaster work, had been advertised during the 1760s and have been discussed above (pages 55–6). Advertisers of the 1780s and 1790s described their architectural moldings straightforwardly as "moulding" or "carved work," "oakleaf," "dental" (dentil—small square blocks used in series), and "dental work" borders.[13] Command of a relatively sophisticated architectural vocabulary for a seller of wallpapers is suggested by the inclusion of the phrase "Vetruvian Scrole" on an invoice of papers sent during 1799 from England to Prestwould in Virginia *(figure 7-10, item 9).*[14] The building trades regularly alluded to the Roman architect Vitruvius Pollio, active in the first century B.C., to describe a classical wavelike ornament applied to narrow borders and bandings.

7-1

7-2

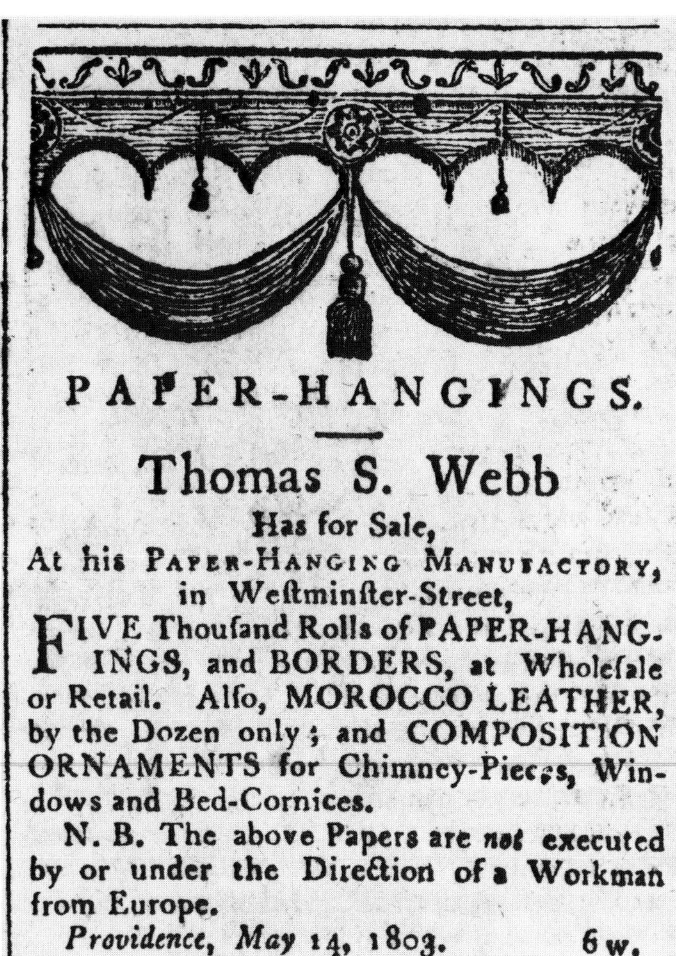

PAPER-HANGINGS.

Thomas S. Webb

Has for Sale,

At his PAPER-HANGING MANUFACTORY,
in Weſtminſter-Street,

FIVE Thouſand Rolls of PAPER-HANGINGS, and BORDERS, at Wholeſale or Retail. Alſo, MOROCCO LEATHER, by the Dozen only; and COMPOSITION ORNAMENTS for Chimney-Pieces, Windows and Bed-Cornices.

N. B. The above Papers are *not* executed by or under the Direction of a Workman from Europe.

Providence, May 14, 1803. 6w.

7-3

Figure 7-1 An American advertiser of the late eighteenth century might have described the block-printed paper hanging on the previous page as a "handsome wainscot or dado pattern." It is a late-eighteenth-century French example. The "moldings" on it are printed in shades of gray-brown on a robin's egg blue ground. The panel would have stood 26 inches high (the paper's width) on a wall. Cooper-Hewitt Museum, 1972-42-177; gift of Josephine Howell

Figure 7-2 In the 1762 edition of *The Gentleman and Cabinet-Maker's Director,* Thomas Chippendale included the page shown above, left, with another page of engraved designs for border ornaments appropriate for paper hangings. Imitative of architectural moldings in a rococo style, they probably inspired some of the "stucco work" wallpaper borders used in America during the mid-eighteenth century. Courtesy of the Henry Francis du Pont Winterthur Museum

Figure 7-3 A paper stainer of Providence, Rhode Island, chose the motif of a wallpaper border for his advertisement in the *Providence Gazette* on June 11, 1803 (above, right). Patterns depicting garlands of fruit or foliage suspended at each end, as well as patterns showing draped folds of cloth looped in the same manner, were most often called "festoon borders" in the wallpaper trade, although "swag borders" was also used. Courtesy of the New-York Historical Society, New York City

7-4

Figure 7-4 Four lengths of a festoon or swag border survive uncut, side by side, as the paper stainer block-printed them late in the eighteenth century. That paper stainer may have been Appleton Prentiss of Boston, who illustrated a length of paper with four borders very like this on the far right of his billhead *(figure 6-2)*. After a paper hanger had separated such strips, he most probably would have placed them at cornice level, like the festoon border illustrated in *color plate 22*. Printed in orange, white, green, and pink on a gray ground, each border is 5¼ inches wide. Cooper-Hewitt Museum, 1955-24-1; gift of Virginia Hamill

7-5

7-6

Figure 7-5 A simple interpretation of a floral stripe imitating a brocade, this fragment, about 14 inches wide, was found behind paneling in the principal room—a second-floor sitting room—of the Lining house in Charleston, South Carolina. The striping is made of black horizontal lines, and the red flowers with green leaves are block-printed on a gray ground. It is probable that the paper is French or English and was hung about 1786. Cooper-Hewitt Museum, 1967-76-1; gift of the Preservation Society of Charleston, South Carolina

Figure 7-6 Simple small-scale patterns like these—each oval on the right is only about 1 inch high—are difficult to identify precisely. An English tax stamp, a fragment of which appears on the reversed piece, bottom right, does mark them as English. But one can only approximate a late-eighteenth-century date for their hanging in a small house in Woodbury, Connecticut. The border shown at the top on the right was used with each of the illustrated patterns: on the ground floor, with the little white meandering flowering vine over black dots on a gray ground; in a second-floor chamber with the same pattern overprinted with a blue ground and with a white diamond grid and orange floral medallions. Cooper-Hewitt Museum, 1968-66-2, -3; gift of the Old Woodbury Historical Society

"PLAIN PAPERS"

Wallpaper simply decorated with a single solid shade of ground coloring brushed on the raw hanging paper stock enjoyed great popularity during the late eighteenth century and well into the nineteenth. An invoice of goods sent from London to George Washington in 1763 includes the earliest documentation yet found for the American advent of plain papers. The entry describes "A handsome plain paper for a room 18 by 16 feet and 6 feet pitch above chairboard."[15]

Documents preserved in the papers of other illustrious Americans and of official residences include additional early references to the use of plain papers. In 1765, Benjamin Franklin wrote from London to his wife about decorating their parlor in Philadelphia: "Paint the wainscot a dead white, paper the walls blue, and tack a gilt border round the cornice."[16] The blue paper for those Philadelphia walls was most probably of a solid color, a scheme reflecting current European fashion, which Franklin would have noticed during his stay in London. Plain blue paper hung with a narrow stripe of gilt leather in the ballroom of the Governor's Palace in Williamsburg in 1771 was almost certainly sent from England.[17]

These plain papers, originally imported, were soon being made domestically. By 1780 Daniel Leeson of New York was advertising "plain and printed Paper Hangings" of his own manufacture,[18] while in the same year another New Yorker, William Mooney, offered imported paper hangings in plain colors.[19]

In 1790, Thomas Jefferson ordered plain papers from Paris. He specified that he wanted both "plain sky blue" and "plain pea-green," each with festoons, corners, and edgings. For the sky blue paper, Jefferson wrote a quite detailed description of the trimmings, indicating how he planned to use them:

> 22 rouleaux of plain sky blue paper for papering a room. 4 rouleaux of festoons to place next below the cornice all around the room. [For festoon borders, see *figures 7-3, 7-4.*] 8 pr. of corner papers. These are stamped with the representation of curtains hanging in furbelo, to ornament the corners of the room.
>
> 300 yds of edging paper. One breadth of paper contains perhaps a dozen breadths of edging: therefore 300/12 yards of paper only are wanting.[20] [For examples of narrow edging papers see the narrowest border papers in the Phelps–Hatheway house rooms illustrated in *figures 4-3 through 4-5* and in *color plates 9 through 11.*]

7-7

7-8

Figure 7-7 One reason for the popularity of dark ground colors was their ability to conceal dirt. In 1798, this pattern was printed in pastel shades on a dark ground by the Parisian manufacturer Robert, from whom Thomas Jefferson had earlier purchased papers *(see figure 4-1).* This ordinary little paper can be documented so precisely because it is one of the hundreds of small-scale repeating patterns whose designs were registered with an agency of the French government to protect them from plagiarism. Actual specimens of all registered patterns, bearing dates and the manufacturers' names, were bound as official documents in volumes of the 1790s, which are now preserved at the Bibliothèque Nationale in Paris. Since these volumes include ordinary patterns in addition to high-style examples, they are potentially useful tools for establishing the dates and sources of specific, relatively undistinguished, patterns found in American houses. Courtesy of the Bibliothèque Nationale, Paris

Figure 7-8 In eighteenth-century terms, this is a "sprigg" pattern, among the most popular of wallpaper types. This one is French, of about 1780, but its American and English counterparts were plentiful. Each sprig, printed in pink, green, and white on a pale neutral ground, is about 1½ inches high. The "pin ground"—the background full of dots—is of the type advertised as effective camouflage for fly spots. Cooper-Hewitt Museum, 1931-45-47; gift of Sarah and Eleanor Hewitt

Again in 1784 the subject of plain papers appeared in the writings of George Washington. In planning to decorate the ballroom at Mount Vernon, he wrote to Clement Biddle on January 17, 1784:

> I have seen rooms with gilded borders; made, I believe, of papier Mache fastened on with Brads or Cement round the Doors and Window Casings, Surbase, &ca.; and which gives a plain blew, or green paper a rich and handsome look. Is there any to be had in Philadelphia? And at what price? Is there any plain blew and green Paper to be had also? The price (by the yd. and width)

Through the spring, letters to Biddle included requests for prices of the plain blue and green papers, as well as for plain yellow papers, and for Mache and gilded borders. In June, Washington ordered a gilded border: "I shall be obliged to you for sending me 70 yds. of gilded border for papered Rooms . . . that which is most light and Airy I should prefer."[21] Fragments of wallpaper found in the ballroom at Mount Vernon before restoration make clear that, after all the correspondence, Washington found green to be the color of plain paper most to his liking. The fragments are plain green paper with a narrow border imitating an architectural molding, combining a band of egg-and-leaf ornament with a band of beading. Printed in shades of gray, this architectural border was apparently substituted for the gilded border Washington had first ordered.

Blue and green were the most popular colors in plain papers not only for the founding fathers but also for the more ordinary eighteenth-century populace. They were the colors standardly advertised for plain papers in American newspapers during the 1780s and 1790s. The blues were occasionally described as "sky" or "verditer"—a color derived from copper carbonate pigments. A few other colors were specifically mentioned among plain papers offered for sale by American manufactories and dealers, but they were mentioned only infrequently. Among these, straw or buff were advertised intermittently, but with nothing like the consistency with which green and blue were featured in the dozens of advertisements available for study. In these ads of the 1780s and 1790s plain papers in other colors were offered, but each in only one or two. Those other colors include "Leds," French gray, pink, salmon, and yellow as well as black and white. Rather than printing lists of colors, manufacturers more frequently offered "Plain papers of different colors" or "every kind of fashionable Plain Paper." During the 1790s, William Poyntell offered to his Philadelphia customers "Plain Grounds made to any Colour or Shade," and

Mackay and Dixey, New Jersey manufacturers, offered "Plain paper of any colour that may be desired."[22] Once a customer had decided against the popular favorites, blue or green, the likelihood of his choosing almost any other color for his plain papers seems to have been about equal.

The fashion for plain colored wallpapers had a practical side. While the decorative effect may have been much like painting, paper had an advantage. It hid the cracks in plaster work.

BORDERS

Because they were such an integral part of wallpapering schemes using plain papers, and because they so frequently took the form of imitations of architectural moldings, border papers have already figured heavily in the discussion of international wallpaper styles. Separating them out here will require a little repetition in order to do justice to the importance of their role in eighteenth-century interior decoration.

Borders were of greatest visual importance, of course, when used with plain papers. Jefferson's order of 1790 has already indicated that several different kinds of borders were needed for the proper embellishment of a wall covered with plain paper. The richness of borders was apparently calculated to contrast with plain papers in an effective decorative scheme. "Rich" and "elegant" were the words most often used to describe borders intended for use with solid-colored papers.

Walls painted in plain colors were also embellished with wallpaper borders. In 1792 Henry Remsen wrote from New York to Jefferson:

> It has become the prevailing fashion to paint dining rooms and large halls, either green, light blue, or yellow, which looks extremely well. A border of paper or paint of a different colour or colours is added.[23]

Borders had been used with repeating patterns as well, right through the eighteenth century. Early documentation for their use is provided in Thomas Hancock's order of 1737/8 to London for wallpapers. That order included a description of some papers that had been recently hung in Boston, and Hancock made special note of the bordering: "At the top and bottom was a narrow Border of about 2 inches wide which would have to mine."[24]

During the 1760s, as noted previously, most references to borders specified that they were "Stucco" or "Stoco" borders, or were made of *papier-mâché.* In a Philadelphia advertisement of 1769, Plunket Fleeson described borders of this type in some

22

Color plate 22 The watercolor "A Piano Recital at Count Rumford's, Concord, New Hampshire," was painted about 1800. It probably depicts the daughter of Benjamin Thompson with her friends and family. Thompson, a notorious Tory sympathizer during the Revolution, acquired the title "Count Rumford" when knighted by George III. An expatriate, he probably never saw this room in his American home. In the Concord room the walls were embellished with wide festoon borders at cornice level, and narrow edgings at chair-rail level and around the door. The walls were probably papered rather than painted, in the green that was, aside from blue, the most popular color for plain papers during this era. The borders were almost certainly paper ones. National Gallery of Art, Washington; gift of Edgar William and Bernice Chrysler Garbisch

23

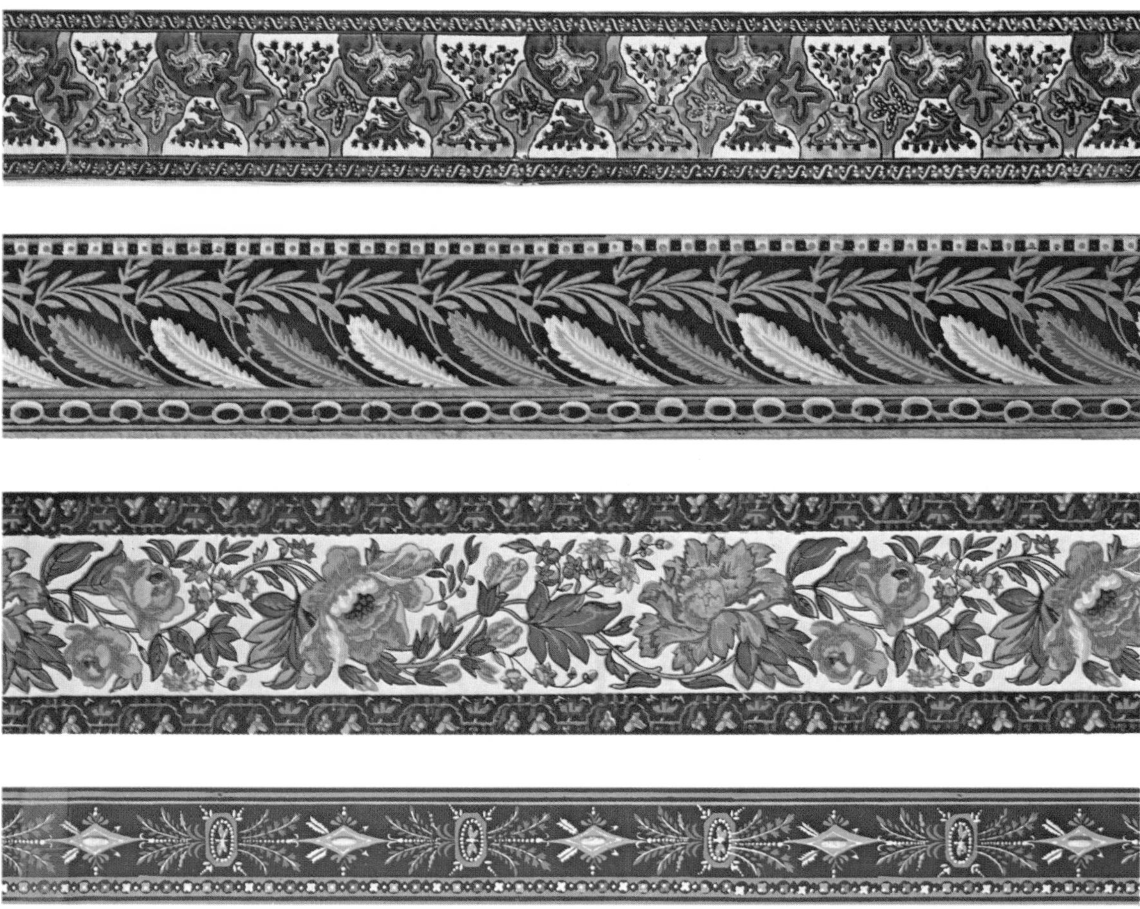

24

Color plate 23 (opposite) Five wallpaper borders of the late eighteenth and early nineteenth centuries—showing part of the wide range of patterns that paper stainers derived from architectural sources. From top to bottom appear a bead and acanthus border, an elaborated bead and reel pattern with foliage incorporated in the oversized beads, an imitation of a molding with bell flowers, and an egg and leaf molding—all based on classical models—as well as an example of eighteenth-century "improvement" and reordering of Gothic motifs. Pointed ovals and quatrefoils are filled with leafy ornament derived from classical sources. The "Gothick" border, bottom, is 7 inches wide, and the narrowest banding, second from top, is 2⅝ inches wide. Cooper-Hewitt Museum, 1972-42-21, -23, -32, -17, -82; gifts of Josephine Howell

Color plate 24 "Rich and elegant" eighteenth- and early-nineteenth-century border patterns shown here are closely related to textile patterns of the period. What was perhaps described as a "patch border" in its day appears at the top, and, below it, a feathery leaf motif, a chintz pattern, and a narrow edging, with beading more closely akin to jewelry than to architectural ornament. The strength of these borders, heightened by dark colors and by black grounds in two examples, is representative of the period's preference for strong elements outlining lighter-colored expanses of wall space. The widths of these borders, from top to bottom, are: 4⅛, 5, 6, and 2⅝ inches. Cooper-Hewitt Museum, 1972-42-72, -7, -71, -2; gifts of Josephine Howell

25

Color plate 25 During the late eighteenth century, floral stripes based on textile designs were very popular for wallpapers. The stripes formed here of block-printed dots and dashes imitate raised woven effects in brocades. This French pattern of about 1780 is about 23 inches wide and has been attributed to the Réveillon factory. Cooper-Hewitt Museum, 1931-45-25; gift of Eleanor and Sarah Hewitt

26

Color plate 26 Geometric patterns were also popular in eighteenth-century America. In this French block-printed sample, the distance between the centers of the stars is about 5⅜ inches. It dates from about the turn of the nineteenth century. Cooper-Hewitt Museum, 1949-144-2; purchased in memory of Mrs. Gustav E. Kissel

27

28

Color plate 27 (opposite) A French wallpaper panel of the late eighteenth century, this was printed from wood blocks in distemper colors. It was intended for use as an overdoor or fire board and is about the size of many easel paintings—27½ inches wide and 30½ inches high. Small scenes like this were forerunners of the much larger-scaled early-nineteenth-century French landscape panoramas that covered whole walls. They also served as a source of motifs that were repeated on patterns like the one illustrated in *figure 12-8.* Fishermen with or without their ladies, but nearly always in proximity to a blasted tree, appear in a number of wallpapers that have survived from the early nineteenth century. Cooper-Hewitt Museum, 1931-45-83; gift of Eleanor and Sarah Hewitt

Color plate 28 A block-printed pattern in the style of Réveillon covers this cardboard box of about 1780–5. Printed pages from a late-eighteenth-century book line the inside. Bits of unfaded wallpaper preserving the bright-blue ground color can be seen where the lid has protected them. The box is 16¼ inches long and 10¼ inches wide. Cooper-Hewitt Museum; gift of Eleanor and Sarah Hewitt

Color plate 29 (overleaf) A small panel of wallpaper made about 1790 in or near Lyon, France, is carefully bordered with imitation *cyma reversa* molding along the two sides, a border of festooned flowers at the top, and a border of butterflies at the bottom. It is 12½ inches wide and 22 inches high; it might have been used to decorate a fire screen or a small fire board, or framed and hung on a wall. Cooper-Hewitt Museum, 1925-1-369; gift of The Council

29

detail as "Paper Mache, or raised paper mouldings for hangings, in imitation of carving, either coloured or gilt."[25] The S and C curves and the scrollwork of rococo ornament adorned many of these mid-century borders with designs derived from sources like that illustrated in *figure 7-2.*

Most often, it was intended that the borders match the colors of papers, plain or figured, with which they were to be used, as well as the patterns. Advertisements typically listed "Paper hanging, with matched borders, of the newest fashions" or, more modestly, a "general assortment of Paper Hangings with borders suited to the same."[26] A Boston paper stainer in 1792 described "paper hangings . . . with festoon, dental, patch and narrow borders suited to each kind . . . borders suited to the fancy of any person."[27] "The fancy of any person" in some cases may have created "matches" between papers and borders that elude modern comprehension. However, for the most part, the intention was to "match" the borders to papers in the same sense we understand the word—or to create pleasing contrasts.

When borders that appear to be mismatched to papers turn up in old houses, there may be explanations besides those of aberrant eighteenth-century tastes. In some cases, mismatched borders may have been used simply because more appropriate ones were not locally available, not necessarily because of any dicta of taste now obscure to us. Another reason borders may now seem mismatched may result from their becoming discolored over the years in ways different from the discoloration of the "fill pattern"—as the trade often designates the principal wall-covering pattern. This may be because different pigments that aged differently were present in colors that looked the same when hung in the eighteenth or nineteenth century, or because different chemicals were present over or under different areas of the papers on the walls. Sometimes a modern restorationist's instinct that the border does not "match" the fill paper is the first clue to discovery that in fact a border now finishing the walls was hung at a later date than the fill pattern. Sometimes the original border survives in place under one added years later, but still so long ago that the newer border also looks and, in fact, is very old.

Wider borders seem to have been preferred for horizontals—as friezes and hung just above chair rails—with narrower borders or "edgings" running vertically, or outlining doors, windows, fireplaces, pilasters, and sometimes the corners of rooms. Border motifs were often unmistakably vertical or horizontal in orientation, and special corner elements were printed to ornament the junctures where the two met. The skillful use of eighteenth-century borders to define wall areas and outline all the interruptions in walls is most clearly illustrated with patterned papers in the Phelps–

Hatheway house, as shown in *color plates 10 and 11* and in *figures 4-3 through 4-5,* and for borders used with plain walls, in the little watercolor showing a piano recital, which is illustrated as *color plate 22.*

Festoon borders *(figures 7-3 and 7-4; color plate 22)* were mentioned even more frequently than the architectural types in late-eighteenth-century references. Most often characterized as "elegant," "rich," or "in the latest taste," festoons were sometimes described more precisely during this period. A Philadelphia advertiser of 1789 offered "Chintz festoon borders," and New York manufacturers a decade later mentioned "feather Festoon" as well as "Plain Festoon" borders. William May of Boston described in 1791 his "festoon borders from 20 inches to 1 inch in width," and William Doyle, another Bostonian, advertised in 1796 "Elegant wide and narrow Festoon frosted Borders, on black and other grounds."[28] Two of Doyle's adjectives, "frosted" and "black" merit elaboration: Borders with black grounds and colors more vivid than the paler tints usually chosen for fill papers were used to outline and thereby to sharply define wall areas and were fashionable during the late eighteenth century and well into the nineteenth. "Frosted" probably refers to the application of particles of mica to give sparkling accents to elements of the patterning. We know about such use of mica from printed references like this advertisement and from Dossie's *The Handmaid to the Arts,* where he described spangles as: "that kind of talc called isinglas [mica], which, being reduced to a gross flaky powder, has a great resemblance to thin silver scales or powder."[29] Like flock, this talc was scattered over areas prepared with varnish, and was used both for grounds and for applied patterns. Very few papers have survived in American houses that actually preserve the glittering materials. One of the few extant examples could in fact be described as a "Festoon frosted border," to use Doyle's phrase from his 1796 advertisement. It is a border found in the Webb house in Wethersfield, Connecticut, in a pattern that duplicates the festoon illustrated in *figure 7-4.* In the Webb house example, which is printed

Figure 7-9 Printed wallpapers imitating marble, like the one on the right, are less familiar to collectors of antiques than are the "marbled papers" produced by floating oil-based colors on water. Numbers of "marbled papers" have survived as book covers and book linings, but examples of the paper stainers' printed imitations of marble are rarer. With its dark ground, good camouflage for scuffs, it is of a type often used on dados and in entries and halls. In the late-eighteenth-century French wallpaper shown here, white, gray, and black veining has been block-printed on a green ground. The sample is 23 inches wide. Cooper-Hewitt Museum, 1972-42-207; gift of Josephine Howell

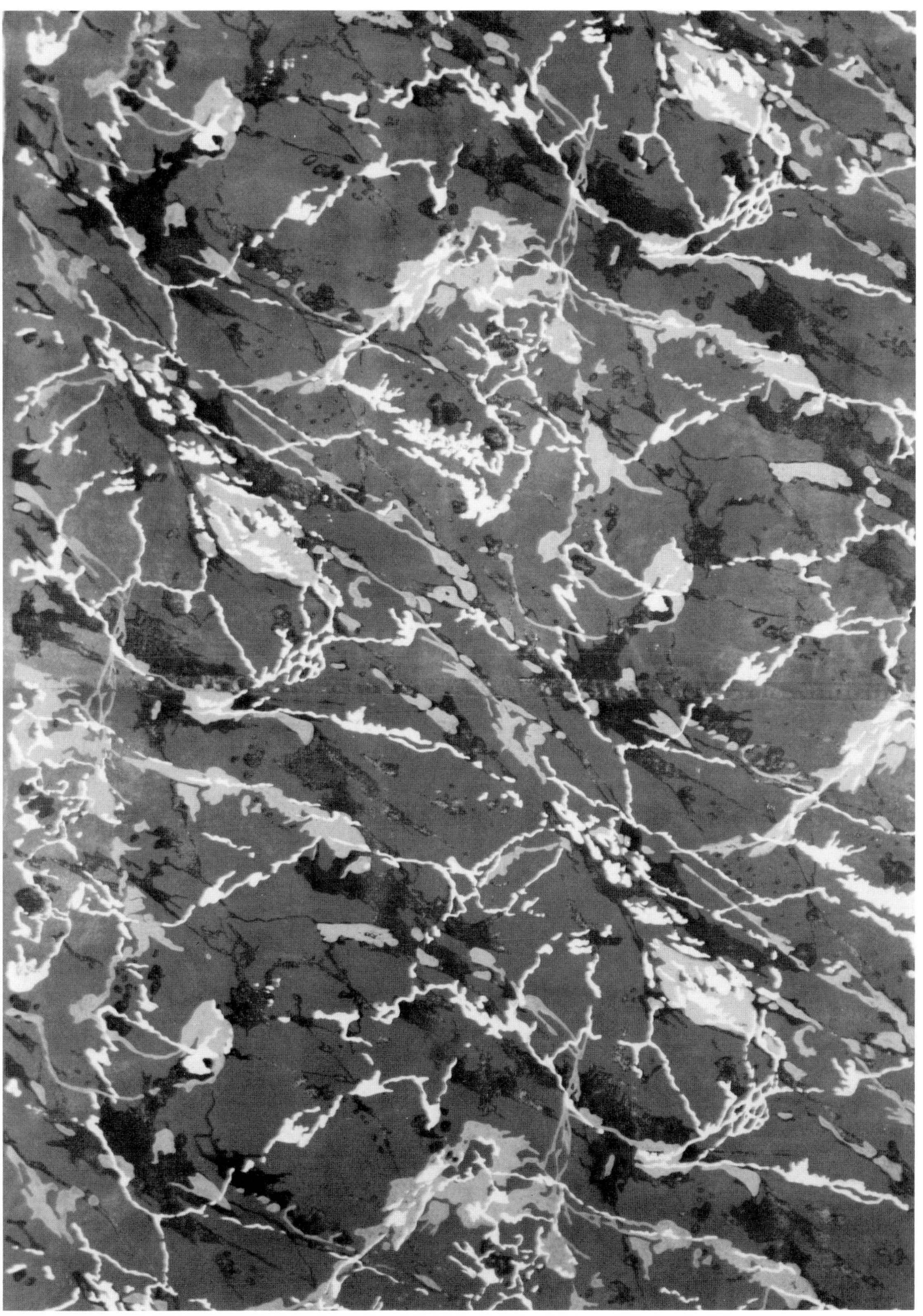

7-9

on a blue ground, mica was added over the patterned areas. Another Boston advertisement of the 1790s, that of Prentiss and May, described what must have been similar borders not only as "frosted" but also as "spangled."[30]

In candlelit rooms, spangles and frosting as well as gilding were valued for the way they picked up the flickering light, increasing the amount of light in a room and making a surface seem almost animated. Several references to gilt borders have already been mentioned in the course of the discussion of plain papers, and many others can be found in sources from the 1760s and 1770s.

Wallpaper sellers of the 1790s frequently listed "patch" borders among their stock. It is unclear exactly which of the several kinds of borders made up of variegated bits of patterning was in the mind's eye of one of those advertisers. Since some surviving geometric border patterns could be described as made up of "patches" of color *(color plate 23),* it is tempting to speculate that the word may have been applied to geometric patterns. What was meant by "race" borders is even less clear.

Floral borders of several kinds were popular; some were called "garland" and "rose." Many of the "chintz" or "chinch" borders advertised and listed in bills of sale were floral-patterned as well, incorporating exotic blossoms derived from East Indian textiles as well as from Western interpretations of such motifs as popularized in publications like those of Pillement (see page 103).

Fruit motifs sometimes appeared on borders in realistic treatments, and were sometimes rendered as if carved in architectural ornament. "Best fruit border," an entry from a Boston invoice of 1794, and "lemon" borders, a listing in a New York advertisement of 1799, are among the many ways they were described in contemporary sources. "Fantail," "canopy," and "feather" are additional words used to describe borders during the 1790s.[31]

Borders had long been valued in the wallpaper trade, as paperers' and painters' manuals attest, because they functioned as practical aids in keeping the fill paper on

Figure 7-10 An invoice of paper hangings shipped from London to Lady Skipwith of Virginia in 1799 includes descriptions of ordinary patterns: a stripe (Item 9), a plaid (Item 7), and a variety of patterns depicting flowers, leaves, and fruits. Such fascinating pattern names as "Worm and pin on sea green" (Item 10) catch the eye with their awkward attempts to characterize in a few words the kinds of motifs familiar enough but requiring endless multiplications of words to describe. The ordered arrangement of the list emphasizes the importance of borders indicating that they were provided quite specifically to be hung with particular papers. A wide variety of the popular types familiar from other sources of the period appear here: Festoon borders, narrow and wide (Item 10), Laurel (Item 1), Leaf (Item 3), Oak Leaf (Item 8), and "Vetruvian Scrole" (Item 9), suggesting imitation of architectural moldings, as well as borders adorned with a variety of other motifs. Courtesy of the Earl G. Swem Library, the College of William and Mary in Virginia

Mess.rs Dawes Stephenson & Co. London Aug.t 3. 1799

Bo.t of James Duppa
No. 34 Old Broad Street

SPS
No 6

No 1	11 ps Wt Sattin Grass purples &c	7/	3.17.–
	16 Doz Laurel Border purple & Yellw	3/	2.8.–
2	11 ps Angle Leaf Greens	5/	2.15.–
	12 Doz Reed & Ribbon Green	2/	1.4.–
3	11 ps Angle & Worm purples &c	5/	2.15.–
	10 Doz Leaf Border	1/6	.15.–
4	13 ps Cloud & Mitre on Grey	4/	2.12.–
	16 Doz Oval Chain on Do	/9	.12.–
5	11 ps Feather, Leads & Greens	5/	2.15.–
	10 Doz Chain border on Green	1/6	.15.–
6	11 ps Ivy Leaf on Drab	4/	2.4.–
	10 Doz Star border on Drab	1/6	.15.–
7	11 ps India plaid on Olymp.n	5/	2.15.–
	16 Doz Diad. Border on Etruscan	1/6	1.4.–
8	11 ps Leaf & Sprig reds &c	5/	2.15.–
	12 Doz Narr.w Oak Leaf border	1/6	.18.–
9	11 ps Bengall Stripe on Buff	5/	2.15.–
	12 Doz Vetruvian Scrole	2/	1.4.–
10	30 ps Worm & Pin on Sea Green	5/6	8.5.–
	18 Doz.n Narr.w Festoon	1/	.18.–
	8 Doz Broad Do	1/6	.12.–
11	11 ps Cloud, purples &c	5/	2.15.–
	12 Doz Baskett border	3/	1.16.–
	Case		.10.–
	Bond, Debenture &c		.16.–
	Cartage		.2.9
			50.12.9
	Cr		
	By Drawback on 1092 Square Yds @ 1¾		7.19.3
		£	42.13.6
	7½ p. Ct. Disct. for Money		3.4.–
		£	39.9.6

7-10

the walls. Pasted along the tops of the strips, and running continuously over those separate pieces of paper, borders helped to hold them together at the seams, right on the ends, where any curling back or peeling was most likely to begin. Borders at the tops and bottoms of the paper lengths covered any irregularities that might have crept in when cutting the lengths of paper. When tacks were used to fix the strips to the wall, borders covered the tack heads. Borders in general contributed to the neatness of finish so much admired all through the eighteenth century.

In the 1780s and 1790s, however, there was a significant increase in the use of wallpaper borders in America. This is documented in wallpapered rooms that survive from the period and in the increased quantity of verbiage in wallpaper advertisements devoted to descriptions of borders. This late-eighteenth-century fashion for borders can be characterized as one manifestation of a more general taste for clearly articulating, marking off, and separating geometric forms in neoclassical design, both in architectural design and the design of furniture. Paper hangers used borders to outline and define areas of wall space in much the same way that skilled cabinet makers used inlaid stringings and bandings of wood in contrasting colors to decorate the surfaces of furniture. Working in the Federal style, America's version of neoclassicism, furniture designers of this period used the wooden equivalents of borders to outline drawers, thereby visually dividing the façade of a chest of drawers into a series of discrete horizontal rectangles, or they ran a line of blond wood around the edges of a dark mahogany tabletop, reinforcing the visual strength of its geometric form. Wallpaper borders were used to horizontally mark off dado from side wall paper, and side wall from frieze; then narrower borders were used to draw vertical lines around those wall spaces so they read as discrete, flat-surfaced geometric shapes making up a wall.

ORDINARY REPEATING PATTERNS AND "COMMON" PAPERS

In addition to wallpapers that bore distinctive marks of their period, including the "architecture papers," plain papers, and borders discussed above, eighteenth-century paper stainers of England, France, and America printed patterns that have a relatively timeless, non-fashion-conscious quality. Often monochromatic or incorporating only a few colors, such standard wallpaper types as stripes, checks, simple geometric configurations, imitations of stone or wood, reproductions of textile weaves, and ordinary little spotted flower patterns were made as cheap papers during the eighteenth century, and they have remained in production to this day. Stylistic quirks in the manner of design, choice of palette or of distinctive motifs, and marks of the ways they were made often give clues to the age of such papers, but the ordinary simplicity of their look kept them in use for long periods, in some cases rendering them nearly timeless.

However, study of an eighteenth-century list of "The most general names the papers are known by" makes clear that only a very few items then deemed appropriate for this category because they were standard features of the wallpaper trade have retained that status. Such a list was published in England in 1770 in a little handbook called *The Compleat Appraiser.* The list seems to have served in America as well during the 1770s: An upholsterer of Williamsburg, Virginia, apparently relied on this very list when he composed a wallpaper advertisement for the *Williamsburg Gazette* in 1771. The list of those "most general names" in the very order they had been given in *The Compleat Appraiser* appeared with the omission of but one word—"check'd" —in William Bucktrout's advertisement of the paper hangings he imported to Williamsburg. Bucktrout's list read: "embossed, Stucco, Chintz, Striped, Mosaick, Damask, and common."[32]

This book has already considered stucco, chintz, and damask patterns among the high-style papers of the eighteenth century surveyed above. Doubtless, "chintz" was a word also used for extremely simple floral patterns that should be remembered in any review of ordinary eighteenth-century papers, as well as to describe the more elaborate designs derived from exotic Eastern textiles. "Embossed" seems to have referred most usually to flocked papers, as well as to papers with raised patterns achieved by stamping under pressure. This leaves but very few words in the lists of the 1770s to describe what was probably, numerically, the bulk of papers in popular use on American walls: "striped," "mosaick," and "common."

Striped wallpapers included not only those with stripes of solid colors, but also floral stripes and stripes made up of small geometric elements *(figure 7-5 and color plate 25).* Eighteenth-century paper stainers copied many of their striped patterns from textiles. Simpler stripes have been standard offerings of the wallpaper trade throughout its history and continue to be. Devices for applying colored stripes to paper—brushes on wheels or rollers, V-shaped troughs with narrow openings to dispense an inch or half-inch of coloring on the paper surface—were among the first mechanized improvements in traditional methods of decorating wallpaper. Their development by the early nineteenth century marked important progress toward machine printing. Early technological innovation made possible relatively low prices for some striped wallpapers.

Interpreting the word "mosaick" proves more difficult than is the case for "striped." Dossie includes it among other patterns having a "small running figure." Not only does "mosaick" appear on *Compleat Appraiser*'s and the Williamsburg upholsterer's lists of the 1770s, but it is also found in earlier American wallpaper references. In the lists of wallpapers on invoices of goods sent from London to George Washington in 1757, two wallpapers were described as "Mosaic."[33] Since no patterns have been found that specifically imitate mosaic floor or wall patterns with stylized copies of bits of stone, and since a great many geometric patterns do survive from eighteenth-century use, it seems possible that "mosaic, -k" may have been the word used in a general sense to describe patterns made up of small elements, such as the small-scale geometrics that have survived in some numbers *(color plate 26).*

Whether or not "mosaic" is the word an eighteenth-century dealer in paper hangings would have used, he would have stocked among his wares numbers of patterns featuring small abstract geometric shapes in repeating configurations *(color plate 26).* The basic elements were simple triangles, diamonds, squares, ovals, circles, and other only slightly more complex forms. Closely related to the eighteenth-century geometrics are patterns in which dots or simple motifs like stars are spaced at regular intervals—"spotted" or "powdered" in the terms of the trade—over a paper. Still other eighteenth-century papers were decorated with more naturalistic but still non-representational shapes—squiggles, wormlike, or spaghettilike forms on a small scale ("vermiculation" in architectural terms).

Perhaps the cheapest wallpapers would have been lumped under the last word on the list of the "most general names." Advertisers did not belabor explanations of what they meant by "common" papers. Customers would have been attracted to descriptions of what was novel, not to descriptions of the designs to which their eyes had become visually inured and numbed through daily encounter.

If we may presume that the word "common" was used in a general sense to mean all cheap papers, we are also suspicious that there was a more specific meaning attached to the word, a meaning now lost to us. During the eighteenth century "common" may have already acquired the special sense, when referring to wallpapers, that we can infer from context in nineteenth-century references: in these sources, common wallpaper seems to be distinguished from finer grades specifically by a lack of ground coloring.[34]

American advertisers in addition to Mr. Bucktrout of Williamsburg did occasionally include relatively detailed bits of information about their ordinary wallpapers in their newspaper notices. In Boston during the mid-1790s there were dealers offering "striped and neat small patterns on Satin [polished, shiny] ground work . . . large ditto," "striped and small neat figures, on different colored mock sattin Grounds," and "many neat and small figured Papers, lately executed, suitable to the present Tastes."[35] "Flowered papers" including "spriggs" were frequently mentioned in such advertisements. *Figures 7-5 through 7-8* illustrate examples of the endless variety of small-scale floral patterns that such advertisers would have had in stock.

Imitations of marble and other kinds of stone and stone work were another category of patterning that an eighteenth-century householder would expect to find in a wallpaper sales room *(figure 7-9).* A New York dealer of the last decade of the eighteenth century called attention to his specimens within this category, describing "Some very elegant gold and marble" papers.[36] But less elegant versions were on most paper sellers' shelves.

While minor changes in details, colors, proportions, and stylistic mannerisms within these ordinary pattern types are apparent to the careful scrutinizer examining nineteenth-century specimens, they look very like their eighteenth-century ancestors. The striped, geometric (mosaic?), and common papers, the sprigs, pseudo textiles, and pseudo stones formed a kind of international vernacular. Although seldom touted in the eighteenth-century advertisements, they acquired the status of standard fixtures in the wallpaper trade during this period. Manufacturers continued to print papers based on these basic formulas, relatively unchanged, throughout the next century. The simplest, comparatively styleless, examples of such patterns are the ones most frequently encountered when restoring old houses, and they remain the ones that most often elude precise cataloguing.[37]

8-1

Figure 8-1 "Flower Pots for Chimney Boards" appeared frequently in late-eighteenth-century American wallpaper advertisements. One of the type is seen here, framed on three sides with wallpaper bordering, printed in reds, greens, blues, and white on a dark-brown ground. Such "chimney prints" or "fireboard pieces" were mounted on wooden boards, then placed inside the openings of fireplaces when they were not in use. This one measures 35⅜ inches by 54⅝ inches. Courtesy of Old Sturbridge Village, Sturbridge, Massachusetts

8-2

Figure 8-2 Paper hangings were often mounted on fire screens, free-standing pieces of furniture that stood in front of the fireplace while a fire was burning. In describing the three shown in this engraved plate, number CXXIV in the 1755 edition of *The Gentleman and Cabinet-Maker's Director,* Thomas Chippendale noted that fire screens A and B each had two leaves: "The fretwork round the paper of each may be cut through; the other [C] is a Screen upon a pillar and claw, to slide up and down at pleasure." Cooper-Hewitt Museum, 1921-6-D292

8 COLOR SCHEMES AND THE USE OF WALLPAPERS

WHAT COLORS WERE USED

To list the myriad color schemes in eighteenth-century wallpapers available at various price levels would be a never-ending and pointless task. But it may prove helpful to review and bring together some observations about wallpaper colors already touched on in the preceding chapters and to generalize from them.

If, in imagination, one assembles all the known specimens of eighteenth-century English wallpapers, mentally averaging the quality and intensity of their colors, and then does the same for all the French examples, a generalized impression emerges: The palette of the English paper hangings that reached America lacked the brilliance and clarity of the French examples. This difference can be explained in part by the fact that most of the English papers date from the mid-eighteenth century, while most of the French papers came to this country later, after 1780. By this time the French had improved on English printing techniques, and the English paper stainers had made advances of their own. But those English improvements, retarded by government restrictions, never quite equaled the mastery of the French industry as a whole during the eighteenth century, especially in the area of color technique.

The English continued to limit their palettes, or perhaps were limited by their technical skills, and printed in more somber shades than did the French. The French seem to have preferred, and were more skilled in printing, clearer, brighter, more intensely contrasting combinations of colors and they mastered the art of controlling large numbers of colors within a single pattern.

Robert Dossie's *Handmaid* of 1758 documents the fact that the English paper stainers did have a broad spectrum of pigments at their command. His list of "Colors Proper to be Used for Paper Hangings . . . for Common Designs Done With Water Only" is reproduced in this book as Appendix A. It includes pigments for all the primary and secondary colors. Yet many English wallpapers exhibit a taste not for color but for grayness, for grisaille—one sees it in the painted landscape papers *(figures 3-10 and 3-12),* and in the pillar and arch patterns *(figures 3-16; 3-19).* Sometimes the grisaille scheme was enlivened by juxtaposing the grays with a single shade of one solid color, as in the mustard yellow used with the grisaille landscape paintings illustrated in *figure 3-9.*

Even when English paper stainers printed patterns in polychromes, they often overwhelmed the colorful effect by introducing quantities of gray, as in the background of *figure 3-6.* And although the many colors that appear on Dossie's list were available to be combined, the English seem to have preferred relatively monochromatic color schemes. We see this preference in floral patterns like those shown in *color plates 6 through 8* and *figures 3-1, 3-3, and 3-4.* While economics doubtless played some part in this, since the fewer the colors, the cheaper the paper, lower prices do not fully account for the limited nature of eighteenth-century color schemes in English wallpapers used in America. In August of 1795 Lady Jean Skipwith, a recent immigrant from Britain to America, included a sentence in her letter ordering wallpapers that perhaps was typical of a wider taste. She stated: "I am very partial to papers of only one colour, or two at most."[1]

Among the monochromatic color schemes so favored by English paper stainers, blue seems to have been the most popular choice of American customers *(color plate 8, figure 3-1).* Blue was *the* color associated with English wallpapers well before 1700. The renown of the blues produced by English paper stainers was celebrated in the name given to one of London's most prestigious shops offering paper hangings about 1700, the "Blue Paper Warehouse" illustrated in *figure 1-7.* Blue wallpapers were not only the most celebrated of papers made in England; they were also more expensive than papers decorated with other colors. This fact doubtless enhanced their prestige among the fashion-conscious. The pigments from which Dossie tells us English paper stainers derived their blues—Verditer, Indigo, and Prussian blue—rendered them more expensive than most colors. Dossie called Prussian blue "one of the Dearer colors." Verditer is a copper carbonate pigment that was manufactured as a byproduct of silver refining. Indigo, derived from various plants of the genus *Indigofera,* had to be imported from the Far East. The desirability attached to something known to be expensive was doubtless reinforced by the centuries-old associations of blue with the Virgin Mary and with the "heavenly" blues achieved in paintings of the Middle Ages and the Renaissance by using powdered lapis lazuli. Blue had long been a rare and much-prized color, difficult for the artist or artisan to capture.

Evidence that blue coloring in a wallpaper carried with it a price increase in America as well as in England has been cited on the inventory of paper hangings in the manufactory of Appleton Prentiss of Boston (see page 118). On that list, blue papers consistently received valuations higher than those assigned to wallpapers with the same patterns printed in other colors. And again, to review, blue was one of the two most popular colors, green being the other, for the plain-colored papers that enjoyed such popularity around the turn of the nineteenth century.

English wallpapers rose to fame partially on the strength of their blues, only to pale in comparison to the palettes of the French papers, in which a lighter, slightly

8-3

8-4

Figure 8-3, detail, and figure 8-4 It is not the paper on the walls, but the elaborate wallpaper border outlining the underside of the canopy on the bed that is of interest in this chapter. The eighteenth-century bed belonged to Sir Peyton Skipwith of Prestwould, near Clarksville, Virginia. It was probably decorated in 1831–2 with the flocked, feathered, and bejeweled wallpaper border seen here. In those years, the house was elaborately wallpapered from top to bottom (*figures 9-17 and 9-20* and pages 229 and 231). Using wallpaper to ornament a bed reflected current fashion. For instance, in 1812, John Perkins of New Bedford, Massachusetts, advertised "elegant cornice borders for rooms: ditto for windows and bedsteads." Bed in the private collection of a descendant of Sir Peyton Skipwith

greenish-blue often provided the background or the foil for vibrant oranges and a range of crisp, lively colors *(color plate 14)*. The colors made by American paper stainers of the eighteenth century could offer little aesthetic competition for the French or the English products. Although American craftsmen of the late eighteenth century boasted that the colors on their papers were equal to the imports, criticism of the American paper hangings repeatedly mentioned the poor quality of their coloring. One observer, writing at the very beginning of the nineteenth century, summarized a widely held opinion:

> The paper hangings manufactured in America are generally very indifferent and the colours too faint and pale. Those made in France are of a fixed substantial nature and . . . will command a preference both on account of cheapness and the superior liveliness of colour.[2]

Surviving examples of French and American wallpapers of the late eighteenth and early nineteenth centuries substantiate this assessment, one in which other writers of the period concurred. Nevertheless, American paper stainers continued to boast of their colors, advertising their patterns in "two to seventeen colours," or "two to twenty-six colours." Patterns were offered which were both "light and dark figured." William Poyntell could provide Philadelphians of 1791 with "Paper made to any particular desire of Ground and Colours, at 3 days notice."[3] Since a different block was required for the printing of each color, the patterns with the largest numbers of colors were the most expensive. This consideration, rather than the dictates of fashion, determined many color choices.

In addition to the diversity of hues, gold, silver, and mica offered glittering additions to color schemes. Their reflective qualities could serve to incorporate colors in the surrounding room within the visual surface of the paper. Not only were these metallic finishes used for borders, as seen earlier (page 138), but they were also applied to fill papers. William Poyntell offered to Philadelphians in 1797 "A handsome assortment of the most fashionable Silver Grounds London and Paris Papers."[4]

Eighteenth-century householders took into account many factors besides the latest fashions or expense when choosing colors for wallpapers. Furniture was covered to match the colors in paper hangings, and vice versa. An instance of the wallpaper dictating the choice of upholstery is found in the writings of George Washington. In 1757 he ordered from London one dozen "strong Chairs . . . the bottoms . . . of three different colours to suit the paper of three of the bed Chambers."[5] A necessity for matching the colors of fabrics on seating furniture with the wallpaper colors was strongly felt throughout the eighteenth century. In 1787 a New York paper-staining manufactory reminded customers that they could order "any kind of Paper Hanging agreeable to their fancy" in any "colour to suit their furniture."[6]

Although it is impossible to generalize dogmatically, to assert that a wallpaper of a given color scheme dates from a specific time simply because the combination of colors in which it was printed was more popular at one date than at another, colors can help in dating old wallpapers. Information about the chemical makeup of the colors can be useful in establishing dates. Sometimes, the date of the invention or discovery of a pigment is known, so that a researcher can establish a date before which the paper could not have been made. In some cases, ingredients went out of use when cheaper pigments came in, and the presence of purely early coloring materials will suggest an earlier date for a paper. (See Appendix A.)

HOW WALLPAPERS WERE HUNG

The English used tacks to hang wallpaper during the early eighteenth century. A London advertisement of about 1700 instructed the paper hanger to:

> First Cutt your Breadths to your intended heights then tack them at the top and bottom with small tacks and between each Breadth leave a vacancy of about an inch for the borders to Cover, then cut out the borders into the same lengths and tack them straight down over the edges of the Breadths and likewise at the top of the room in imitation of a Cornish and the same (if you please) at the bottom . . . But if you will putt up the same without borders, then cutt one of the Edges of each piece or breadth smooth and even, then tack itt about an Inch over the next breadth and so from one to another. But whether you putt them up with or without Borders gently wett them on the back side with a moist sponge or Cloth which will make them hang the smoother.[7]

Similar methods for hanging paper were apparently followed in this country. A 1741 bill for work done at the Province house in Boston includes a charge explained in this way: "To New Tacking the paper hanging above in the chamber & New papring one roome below stairs. Error Except'd . . . £ 4-10-0."[8] As late as 1784, George Washington wrote, when ordering wallpaper bordering: "I do not know whether it is usual to fasten it on with Brads or Glew; if the former I must beg that as many be sent as will answer the purpose."[9]

The eighteenth-century passion for borders seems to have had its beginnings in the recognition of their usefulness in covering the rough edges, in some cases concealing the tacks used to fasten wallpapers to stretchers or walls. As suggested earlier, borders were used not simply because they were thought decorative; nor were they a refinement used only by the very stylish. They were utilized, whether papers had been hung with tacks or with paste, to cover the edges, to simplify the paper hanger's job by concealing and fastening the cut ends of the paper.

French and English methods for hanging paper were much the same and were doubtless brought to America by immigrant upholsterers and paper stainers. During the 1730s, J. M. Papillon made a series of drawings showing workmen hanging papers. Although no captions for the drawings survive, some of the series clearly show workmen pasting, rather than tacking, sheets of paper, one by one to walls. They show that the little sheets contributed elements to patterns that extended beyond the confines of any one sheet, and that these fill patterns were used in combination with many borders, corners, and ceiling centers. Other Papillon illustrations show workmen pasting sheets of paper to fabric, probably canvas, then pasting the textile-backed paper to walls.[10]

American references indicate that, as in France, papers were in some cases fixed to fabrics before they were hung. The 1770 inventory of the Governor's Palace at Williamsburg, Virginia, seems to imply that such a hanging method was planned. Listed in a storeroom were "2 green Damask Curtains—Oznabrigs [coarse linen] intended to paste the Paper on in the Supper Room."[11] In 1786, Thomas Hurley of Philadelphia advertised wallpaper

> put on . . . at the usual moderate price of 1s6 per piece on plain walls, or 3s for pasting on paper and canvas, which he warrants to execute so as that no damp can long after affect his paper.[12]

Papers have been found pasted to canvas tacked to wooden stretchers. This simplified the task of taking down and moving a paper to other walls. From this expensive and careful refinement, the list of hanging practices descends. An invoice of paper hangings shipped in 1799 from the London shop of James Duppa to Lady Skipwith at her Virginia mansion, Prestwould, was accompanied by a note:

> The process of putting up paper Hangings is to have the wall as smooth as possible and then to be well sized over. The ingredients used for making of paste is flour & water with a small quantity of Allum put in and boiled till quite thick.[13]

Among George Washington's manuscripts, in his own hand, are preserved some of the most famous directions for hanging wallpaper.

> If the walls have been whitewashed over, [hang the paper] with glue. If not, simple paste is sufficient without any other mixture, but in either case the Paste must be made of the finest and best flour and free from lumps. The Paste must be made thick and may be thinned by putting water in it. The Paste is to be put upon the paper and suffered to remain about five minutes, to soak in before it is put up, then with a cloth press it against the wall until all parts stick. If there be rinkles anywhere, put a large piece of paper thereon, and then rub them out with cloth as beforehand.[14]

The instructions sent to Lady Skipwith, and George Washington's more elaborate notes about the practical details of handling the paste on the paper, assume that the walls to be papered would already have been plastered. But although smoothed and plastered walls were recommended as the proper surface on which to paste papers, early paper samples are often found in American houses pasted directly on unfinished boards. Skills of paper hangers and the budgets of houseowners as well as their varying standards of fastidiousness dictated the methods used to hang papers more often than did any printed standard instructions.

The flour and water paste "with a small quantity of Allum put in and boiled till quite thick" recommended to Lady Skipwith was fairly typical of pastes used through the eighteenth and nineteenth centuries. Standardization, however, was hardly a mania of the era. Individual paper hangers devised varieties of concoctions for hanging wallpaper. In his Charleston, South Carolina, advertisement of 1762, Richard Bird, "Upholsterer from London," boasted of "a paste that has a peculiar quality of destroying all vermin in walls."[15] Edward Ryves addressed the subject of wallpaper paste in an advertisement in a Philadelphia newspaper of 1786:

> From the nature of the climate or the strength of the lime [in plaster], the paper often comes off in many places notwithstanding the greatest care in the hanging . . . He has found out a composition he mixes with his paste, which entirely prevents this inconvenience.[16]

Such unspecified "compositions" sometimes defy the best guesses at identification made by restorers of old houses trying in vain to remove papers hung two centuries ago. Molasses has been suggested as a sometime-substitute for wallpaper paste. Unlike the more often encountered wheat paste, some of the glues and other adhesives that have been found holding old wallpapers in place prove to be non-water-soluble. Skillful testing of solvents is required to find a solution for softening these adhesives, and in some cases only laboratory analysis can determine their makeup.

A poorly mixed or overly watery paste, or the clumsy use of a good paste, could ruin the paper. At Benjamin Franklin's Philadelphia house, redecoration was in progress while Franklin was in London during 1765. His wife wrote to him that in "The Blewroom . . . I think the paper has loste much of the blume by paisteing of it up."[17]

WHERE WALLPAPERS WERE USED

"The houses are seldom without paper tapestries, the vestibule especially being so treated," observed a visitor to Philadelphia during the 1780s.[18] His comments confirm evidence gleaned from advertisements, from bills for the sale and hanging of papers, from newspaper descriptions of houses for sale, and from actual survivals on the walls of old houses that entryways, passages, and halls were the rooms most likely to be papered in eighteenth-century American houses.

In descriptions of the papers they were selling for these most public of domestic rooms, advertisers tended to lavish more care, to go into greater detail more often than they did in their comments about papers for other parts of the house. Papers of the type variously designated "architecture papers," "pillar and arch figures," and "large pillar papers" are consistently recommended for entryways, spaceways, and halls; these have been illustrated in *figures 3-16, 3-17, 3-19 through 3-22,* as well as in the related pattern of *color plate 9 and figure 4-3.* Doubtless akin to those illustrations was the "Large Arch Paper" Cornelius and John Crygier offered their New York customers of 1799 specifically "for Halls."[19] The appearance here of the adjective "large" in association with a pattern recommended for hallways is significant. It is representative of many advertisements in which hall papers are described as "large" and "large figured." These advertisements indicate a preference for large-scaled patterns in American hallways, a preference corroborated by the survival of numbers of large-scaled eighteenth-century wallpaper samples in entry and stair halls. Wallpapers specifically "for staircases" were advertised by William Kidd in Williamsburg in the 1770s and by William Poyntell in Philadelphia in the 1780s.[20] These were probably of the same types deemed appropriate for halls and entries.

In addition to "Large Arch Paper," the Crygier advertisement suggested two other kinds of paper "for Halls": "Stone" and "Black and diamond." This mention of

"Stone" paper is an early instance of the association with halls of patterns imitating stone and brick work, an association that was to become very commonplace during the nineteenth century *(figures 12-32 through 12-34).* "Brick paper" was ordered by Thomas Jefferson in 1790, and described straightforwardly in one of his several letters detailing the kinds of wallpapers for which an agent in Paris was to shop: "This resembles brick work."[21] The phrase "Black and diamond," which the Crygiers used to describe another kind of paper specifically for halls, suggests the kind of geometric pattern illustrated in *color plate 62.* Numbers of papers featuring bold geometric patterning have been found in late-eighteenth- and early-nineteenth-century hallways in America.

Advertisers also specified that they sold papers appropriate for dining rooms, "lower rooms," parlors, and drawing rooms, though they rarely described these papers. Zecheriah Mills of Hartford did specify in 1797 that "a great variety of small figures" were "for lower rooms and chambers [bedrooms]" in contrast to the large figures he suggested for halls.

In addition to printing different categories of patterns for specific rooms, eighteenth-century paper stainers also printed papers intended for specialized placement on walls, for use on objects other than walls, and for ceilings. Again, an invoice listing wallpapers sent in 1757 from London to George Washington yields an unusually early bit of documentation for an interesting aspect of American wallpaper use. Washington's wallpapers included "A set of best painted ornaments for a Ceiling, Qty 32 Square yds" and "A set of Papier Mache for ditto."[22] These *papier-mâché* ornaments probably imitated the carved and plaster-work ceiling rosettes familiar in English Palladian country houses of the eighteenth century. The motifs on these rosettes most often were derived from classical architectural details and from rococo designs. Nearly a decade after Washington's ceiling ornaments arrived from England, a Charlestonian was offering "machee Ornaments for ceilings &c, to imitate stoco work."[23] In 1762, a New York dealer advertised with his "great variety of paperhangings" that he had "Bass Relievo for Ceilings."[24] Apparently ceiling ornaments fell into a standard category, for which potential customers, in the advertisers' eyes, had little need of description. Ceiling papers and the services of hangers willing to affix them to ceilings were advertised, though not featured, right through the eighteenth century.

Wallpaper pictures of relatively small size—about the size of easel paintings—were made by French paper stainers to be placed in the space above doorways. Known as *dessus-de-porte,* these often charming pictures were among the forerunners of much larger wall-sized "landscape papers" and "views" that became the most celebrated features of the nineteenth-century French wallpaper trade. French immigrant craftsmen introduced the *dessus-de-porte* to Americans. For instance, Francis Delorme, "lately from Paris," advertised during 1796 in Charleston, South Carolina, "Landscapes for Chimney Pieces, and to place over doors."[25]

However, there are few such advertisements or survivals extant in America indicating that these wallpaper pictures were placed over doors. Indeed, few American houses boasted rooms with ceiling heights that could have accommodated such decorations in the space between the door frame and the ceiling. Instead, these specialized products of paper stainers were used here to decorate overmantles—"for chimney pieces" to use the words of the Charleston advertisement—and as "fire boards." These were wooden boards cut to fit into a fireplace opening. They served as covers when no fires were burning. Fire boards blocked drafts, hid unsightly leavings from fires, and filled the dark and dirty-looking space during warm seasons.

Also called "chimney boards," they were usually horizontal rectangles finished with "frames" of border paper pasted around the edges. Vases filled with flowers were a favored motif for the prints that paper stainers produced as decorations for fire boards, like that illustrated in *figure 8-1.* Since many eighteenth-century Americans placed real vases or pots of flowers in unused fireplaces during the summer months, it seemed logical to print colorful and more lasting replacements for these popular ornaments.[26] As early as 1766 John Blott, upholsterer and paper hanger, advertised in Charleston "Neat flower pieces for Chimneys."[27] There were many other styles offered as well. In 1771 "Mock India Chimney boards for rooms and chambers" were available from John Mason in Philadelphia.[28] In 1796 William Poyntell offered Philadelphians "Borders, Landscapes, and Chinese pieces for ornamenting Breast Works and Chimney Boards."[29] Some advertisers featured imported fire board prints; others, their own products. The merchants at # 40 Cornhill, Boston, who sold "Beautiful Landscapes, Frontispieces, and Flower-Pots, for Chimney-boards" in 1792 noted that these had been imported from France.[30] Joseph Hovey, also of Boston, sold "beautiful Flower Pots for chimney boards" made at his own factory during 1790.[31] "Antient Statues" were also advertised as motifs available on fire board prints offered for sale during the eighteenth century.[32]

Screens and fire screens were frequently covered with paper hangings during this period. The 1755 edition of Chippendale's *Director* provided models for elegant fire screens incorporating papers with Chinoiserie motifs. In 1759 "2 neat Maha^{y} Pillar and Claw fire Screens, India Paper on both sides" were sent from London to George

Washington.[33] They may have looked very much like Chippendale's design "C" as illustrated in *figure 8-2.* "India Paper" here, as in other eighteenth-century documents as well, probably designated a Chinese paper or a Western example of a paper decorated with Chinoiserie motifs. A Boston advertisement of 1767 announcing the arrival from London of "fine India Figures for Screens" suggests that the fashion for decorating fire screens with Chinese paper remained stylish during the following decade.[34]

Upholsterers during the 1760s and 1770s frequently offered to cover screens with paper hangings—large folding screens as well as fire screens. John Blott of Charleston advertised in 1771 that "he makes and sells . . . six and four leaved screens."[35] As early as 1756 another Charlestonian, one Booden, upholsterer from London, noted in the newspaper: "I make and sell all sorts and signs of paper and other screens and new cover old ones."[36] Evidence that old screens were frequently recovered is provided by eighteenth- and early-nineteenth-century examples that sometimes preserve six and eight successive layers of repapering.

A more unusual use for wallpaper was as a covering for valances. In Charleston newspapers during 1793 and 1794, Francis Delorme referred to what decorators now call valances as "cornishes" (i.e., cornices, still a word for horizontal moldings or frames used to conceal curtain rods). Delorme advertised "He also makes and sells bed and window cornishes, covered with colored paper, in the neatest and most elegant taste."[37] A simple wooden valance covered with a wallpaper that probably dates from the turn of the nineteenth century was found in a house in King and Queen County, Virginia; it is now in the Colonial Williamsburg Collections.

In addition to the more decorative uses of paper hangings on screens, chimney boards, and cornices, they were put to other more functional and improvisational uses off the wall, uses that did not always require an upholsterer's skills. Americans used paper hangings during the eighteenth century to line trunks and boxes as well as to cover them. They also covered books and ledgers and lined drawers with wallpaper. Eighteenth-century wallpaper samples frequently turn up in very unlikely places, such as in the archives of government records offices, where they were used to bind official documents.

WHO BOUGHT WALLPAPERS

The fact that this book incorporates a great many wallpaper references that survive among the reverentially preserved documents of rich and illustrious early Americans may have skewed the reader's impression of just who bought wallpaper in the eighteenth century. The two most numerous sources for documenting the history of wallpaper are papers, including the business records and correspondence of important people, and published advertisements. The nature of these sources leaves uncertainty when attempting to assess the popularity of wallpaper with ordinary people or estimating just how often wallpaper was used in eighteenth-century American houses. It is uncomfortable to be forced to generalize from soft rather than hard facts, but most of the case for the popularity of wallpaper as a middle-class, widespread feature of interior decoration comes from circumstantial evidence.

In contrast, there are many more sources available to the student of eighteenth-century furniture history in America. Researching wallpaper, one can make but little use of some of the furniture historian's richest sources—inventories. They reveal the net worth of the deceased and also itemize the kinds of property that made up that wealth—land, buildings, livestock, clothing, jewelry, furnishings. And there are many of these inventories, so a researcher can use them to form an understanding of comparative wealth among members of a community. One can assess these material, economic factors in light of the documentation of an individual's political position, church membership, social affiliations, and other information available from other kinds of eighteenth-century records. The social historian uses inventories to gain a broad sense of the circumstances of material life during periods distant in time. The furniture historian uses inventories to understand the context within which a given furniture form would have been used, to assess the rarity of that form, and to gain an understanding of which possessions constituted signs of wealth, luxury, and rank—status symbols. Analysis of large numbers of inventories can provide relatively reliable answers to such questions as "Would the majority of householders in Boston in 1775 have been likely to own a bedstead, or an upholstered chair?" Unfortunately, inventories do not say whether that same group would have papered the walls of their homes.

However, in a few exceptional cases, wallpapers are noted on inventories. For instance, because they had not been hung, wallpapers were listed in the inventory of Joseph Weld, currier, of Roxbury, Massachusetts, a man of modest estate who died in 1766. "Four Roles Paper and Border" were noted "In ye Southwest Chamber." An evaluation of 14 shillings was given to the papers. To gain some perspective on the relative value of that wallpaper, we can compare it to other items in Weld's estate. The most expensive furnishings listed were "one Bed Bedsted, and its Appurtenan^{ces}," valued at 60 shillings. His silver watch, Weld's only piece of jewelry or of silver, was valued at 100 shillings. The total value of his real estate came to £686.13.4. Weld's estate fell within a middle range when compared with the estates of his neighbors who died between 1750 and 1775.[38]

Such evidence for the presence of wallpaper in the house of an ordinary man provides a counterbalance to the descriptions of wallpapers gleaned from the archives of George Washington and Benjamin Franklin.

More weight is added to the case for the widespread eighteenth-century use of wallpaper in America by the fact that paper hangings were advertised repeatedly in newspapers of all the eighteenth-century American cities. Merchants do not continue to stock and advertise that for which there is only a very limited market. Therefore, the simple fact that these notices appeared frequently in daily newspapers suggests that wallpapers were used by the middle class and used more frequently than the restored house museums and museum period rooms might lead us to believe.

American dealers not only advertised the relative cheapness of domestically produced wallpapers as opposed to foreign ones; they also stressed the economy of paper as opposed to other ways of finishing interior walls. While the claims of advertisers are not to be taken too literally, some of them can be accepted as giving an idea of the relative expense involved in hanging paper on the wall. A Charleston paper hanger insisted in 1765: "The expense of papering a room does not amount to more than a middling set of prints."[39] Imported papers were sometimes advertised as "cheap."[40] William Poyntell of Philadelphia said of his London imports in April of 1783: "The low Prices at which they will be sold will make Papering as cheap as Whitewashing." By August, he had amended his sales pitch and was advertising that at his prices, papering was *cheaper* than whitewashing.[41] In 1786, Joseph Dickinson, a Philadelphia manufacturer, suggested that his customers hang papers with "dark grounds, which the smoke will not considerable effect in the course of twenty years, at such low prices will eventually be found cheaper than whitewash."[42] And in the same vein, an advertisement of 1791 for papers made in Boston but sold in New York stated:

> It is found, by experience, that papering of rooms, both walls and ceilings, as well as entries, with this cheap paper, is far less costly, and much handsomer, than white washing.[43]

An additional inducement for papering rather than painting the walls of a newly built house in the eighteenth century was the relatively cheaper cost of finishing a wall to receive paper as opposed to preparing it to be painted. Evidence for this survives in the form of a published "List of the Different Kinds of Masons', Bricklayers', and Plasterers' Work" dated 1801, preserved in the papers of architect John J. McComb at the New-York Historical Society. Using what were apparently current London prices, McComb filled in the figures on the printed list. He noted that to lathe and plaster "two coats and trowelled for papering" cost 3 shillings per yard as opposed to 3 shillings 9 pence to lathe and plaster "2 coats floated and finished for painting." The list includes the entries: "Plastering on brick, 2 coats trowelled off for paper," 1 shilling 3 pence a yard, as against "set in white," 2 shillings 5 pence, and "finished for painting" at 2 shillings 8 pence per yard. This saving must have encouraged some who built houses to hang their walls with paper.[44]

Substantiating the evidence that the cost of papering walls compared favorably to other ways of finishing walls are the actual prices for wallpapers that advertisers frequently published. From the 1730s through the 1760s, the prices were sometimes given by the yard, as on a Boston trade card from that period, giving a range "from 2d [pence] to 6d per yd and Upwards."[45] More frequently, the price was given by the piece, each piece being twelve yards. Several advertisements from Boston during the 1760s gave "20s [shillings] and half a dollar a piece" as the lowest prices for paper hangings.[46] By 1771, the prices of the least expensive papers and even those imported from London had dropped. In that year a Salem, Massachusetts, advertisement offered such papers "from 3s to 4s a piece."[47] Prices of the 1780s ranged in New York "from 6s. to 22s.6d. per piece" for English and American wallpapers,[48] and in Philadelphia "from 2s.6 per piece" to "higher prices in proportion,"[49] for papers of local manufacture. In 1786, Joseph Dickinson announced to his Philadelphia customers: "As importation, evidently, hath nearly drawn all the specie from America, to facilitate business, I will take produce and every current money."[50] In the 1790s, with increased competition between growing numbers of manufacturers, prices appear to have dropped again, as advertisements stated "the lowest are at 1s.9d. per piece,"[51] and included price ranges from "2s to 20s the Piece" or "1/4 to 20s per roll."[52] In 1797, a Boston manufacturer announced that his prices were "from 30 Cents to 12 Dollars per roll" for French, English, and American paper hangings.[53]

In the middle of this period, during the 1770s, a typical daily wage for a skilled craftsman was about seven shillings six pence, or one dollar.[54] At prices as low as 3 shillings for 12 yards of paper hangings, they were within the reach of a great many Americans.

The actual numbers of paper hangings offered in eighteenth-century advertisements add further evidence that quantities of wallpapers were stocked and presumably sold. Dealers like William Poyntell of Philadelphia, who in 1783 kept a stock of 4,000 pieces, repeatedly announced arrivals of newly imported papers, sometimes specifying that the shipments included several thousand additional pieces, like the 2,500 Poyntell advertised that same year.[55] In 1787, responding to the boasts of a rival whose advertisements had called attention to the large numbers of wallpapers he had recently sold in Philadelphia, Joseph Dickinson published his own claims: Dickinson reported "he cannot but smile to find 3000 [pieces of paper hangings] should want so much advertising, especially as they are London made, superb, and cheap; when I, though in a remote part of the town, have sold three times that quantity."[56] Dickinson stood ready to supply "merchants and storekeepers" with "as many thousand pieces as may be required."[57]

One of the several paper-hanging manufacturers in Philadelphia was reported to have produced between the autumn of 1789 and July of 1790 more than 10,000 pieces and, with planned improvements, was expecting to make 20,000 to 30,000 pieces in the following year.[58] Announcements of arrivals of imports gave figures in the tens of thousands of pieces, and assured customers of equally large supplies of papers kept constantly in stock. These announcements were scattered through New York and Philadelphia newspapers of the 1790s. The figures imply steady consumption of large quantities of paper hangings during the late eighteenth century in America. Added to other kinds of statements from the newspapers, to information about the numbers of craftsmen printing wallpapers in this country gleaned from city directories and tradecards, and to the many bits of wallpaper that survive in old houses, they give us good reason to conclude that during this period, wallpaper was popular in houses of those well below the exclusive circles of the rich, and far beyond the major cities.

At Andrew Jackson's Hermitage, near Nashville, Tennessee, the bedroom of his granddaughter Rachel retains French wallpapers and borders hung during the 1830s. The wall patterning is yellow, and the flocked borders, in shades of gold and brown, are accented with bright turquoise blue flowers. The furniture was placed in this room during the 1850s. Courtesy of the Ladies' Hermitage Association, Hermitage, Tennessee

PART TWO

INTRODUCTION

Wallpapers of the first half of the nineteenth century show an intensification of the tendencies that had surfaced just after the Revolution. Neoclassical patterns *(color plate 9)* are rendered more flamboyant *(color plate 41).* Landscape vignettes *(color plate 27)* are blown up to mural-like productions covering whole walls *(color plates 31 through 37).* Patterning derived from textiles *(color plate 25)* is supplemented by walls swathed in imitation drapery *(color plates 48, 49, figures 12-1, 12-2),* and by eye-fooling pseudo-textiles, fastidiously depicting the gleam of watered silk *(figure 15-4)* or other special woven effects, some created with such compulsive attention to detail that a thread count might be taken. In addition to floral patterns, there are wallpapers lavished with flowers so meticulously detailed and intensely colored that they seem real *(color plate 70).*

The French excelled in making these papers, and Americans indulged an appetite for them, which had been whetted prior to the turn of the nineteenth century. At the same time, they bought wallpapers made by American paper stainers, whose numbers multiplied during the first forty years of the century. Within the wallpaper trade, both fashion and business trends developed during this period along paths laid out before 1800. Eighteenth-century styles persisted into the nineteenth century in pattern formulas like that of the Washington Memorial paper, in the use of plain papers with borders, and, as will be shown, in numbers of simple repeating patterns. Styles were not simply started and stopped on signal. Some appeared only briefly; others continued at various levels of stylishness for long periods.

To identify these early nineteenth-century styles, it is still necessary to rely on many sources similar to those used for the earlier periods—surviving wallpapers, sometimes in the form of dingy scraps in old houses; advertisements; bills; comments in correspondence, diaries, and travelers' accounts. Fortunately, larger numbers of such sources are available. In addition, with the nineteenth century comes an over-abundance of published testimony to wallpaper's growing popularity. However, that testimony is rarely accompanied by illustrations until quite late in the century. An English writer influential in America, John Claudius Loudon, illustrates practically every other furnishing detail and encourages the use of wallpaper in his book of 1836, *An Encyclopaedia of Cottage, Farm and Villa Architecture and Furniture.* However, Loudon omits pictures of wallpaper patterns, explaining:

> As the fashions of most of these papers change as frequently as those of printed cottons, it would serve little purpose to offer designs of them.[1]

An outpouring of published advice to the homemaker extolling the virtues of wallpaper and detailing just how it should be used appears as part of a new literature —or sub-literature—of home decorating books addressed to middle-class women. The category is a nineteenth-century creation. In the eighteenth century, books that illustrated furniture styles and suggested schemes for room decoration usually took the form of expensively engraved limited editions intended for men—not only for the cabinet makers and upholsterers who outfitted the homes of the fashionable but also for their gentleman clients. The lists of pre-publication subscribers to such celebrated works as Thomas Chippendale's *Gentleman and Cabinet-Maker's Director* of 1754 included peers of the realm along with the leading artisans of London.

With the nineteenth century comes a dramatic shift in the audience for literature illustrating and prescribing styles and standards for furnishings: It is a shift from an aristocratic to a middle-class audience, and from a male audience to one that included and eventually was dominated by females. The literature addressed to this new, much larger, readership reflects major changes that accompanied industrialization. By the middle of the nineteenth century, factories had generated enough new wealth to create a consumer class able to purchase factory-made furnishings that catered to more than utilitarian needs: they displayed various forms and numerous styles of decoration. In the 1840s, proto-consumer guides, concerned with matters of taste and style, and sometimes recommending specific manufacturers, began to appear in Britain and America alike. Especially noteworthy among them are illustrated chapters on furnishing, including wallpapering, that appeared in Thomas Webster and Mrs. Paine's *The American Family Encyclopaedia of Useful Knowledge,* an English book of 1845 republished in America, and in A. J. Downing's *The Architecture of Country Houses,* published in New York in 1850.

Underlying both the popularity of such books and their special appeal to women was a relatively new view of the working relationships of men and women, a new conception of separate spheres in which each of the sexes had distinct functions. If we are to arrive at an understanding of the nineteenth-century's willingness to accept the notion that moral criteria should be applied to the choice of wallpapers, if we are to imagine how readers could have been persuaded to wade through pages of prose insisting that moral principles underlay the aesthetic choices necessary for furnishing a house, we must consider this home-decorating literature as but one manifestation of the nineteenth-century rearrangement of basic patterns of living.

Recent scholarship concerned with modernization has detailed the process by which demarcation lines between the public sphere of men and the private, domestic sphere of women were sharply drawn. The tasks assumed by men and women had always been differentiated; what was startlingly new by the early nineteenth century was the physical separation of the majority of men from women when performing their tasks, as well as the fact that men took their activities while earning a livelihood away from home, away from the realm in which women could participate. Home ceased to be a center of production and became instead the primary place of consumption. Laborers were drawn off the land, away from home, and into the factory. Working in the new world of commerce and industry often transformed farmers and craftsmen into middle-class urban consumers, and almost inevitably their wives were left at home with responsibility not only for child-rearing and housekeeping but also for furnishing the scene of those activities.

As men were co-opted by the developing industrial order into a scheme requiring that they leave home to work, the concept, the idea, of "Home" assumed greater and greater importance for them. The rise of urbanization provoked in many souls a reactionary need to stay in touch not only with nature but also with what was now conceived to be a "natural" scale of domestic life. Unprincipled competition for jobs and profit was understood to dominate the man's public world, a world that was frequently characterized as a ruthless fray—amoral, impersonal, and unfeeling. In contrast, home was appreciated as a refuge for the man, a place where he could come to find the love, religion, and morality he had left behind there in the special care of his women.

The new decorating literature promised women that it would help them in their primary task of serving others, by showing and telling them how to make home a pleasanter, more beautiful refuge into which men could retreat from the public world. This guidance began to appear in relatively cheap books like those already mentioned, and in journals like *Godey's Ladies' Magazine,* which ran a series of furniture plates in 1849. It had many functions. One was a straightforward service to people who had only recently acquired the power to buy. It showed them what was available on the market and told them how to use objects that the genteel deemed necessary for polite living. But more important, this literature justified consumerism. It transformed this intrinsically passive phenomenon into an active social principle by stressing that the new role of women as consumers was a positive, morally decisive calling.

Illustrations of rooms and wallpaper samples dating from the 1840s and 1850s *(color plates 69 through 77, figures 15-4 through 15-8)* show us that the working notion of "pretty" adornments for this refuge involved realistic depictions of natural objects, especially of flowers and fruits. It also included sinuous, full-bodied curves that were readily associated with the female form. It featured a profusion of the kind of delicate detail that for centuries had been produced only by the laborious attention of an individual craftsman working for an important and wealthy individual, since only such people, and the Church, could afford to own highly ornamented objects. Bright, delicate, easily soiled colors used with a quantity of white dominated the decorative schemes of mid-century wallpapers. All this can be read as expressive of the nostalgia for nature, for the soft and feminine, for individual expression, and for the vivid coloring of the outdoors—in sum, it all contrasted sharply with the impersonal quality of life among large numbers of strangers, with the hard-edged mechanical forms of the factory, with the dirt of cities, with the grime surrounding the urban poor, and, perhaps most of all, with the toughness and sharpness men were being forced to cultivate in the commercial world. The elaborate ornamental vocabulary nurtured through the middle years of the nineteenth century offered an intensified rendition of nature and of qualities associated with femininity from which men were separated during most of the waking hours of their week. Owning objects cast in these elaborate forms clearly demonstrated to the neighbors that a man was possessed of purchasing power. It was obvious to the most casual observer that such furnishings could not have been made by their owners.

The literature of home decorating told women that furnishings were tools of influence crucial to motherhood's all-important task of molding the minds of growing children. The influence exerted by furnishings could and should be moral, it insisted. Deriving the seeds of many of their ideas from the great English moralist, social philosopher, and critic of art John Ruskin (1819–1900), writers of these books tried to explain how moral criteria could be applied to the judgments of tables, chairs, or wallpapers. They insisted that moral virtues like "honesty" in the expression of structure, or "straightforwardness" in the treatment and revelation of the materials from which an object was made, would be perceived by and would act upon the minds and souls of those who lived with true and beautiful things.

Arguments calculated to persuade mothers that there was a potential for forming character latent in material things appear again and again in the decorating books of the nineteenth century. Authors contended that in the homes of boys it was especially important to manipulate the power of wordless influence imbedded in the designs of household furnishings, since the boys would someday be the men who ran the nation. Like many other kinds of how-to books, manuals of domestic economy and decorating books argued that a woman's power through such influence over future civic leaders and prospective presidents was awesome, if not always obvious.

They held that women enjoyed by means of this power a form of participation in the political process more crucial to the national welfare than a mere vote.

While decorating books were aimed more and more pointedly at women, books about architecture continued to be written by men and addressed to men. The exterior aspect of the home continued to hold men's interest: that was the public face. But even so, in basically architectural books like Downing's, furnishings began to find a larger place than they had in the eighteenth century. In *The Architecture of Country Houses,* Downing emphasized the affecting qualities of material objects. The influence of John Ruskin is readily apparent. Downing relied on the symbolic connotations of things to inculcate moral virtue, explaining how this should happen:

> Everything in Architecture that can suggest or be made a symbol of social or domestic virtues, adds to its beauty, and exalts its character. Every material object that becomes the type of the spiritual, moral, or intellectual nature of man, becomes at once beautiful, because it is suggestive of the beautiful in human nature.[2]

Convinced of the power of material things to influence human character, and to reveal it, possessed of money to buy examples of "the industrial arts"—manufactured objects of every description—Americans of the second half of the nineteenth century purchased quantities of wallpaper. And they had to make their choices from a bewildering multitude of styles.

Throughout the nineteenth century, wallpaper dealers boasted of the range offered at any one time, in patterns they labeled "Renaissance," "Tudor," or "Elizabethan"; "Modern French," "Louis Quatorze," "Louis Quinze," "Louis Seize," or "Rococo"; "Roman," "Greek," "Neo-Grec," "Empire," "Gothic," or "Romanesque"; "Indian," "Chinese," "Japanese," "Moorish," or "Egyptian." They often used one of these labels even though elements of several styles appeared in combination within a given pattern. Or they were quite capable of making up other even more exotic names, such as "Hindoo" and "Anglo-Jap." Celebrating itself as the age that was heir to all the ages, the nineteenth century gloried in the power bestowed upon it by scholarship, exploration, and publication to pick and choose from the arts of all the cultures on the face of the globe throughout time. With Owen Jones's illustrated *Grammar of Ornament* (1856) in hand and with such an attitude about the world, wallpaper designers selected and combined motifs with abandon and relish. Given this situation, it would be pointless to devote the remainder of this book to chronicling, naming, and falsely forcing into sequence all the ornamental styles that were simultaneously popular in the wallpapers of the nineteenth century.

30

Color plate 30 Thirteen layers of paper removed from a wall, then steamed apart, are here arranged in the sequence they were hung, one covering another, in the Nathan Beers house in Fairfield, Connecticut. The earliest pattern, shown on the far right, is an *ombré* of the 1820s. Following the chronological succession forward through time from right to left, layers of machine-printed wallpaper succeeded one another: from rococo scrolls through vignetted scenes, realistically rendered vegetation, and abstracted foliage to a geometric stripe and then a return to scrolls. The cheaply machine-printed stripe with grapes, which appears on top, was probably pasted over its predecessors about the turn of the twentieth century, or even as late as the 1920s. Such accretions of wallpapers are not uncommon in old houses. Examples like this document the range of inexpensive papers chosen by individuals who lived in a particular house. Collections of such documentation suggest trends in the use of ordinary wallpapers through the nineteenth century. Cooper-Hewitt Museum, 1960-211-1 A-M; gift of Edith L. R. Fisher

31

Color plates 31 and 32 *"Les Sauvages de la Mer du Pacifique,"* or *"Les Voyages du Capitaine Cook,"* was the first of the great panoramic papers of Joseph Dufour. Designed and printed in Macon between 1804 and 1806, before Dufour moved the firm to Paris, it is topographical in its attention to the vegetation and to the costumes, dwellings, and customs of the people encountered by the British naval captain James Cook, who was killed by the natives of Hawaii in 1779. All 20 panels of the paper are shown here. Each is 8 feet 3 inches tall. Courtesy of the M. H. de Young Memorial Museum, The Fine Arts Museums, San Francisco, California

2

33

Color plate 33 (above) Niagara Falls as rendered in *Les Vues de l'Amérique du Nord,* a scenic wallpaper printed by Zuber of Rixheim, in Alsace, is illustrated as photographed in the diplomatic reception rooms at the White House. During the nineteenth century this set of paper was hung in the Stoner house in Thurmont, Maryland. In 1961, while the house was being demolished, the panoramic wallpapers were removed, then rehung in Washington. Zuber first printed the paper in 1834. Courtesy of the White House

Color plate 34 (opposite) A portion of "Monuments of Paris," another Dufour paper, is shown as installed at the Winterthur Museum—2,062 blocks and 80 colors were used to print the thirty lengths that make up a full set. In it, the major buildings of Paris are depicted as if they had been removed from their original settings and lined up along the Seine. Here, to the left of the clock, appears the column in the Place Vendôme. Just before the corner is Val-de-Grâce with its dome. At the far right, the east front of the Louvre. Dufour introduced the paper about 1814. Courtesy of the Henry Francis du Pont Winterthur Museum

Color plate 35 Zuber introduced this *"Décor Chinois"* in 1832. It is shown here as illustrated in Zuber's 1975 catalogue of scenics in production. The design by Eugène Ehrmann and Hermann Zipelius relies on Chinese precedents for motifs and composition *(color plates 15 through 17),* but the French factory used the *ombré* technique, which had been its specialty since 1820 (page 273), to achieve delicate blended color effects rather than attempting to imitate the linearity of the painted Chinese models. Courtesy of Zuber et Cie.

35

36

Color plate 36 A scenic paper by an unknown maker, this has been called *"Les Bords de la Rivière"* in twentieth-century publications. On a bill of sale of 1835 a New York wallpaper dealer called a duplicate paper "Roman Ruins." This example survives where family tradition records it was hung about 1830 in the dining room of Shandy Hall, built at Unionville, Ohio, in 1815. In the relatively low-ceiled room, portions of the sky, not needed above, were used for a dado. Courtesy of the Western Reserve Historical Society

Color plate 37 A detail from an early-nineteenth-century scenic made by a French firm, as yet unidentified, shows the Bay of Naples with Vesuvius in the distance. Duplicate sets were hung at Gay Mont in Rappahannock Academy, Virginia, and at Hurricane Hall in Lexington, Kentucky. This fully colored view has been confused at times with the better-known monochromatic Dufour scenic *"Les Vues d'Italie" (figure 9-12).* Cooper-Hewitt Museum, 1974-80-1; gift of Mrs. Edward L. Nesbitt

37

A historical survey of nineteenth-century wallpaper in America must emphasize three points. The first is that there was a shift from French domination of the American wallpaper trade through the first seventy years of the century to English domination during the remainder of the period. The second is that the trade's relatively late shift to machine production—late compared to an industrial revolution well under way in the textile industry in the eighteenth century—had enormous impact on the design and the use of wallpaper. The third is that the introduction of critical theorizing about design was reflected in abstract wallpaper patterns produced during the late nineteenth century. From the mid-century on, English design critics, and later their American disciples, taught an ever-widening audience to differentiate realistic from abstract wallpaper patterns and to prefer the abstract. Since the French excelled in realism, in illusionistic imitations—whether of landscape scenes or of textiles—the call for abstraction encouraged the shift away from French styles. In addition, abstraction was often accompanied by simplification of patterning, and this in turn was more congenial to methods of machine production that the English were developing. Both mechanization and the plea for abstract patterning conspired to favor the change from French to English leadership in the American wallpaper trade.

In the early nineteenth century, the dominance of French wallpaper styles paralleled a trend apparent in all the other decorative arts in this country. Thomas Jefferson and the others who set the national style were Francophiles. American gratitude for French support during the Revolution and the War of 1812 was a potent force in the arts of design. Embargoes also played a role, keeping English goods out during that war. Lafayette's tour of America in 1824–5 rekindled enthusiasm for France and things French. Into the mid-century, Americans who papered their walls aspired to do so with elaborate French papers or with American patterns that looked like them.

But in the 1870s stylish Americans discovered English "Art Wallpapers." Not coincidentally, the English were finally enjoying a rising tide of popularity at this time. Events had also lowered the American opinion of the French. France had violated the Monroe Doctrine when she set up a Hapsburg empire in Mexico during the early 1860s. Communes, strikes, and riots in France frightened Americans, who were proud of their capitalists' accomplishment and of the industrial growth and prosperity that had accompanied it. The ordered country house life of the English gentry and nobility had more fashionable appeal. Theirs was the style to emulate by the 1870s. Americans like the millionaire's daughter Jennie Jerome, who married

Lord Randolph Churchill in 1874, affected the public consciousness. Fashion and style in interior decoration, as always, reflected trends in other areas of national life.

The second and third points introduced here—the effects of industrialization and the introduction of self-conscious theorizing to the process of designing wallpaper —emphasize changes that constituted a "paper revolution." One aspect of this was technological, and it was the more successful. Technological innovations widely utilized after 1840 within the craft that fast became an industry made wallpaper an important feature of very ordinary nineteenth-century interiors as well as of those of the rich. In the 1880s, writing in *The Decorator and Furnisher,* one Marion Foster Washburne summed up the popular opinion of the public to which an eager wallpaper industry catered. In an article called "About Walls and Wallpapers," Washburne proclaimed: "Next to whiteness, bareness is objectionable."[3] The wallpaper industry had triumphed in making its product the universally preferred wall finish.

The other aspect of this "paper revolution" enjoyed briefer and more superficial success. It was the attempt begun by English art critics and avant-garde designers to revolutionize public taste in choosing wallpaper. They tried to popularize abstract flat patterning, to create a market for simple geometric designs, and to end the production of realistic picture papers. In America, critical opinion about wallpaper design was rather late in reflecting the body of English critical reactions to what were called the "shams" and "vicious overelaborations" of furnishings of every description, especially those produced by machine, during the middle of the century.

"A surfeit of means is a danger to design," a German architect had warned in 1851 as he reviewed the "Industrial Arts" displayed at London's Crystal Palace Exhibition.[4] Objects shown there betrayed signs that their designers had indulged a pride in ingeniousness that seemed to blind aesthetic sensibilities. Similar pride apparently affected wallpaper manufacturers in America. Mid-century papers seem to be the products of an industry so consumed with delight over its new-found powers to use machines capable of making elaborate designs inexpensively in great quantity and variety that it had neglected the possibilities inherent in more rigorous formal order.

Well before the general outcry that greeted displays at the Crystal Palace in 1851, English critics, horrified by the cheap papers emerging from the machines, inaugurated their campaign to educate English industrial designers and to persuade buyers that the best wallpaper patterns should be relatively simple. During the 1840s, they decreed that for wallpaper designs conventionalization—stylized abstraction, conforming to the conventions of a given body of design, be it Greek, Gothic, Japanese, or some other—was preferable to realistic imitation of nature. In praising simple patterns, critics sometimes called them "chaste." They insisted that wallpapers be "honest," that is, true to their two-dimensional character, incorporating no tricks of shading or perspective drawing to fool the eye into believing that there were real roses strewn across a wall or blocks of marble finishing a room that was in fact wallpapered. For them, it was "wrong" and "deceptive" to create the appearance of depth, to depict on the paper's surface a vista that seemed to recede into an illusionary distance, to do anything, in fact, that might seem to pierce the actual flat plane forming the side of a room.

With the introduction of these moral terms into nineteenth-century criticism of wallpaper, we have come back to the source of the special appeal the rhetoric of the decorating literature held for women. Charged with safeguarding moral values, women responded to the critics' teachings about just what constituted morally uplifting patterning. If even a properly chosen wallpaper might help their children to appreciate the quality of honesty, mothers wanted to know which among the thousands of available patterns might be deemed the most "honest" of all.

"To suggest some fixed principles of taste," Charles Locke Eastlake summarized the current rules of the reformers in a book called *Hints on Household Taste,* which was published in London in 1868 and first published in the United States in 1872. Although he objected to realistic patterns, Eastlake himself understood their appeal. He dismissed them rudely from the homes of the tasteful, explaining:

> We require no small amount of art instruction and experience to see *why* the direct imitation of natural objects is wrong in ornamental design. The *quasi*-fidelity with which the forms of a rose, or a bunch of ribbons, or a ruined castle, can be reproduced on carpets, crockery, and wallpapers will always hold a certain kind of charm for the uneducated eye . . . ingenious, amusing, attractive for the moment, but [not] within the legitimate province of art.[5]

Six years after Eastlake's book first appeared in England, the French wallpaper manufacturers Desfossé et Karth gave voice to the view of the paper stainer's art that continued to prevail across the Channel. In a London trade journal, *The Furniture Gazette,* on the very pages that regularly published the opinions of the moralizing English zealots, Desfossé et Karth noted in 1874:

> Wall-paper, being before all else an art of falsification, should never give the lie to its first destination. Thus one of the most profitable of its branches is the exact imitation of stuffs [textile fabrics] and even of their floss.[6]

Throughout the following survey, it should become apparent that French and French-inspired patterns, which defied all the English critical rules, like the realistic roses of 1855 shown in *color plate 70,* were constantly vying with the patterns Eastlake would have judged "correct," like those of *figures 16-2, 16-4, 16-9, and color plates 81 through 94.* However, in the wake of the popularity of Eastlake's book during the 1870s and 1880s, the American wallpaper industry aspired to elevate wallpaper to Eastlake's "legitimate province of art." It was a province where the decorative arts were to abide by rules very different from those applicable to the fine or high arts. The revolutionary English standards for wallpaper seemed likely to score a permanent triumph with the purchasing public during the 1880s. Even the cheapest American machine-made wallpapers often reflected the reforming principles. Because such papers traveled widely over the American continent, they were important carriers of the "look" and provided popular demonstrations of the principles inspired by English reformers.

But the abstract patterns that satisfied the English critics won no more permanent a position in the marketplace than did any other passing fashion. In part, this was precisely because they had become overly fashionable. The public had simply heard too much earnest moralizing about the whole subject of interior decoration, an area in which it quite understandably missed its freedom to indulge in idiosyncratic fancies. Designers aspiring to elegance and grandeur now began to use a new vocabulary of critical assessment as the ideals of architectural Beaux Arts classicism gained popular currency during the last decades of the century. This was, indeed, the major artistic counterattack of the great French École des Beaux Arts. It took hold because the English rhetoric had become the cant of a fad, and reforming theory severely demeaned. Therefore, when reformers condemned designs that did not conform to their narrow formulae, they now met resistance from sophisticated designers who were appreciative of qualities beyond those recognized by so myopic a canon.

Finally, the demands of nineteenth-century consumers for something new had to prevail. So wallpaper manufacturers of the 1890s lost their English-derived "art consciousness" and resumed production of many patterns reminiscent of those which had evoked such negative reactions in the mid-century. Catering to a popular craving for novelty, American manufacturers introduced new patterns every year—"ingenious, amusing, attractive for the moment," and as profitable as could be.

9 FRENCH SCENIC WALLPAPERS

PRECEDENTS AND EARLY EXAMPLES

In January of 1802, William Mooney placed an advertisement in New York's *American Citizen and General Advertiser* for paper hangings, including:

> . . . one sett for a large drawing room, composed of 48 sheets done in distemper, representing one of the most enchanting landscapes which can be imagined all in the most natural colors agreeable to nature, from the stone on the river's bank to the golden tinged cloud in the air.

Mooney could only have been describing one of the earliest of the French *paysages panoramiques, papiers peint panoramiques,* or *tableaux panoramiques*—scenic wallpapers as the trade now calls them. To Americans of the nineteenth century they were "landscape papers," "landscape views," "long-strip landscapes," "scenery papers," or simply "views."

They were colossally large prints produced in multiple, and they were indeed something quite new in wallpaper when one example could so move William Mooney to poetic utterance on a mundane advertising sheet. The paper stainers of early nineteenth-century France had only recently begun to use their wood blocks and distemper colors to create these *tours-de-force* of their craft. Their accomplishment was formidable in terms of sheer size. The *paysages* were designed to cover the walls of large rooms from chair rail to ceiling, with continuous, non-repeating panoramic vistas. Most of them were printed in sets of between twenty and thirty lengths, each contributing elements to the scene. Each length was in turn made up of joined sheets of handmade paper. These were finer and stronger than machine-made continuous papers, which became available around 1820, so manufacturers continued to prefer the handmade as the base on which to print their fine scenics.[1] Each length in a scenic "set" or "collection" was about 20 inches wide and between 8 and 10 feet long. The upper portion of each had a coating of color—a "sky"—which could be ordered with or without clouds, and could easily be cut to fit the varying heights of walls without interfering with the most densely decorated portion of the paper. A set of forty-eight lengths, which William Mooney advertised, would have constituted an unusually large "view." Perhaps Mooney had combined two sets of landscape papers, a frequent practice *(figure 9-2).*

These papers were remarkable not only because of their large scale and complicated overall composition, but also because, at the other extreme, in their smallest details they were filled with intricate and intriguing touches. They imitated the minutiae of nature as faithfully as the printing technique could allow. Costumes were lavishly detailed and the niceties of architectural ornament were carefully delineated. Such rendering of detail was no mean accomplishment in this medium, since each element had to be built up by superimposing one atop another, flat, opaque colors, printed from carved wood blocks.

To depict any one of these elements, first a generalized silhouette would be printed from the raised surface of one block. To print a leaf, for instance, the rough silhouette would be stamped in, say, medium green. After it had dried, a darker green would be added from a block carved to provide the shapes of shaded areas. Highlights would be added in a lighter green, using another block. Then linear veining might be added in yet another darker color, from yet another block. When one considers how complicated the process of representing a detail as simple as a leaf was, and then takes into account how many leaves and how many more elements were required to produce a view of a park or of a river scene, one can begin to appreciate the complexities involved in making a full set of scenic wallpaper.

In fact, thousands of blocks sometimes had to be carved to produce a fully colored example of these great papers, which aimed to imitate the hues of nature as closely as possible. An exhibitor at London's Crystal Palace Exhibition of 1851, Étienne Delicourt, boasted that he had used four thousand blocks in making his showpiece, "*La Grande Chasse.*"[2] A manufacturer could greatly reduce the number of blocks to be carved if the scene was planned as one to be executed in a monochromatic color scheme. Such schemes were popular in shades of gray—*en grisaille*—and in shades of brown—*en bistre, en chocolat, en nanquin*—and some were printed in mauve tones. For the monochromatic papers, the reduction of the numbers of printing blocks and the resulting simplifications of production permitted the factory to charge lower prices. However, if they printed monochromatic scenes on top of sky blue grounds, they could achieve colorful effects that were almost naturalistic.

Once the blocks had been cut, a scenic could be kept in production, as a book is kept in print, for many years. When a wallpaper manufacturer retired from business or died, successors often bought the wood blocks and continued to print a scenic designed a decade or a quarter century earlier.[3] Today, original nineteenth-century wood blocks, sometimes recarved or much repaired, and sometimes supplemented by newly carved blocks, are still used by some manufacturers to print scenes first made in the early nineteenth century.

The task of carving hundreds and sometimes thousands of wood blocks was in itself a formidable one. However, before printing could begin, workers had also to perfect the chemistry of color preparation, often for as many as a hundred colors. The brushing of ground color onto the hanging paper stock was an important part of the preparation for printing.[4] Some manufacturers, notably the Zuber factory, carefully shaded the ground color, producing what was called an *ombré* effect (see pages 190ff. for more about the factory). At Zuber this effect was usually achieved by shading the ground from stronger blue near the top of a paper's length to paler tints near the horizon line of a scene, and often incorporating shades of golden yellow and pink. Before actual printing could begin, manufacturers also had to plot the sequence for using each of the blocks, and plan for drying time between impressions, so that colors could be added on top of other colors.

The purple prose of William Mooney's advertisement suggests that the results of all this labor did not go unappreciated. His rapturous description of a scenic wallpaper provides an early instance of the kind of enthusiasm these French papers evoked in America. By 1802, when Mooney's advertisement appeared, French *paysages panoramiques* were well on their way to arousing a greater interest than had been commanded by any other kind of wallpaper in this country. During the first half of the nineteenth century, these papers made the most conspicuous contribution to the history of wallpaper used in America.

The popularity of scenic papers derived from the fact that they were the most seductively realistic of all wallpapers and created the most convincing illusions of space. Rather than reemphasizing the actual physical presence of the surface of a wall, they exploited the possibilities of visually obliterating all sense of that wall's confines. They blasted away the reality of being shut inside and offered access through illusion to the vastness of all outdoors, catapulting the viewer into some remote corner of the earth. They delivered American provincials to the glamorous tourist meccas of Europe and soothed the nerves of city-dwellers with vistas of serene countryside and garden. All this was utterly counter to those obsessive English rules that demanded abstraction and flat patterns.

English critics might dismiss scenics as misguided, if impressive, examples of technical ingenuity, but through the first half of the nineteenth century Americans felt they were getting "Art" when they bought these papers. That reaction to scenics is reflected in the diary entry of Sidney George Fisher, a young Philadelphia bachelor who in 1847 accompanied relatives to the New York emporium of George Platt, then the city's leading decorator. Fisher recounted his visit:

> Saw quantities of elegant things, furniture, mirrors, picture frames, paper hangings, and so on. Had no idea before of the beauty of French paper. The various patterns for drawing and dining rooms, halls and libraries, were really works of art.[5]

Indeed, the French manufacturers were attempting to mass-produce works of important artists, to bring them within the means of a large group of consumers. They did not think of the papers as "mere" decorations, subject to a separate set of design rules appropriate for the "lesser arts." Rather, their appreciators judged them by critical criteria that valued faithful imitation of the appearance of objects in the natural world. Accusations of crudity in the block-printed rendering of the artist's original paintings might have carried some weight with their French producers. However, English outbursts claiming that realistic, illusionistic effects were inappropriate to wallpaper were meaningless within the French understanding of wallpaper making as "before all else an art of falsification." Through the first half of the nineteenth century English objections seem to have been equally meaningless to Americans, who responded enthusiastically to the products of this French view of wallpaper by buying scenics in great quantity.

Precedents for these panoramic landscape views, which became the pride of the French wallpaper industry, and one of its most profitable exports to America, are plentiful. The sources of separate elements in the scenic formula—the largeness of scale, the derivation from works of fine art, the non-repeating continuity of scene, as well as their most popular subject matter—can be traced not only to earlier wallpapers but also to other eighteenth-century phenomena.

The idea of replicating major works by fine artists had occurred to several wallpaper manufacturers in France well before the turn of the nineteenth century. During the 1780s, Arthur et Robert, the Parisian firm from which Thomas Jefferson had purchased papers *(figure 4-1),* reproduced paintings by Louis Michel van Loo (1707–1771) and Hubert Robert (1733–1808) in sizes appropriate for use over doorways, sizes that approximated average-sized easel paintings.[6] Their polychromed subjects presaged those of many later panoramic papers—populated landscapes of the waterside, with Italianate architectural elements. The "very handsome chimney landscapes as ever imported into this country" and "variety of landscape pieces ornamented for firebords" advertised by Charleston and Philadelphia wallpaper dealers of the 1790s were probably of this type.[7]

The London wallpaper makers who had painted large-scale scenes for the Jeremiah Lee mansion *(figures 3-10, 3-12)* and for the Stephen Van Rensselaer house *(figure 3-9)* had also chosen views of harborsides and landscapes dotted with classical architectural elements. They too had relied on the works of fine artists. However, they imitated prints after paintings, rather than the paintings themselves. While the scale of these eighteenth-century English painted landscape wallpapers approaches that of French scenic wallpapers of the early nineteenth century, the English papers look more like a series of prints, enlarged and framed, than like mural paintings. More important precedents for continuous, non-repeating panoramic effects lie elsewhere—in Chinese papers, Italian paintings, and "panoramas."

Chinese papers (see Chapter 5), with their continuous, non-repeating scenes showing landscapes and flowering trees, were important antecedents for the nineteenth-century scenic formula. Their boldness of scale and bright freshness of coloring was doubtless suggestive to French manufacturers. However, the shallowness of space in most of these papers—the effect of friezelike procession around the wall, with little suggestion of depth behind the foreground elements—renders their effect on interior spaces very different from that of most of the nineteenth-century French scenics.

In addition, great mural paintings ranging from those of Pompeii through those of the Renaissance should be understood to constitute precedents in the most general sense for these French wallpapers. However, the "panorama," an eighteenth-century product of the combined worlds of art and amusement, is a closer and more suggestive forerunner of the *paysages panoramiques.* Writing in the 1920s, Henri Clouzot cited the "panorama," an amusement popular in London and Paris during the late eighteenth century, as an immediate source for the wallpaper formula. The panorama was invented in 1787 by an Irish painter working in Scotland, Robert Barker (1739–1806). He patented a painting executed on the vast cylindrical surface of the walls of a specially constructed round building. It showed the entire city of Edinburgh and drew admission-paying crowds both in London and Edinburgh. Soon the pleasure-seeking public in every major city was paying to enter rotundas or other round structures in which a painting of a landscape, usually re-creating a major battle, or detailing the natives and the countryside of some exotic land, was displayed continuously along the wall of the circumference. In 1799 an American, Robert Fulton, opened two rotundas in the Jardin d'Apollon in Paris, one showing the city of Paris itself, the other detailing the siege of Toulon.

Characteristics of the "panoramas" recur often in the scenic wallpapers—the frequent choice of battles and military campaigns as subjects, as well as city views and exotic scenery; the combination of close attention to foreground detail with an emphasis on creating a sense of sweeping recession to distant spaces; and the care

9-1

9-2

Figure 9-1 (opposite, top) *"Les Jardins de Bagatelle"* is said to be one of the first scenics made, but it seems unlikely that it dates before 1800. Thirty-eight of the engravings by Louis-Philibert Débucourt (1755–1832) on which it was based were published in 1800, and the full series of 52 was published as *"Modes et Manières du Jour à Paris à la fin du 18ème Siécle et au Commencement du 19ème"* only in 1810. Courtesy of the Metropolitan Museum of Art, New York

Figure 9-2 (opposite, bottom) Between 1800 and 1805, the parlor of the George Shepard house in Bath, Maine, was hung with 12 lengths of *"Les Jardins de Bagatelle"* and 8 strips from another scenic set, not fully identified. Combining elements from different sets has led to confusion in identifying scenes, crediting manufacturers, and attaching dates to scenic papers. Courtesy of the Museum of Fine Arts, Boston

Figures 9-3, 9-4 Two plates from Zuber's *Collection d'Esquisses* show Swiss and Italian scenes. The top plate was labeled *"Paysage Colorie L'Helvétie"* in the album of prints published about 1831–5 by the factory, although the paper it depicts more often appeared as *"La Grande Helvétie"* in factory records. After designs by Mongin, it was introduced in 1813–14. *"Paysage Colorie L'Italie"* was the album title for the plate below. More often called *"Les Vues d'Italie,"* it too was designed by Mongin, and introduced about 1818. Courtesy of the Metropolitan Museum of Art, Elisha Whittelsey Fund, 1957

9-3

9-4

9-5

9-6

Figures 9-5, 9-6 (left) Two plates from the nineteenth-century Zuber *Collection d'Esquisses* illustrate designs after J. Michel Gue—*"Les Vues d'Écosse"* (top) and by Jean-Julien Deltil—*"Les Vues de la Grec Moderne ou Combats des Grecs."* The medieval tale set in Scotland is shown with appropriately Gothic frieze and dado. While the Scottish scenes were printed in monochromes—*"en camayeux"* as captioned in the album—the Greek views were *"colorie."* Both designs were introduced by the Zuber factory about 1827. Courtesy of the Metropolitan Museum of Art, Elisha Whittelsey Fund, 1957

Figure 9-7 (opposite) Zuber showed *"L'Arcadie"* in its *Collection d'Esquisses* with an elaborate surround of pilasters, frieze, and dado derived from classical architectural ornament. The paper, first printed by the factory in 1811–12, is an example of the classicism of Antoine Pierre Mongin—dreamy, pastoral, and generalized. Courtesy of the Metropolitan Museum of Art, Elisha Whittelsey Fund

9-7

with which French manufacturers designed the last panel of their sets of wallpaper to make a perfect joint with the first panel, so that the scene was continuous all around the walls when hung.

Whatever the most influential of its forerunners, the scenic wallpaper formula evolved during the 1790s and came to prominence nearly simultaneously in several early-nineteenth-century French manufactories. No firm evidence has yet come to light revealing conclusively just which of the more than 100 known sets of these papers started the fashion in exactly what year.

Perhaps the *atelier* of François Robert (d. c. 1807) was the first to produce scenic wallpaper in the form that became so popular in America. In 1790, Robert joined the Manufacture Royale of Jean-Jacques Arthur (1760–1794), which Arthur's father had established by 1772, to manufacture paper hangings in Paris. To work out a full-blown scenic they would have had to do little beyond increasing the scale of their wallpaper imitations of the paintings of Van Loo and Hubert Robert, made during the 1780s (see page 184).[8]

Two examples of *"Les Jardins de Bagatelle" (figures 9-1, 9-2),* attributed to Arthur et Robert, are thought to have been hung in America at dates rather early for scenics. "Prior to 1800" is the phrase used by the Falmouth Historical Society to indicate the time that a set of this paper was originally hung in the Joseph Bourne house in West Falmouth, Massachusetts. This is the period in which, according to family tradition, the paper was brought from France by Captain Nathaniel Eldred. Another set of the paper is supposed to have been hung in 1804 in the George Shepard house in Bath, Maine. That set of *"Les Jardins de Bagatelle"* is owned by the Museum of Fine Arts, Boston. These two instances of American use of this particular paper at such early dates bolster published opinion, based for the most part on stylistic analysis, that this is one of the earliest examples of the type.[9] William Mooney's advertisement, quoted at the beginning of this chapter, documents the fact that scenic wallpapers, probably very like *"Les Jardins de Bagatelle,"* were available in New York as early as 1802.

ZUBER

It is probable that the very first of the *paysages panoramiques* was the product of a factory that did not survive long enough to have a company history published during the late nineteenth century, when so many manufacturers in every branch of industry committed their accomplishments to print. Fortunately, the Zuber wallpaper factory in the tiny village of Rixheim, near the textile manufacturing city Mulhouse, in Alsace, did survive (and still does). In 1895 the firm published an autobiographical volume by its founder Jean Zuber (1773–1835) called *Réminiscences et Souvenirs.*[10]

In telling his company's story, Zuber gave an intimate glimpse into its workings at a moment crucial to the future success of scenic wallpapers—that time when he invested a great deal of money in a gamble on what was to become the specialty of his factory. Zuber attached a date of 1803 to the planning of his first major *paysage.* While this published account of the first formal steps toward its production does not trace the history of the scenic wallpaper type to the moment when the idea for scenics was first conceived, Zuber's autobiography does document part of the process by which one of the first commercially and critically successful full-blown *paysages panoramiques* came into being.

In giving the background for the event, Zuber recalled that he had begun working in 1791, when he was eighteen years old, as a salesman who traveled all over Europe for the wallpaper and textile printing factory of Nicholas Dolfus in Mulhouse. By 1798, Zuber had become a co-partner in the firm. In that capacity, he was charged with installing a wallpaper factory in a complex of old buildings—the *commanderie de l'ordre teutonique*—in Rixheim. By 1802, Zuber was sole owner of the factory. He brought his family to live in the grand residence at the center of the complex.

Once in control of the factory's fifty printing tables, he lost little time in launching his bold venture. In his autobiography he recounted his trip to Paris in 1803 where he met the artist Antoine Pierre Mongin (1761–1827), known for his battle scenes. In Paris, Zuber and Mongin planned the production of the factory's first major landscape, *"Les Vues de Suisse."* Zuber's autobiography reveals his sense of risk in committing large sums of money to this venture:

> This enterprise exceeded my means for the moment but I had become audacious and the success made a suitable return to my expectations. Mongin applied himself with dispatch and came to live at Rixheim with his wife at the beginning of 1804 to oversee the execution of his work. . . . The execution of our "Vues de Suisse" was advanced far enough to enable Soperlin [Zuber's brother-in-law] to take samples of it on his sales trip [all over Europe] during the winter of 1805. The first delivery was in the spring and the success was such that the inventory attained a profit of 72,000 francs . . . A first exposition of Industry at Paris awarded us a silver medal for our "Vues de Suisse."[11]

Zuber's reminiscences make clear that the production of each of the scenic papers was a large-scale undertaking requiring years of design and planning, the refined and carefully coordinated skills of highly trained artists and craftsmen, and international sales promotion. His three-year effort to make a single set of the scenic wallpaper was financially justified. Though certainly not the first to make such a wallpaper, his factory brought the craft of making these grand views to heights of technical perfection and popular success. Following its initial triumph of 1805, Zuber's factory was soon to rank, with that of Joseph Dufour (1752–1827), as one of the two most important producers of scenic wallpaper in France.

The Zuber firm, which Jean *père* turned over to his sons in 1835, went on to bring out at least twenty-three *paysages panoramiques* during the nineteenth century. Views of lands far and near, almost topographical in their obsession with details of geography and local monuments, as well as with costume, were among the factory's successful specialties. According to company records, many were derived from published engravings and etchings. Following the precedent of *"Vues de Suisse,"* by 1835 Zuber had brought out views of India (1807), another more extensive view of Switzerland (1813–1818) *(figure 9-3),* views of Italy (1818) *(figure 9-4),* of Scotland (1827) *(figure 9-5),* of modern Greece (1827) *(figure 9-6),* of Brazil (1829), and of North America (1834–1836).[12] This last paper, *"Vues de l'Amérique du Nord" (color plate 33),* became one of the most publicized of wallpapers during the 1960s when Jacqueline Kennedy had a set, which had been taken from the Stoner house in Thurmont, Maryland, put up in the White House. Zuber used its views of Scotland (1827) as background for scenes calculated to appeal to the Romantic tastes of the early nineteenth century. Figures in the foreground illustrated Sir Walter Scott's poem "The Lady of the Lake" (1810) *(figure 9-5),* which thrilled nineteenth-century readers with tales of chivalry and valor and put Scotland on the map as a point of pilgrimage for tourists.[13]

The Zuber factory issued scenic wallpapers in which sales appeal was heightened by the inclusion not only of scenes from romantic literature but historical events as well. In 1848 it reissued the paper first introduced in 1829 as "Views of Brazil," this time inserting episodes illustrating the Spanish conquest of Mexico. Accuracy of detail in presenting the historical setting was apparently not a first priority. In 1852, the factory offered for sale "Views of the American War of Independence." It was made up of the printed scenic first issued in 1834–6 as *"Vues de l'Amérique du Nord"* with additional hand-painted figures engaged in battles. The patriotic subject matter

9-8

Figure 9-8 The busy port of Boston is seen here in a detail from the set of Zuber's "*Vues de l'Amérique du Nord*" now at the White House *(see also color plate 33)*. Although the State House at the top of the hill clearly marks the scene as Boston, the foreground bears a close resemblance to waterside views of a variety of European ports pictured on other scenic papers. Courtesy of the White House

apparently commanded such a profitable American market that in 1927 blocks were carved so that the "War of Independence" paper could be more standardly and cheaply produced by the factory. Zuber continues to offer "War of Independence" in 1980.

The Greek views of 1827 *(figure 9-6),* as originally printed, incorporated battle scenes from the Greek war for independence from the Turks (1810–29). The paper commemorated and glorified recent events of Greek heroism, which were then commanding the sympathy of most of western Europe. Zuber's promotional literature accompanying the paper emphasized that it:

> recalls some of the admirable traits of self-sacrifice and of bravery, with which the modern history of Greece abounds and to which noble foreigners who, having no fear of sacrificing their lives and their dearest concerns, have risen to the aid of these unhappy people, and have taken such a glorious part.[14]

In addition, Zuber's directors would not have failed to recognize the sales potential for a subject that so prominently incorporated major examples of classical architecture. In this panoramic paper, each battle of the valiant contemporary conflict was pitched before an architectural monument of Greek antiquity. The dual appeal of a popular political subject and the models for much of Greek Revival architecture must have been carefully calculated. In American advertisements, this paper was sometimes referred to as the "Battle of Navarino."[15]

Battle scenes are relatively rare among Zuber papers compared with the production of other factories, but the classicism in the Greek views was more closely linked to the classical themes in an important and characteristic group of Zuber papers. However, the Zuber factory usually presented its classical subjects in serene and idyllic settings, far removed from the realities of nineteenth-century battles. The more typical classicism in Zuber's scenic papers has the air of a dilettante's dreamy and generalized appreciation of mythology and ancient forms. A vagueness of subject matter distinguishes Zuber papers, which incorporate casual visual references to the classical world, from the more archaeologically correct classical scenes by the rival Dufour factory. In 1811, Zuber introduced a paper advertising that it was inspired by Salomon Gessner's poem *"L'Arcadie,"* written in 1756. In Zuber's *"L'Arcadie"* *(figure 9-7)* anonymous figures draped in classical garb relax in a garden, at some distance from the viewer. In contrast, Dufour papers often depict specifically identified characters, at large scale, placed close to the viewer in settings, often interiors, that abound with clearly defined classical ornaments. (Compare and contrast *figures*

Figure 9-9 A portion of Zuber's *"Les Jardins Français,"* designed about 1821 by Antoine Pierre Mongin, is mounted on a folding screen six feet high. This scenic went through three "editions": printed figures costumed in the styles of 1830 were substituted in later printings for the figures shown here. Still later, in making a descendant from this paper, the factory eliminated all the printed figures and had artists paint figures dressed in Spanish costumes over the printed settings. Copies of the colorful view shown here have survived from houses in Tennessee, Nova Scotia, Massachusetts, and Virginia (where it is known in three houses). Cooper-Hewitt Museum, 1971-7-3; gift of Channing Hare

9-9

9-7 and 9-15.) Zuber's romanticized and generalized classicism also appears in *"Les Lointains"* of 1825. In it, as in *"L'Arcadie,"* Zuber gives us classical architecture tucked in amongst the greenery of a pleasant garden. In contrast, Dufour's designers, apparently responding with more enthusiasm to the Empire style prevalent in France's capital during the early years of the nineteenth century, give us images of classical forms that reflect hard, studious looking at the ancient models. For instance, in "Monuments of Paris," first produced by Dufour about 1814, the major buildings of Paris based on classical models have been removed from their actual settings, and presented in a friezelike configuration, perfectly frontal, in a straight line across the middle ground *(color plate 34).* Even when classical architecture plays a relatively prominent part in a Zuber scenic, as in "Views of Italy" (1818), its presentation is different from that in Dufour's papers. It is shown within a romantic, naturalistic setting. Where Dufour's versions of classical buildings and figures seem to demand study, in the Zuber scenes these elements keep a more polite distance: present, sufficiently interesting to divert the eye, but not striving to dominate interest so blatantly as do many of Dufour's.

Closely related to Zuber's purely classical garden scenes is a paper issued by the

Alsatian factory in 1821, *"Jardins Français" (figure 9-9).* The architectural ornaments in the garden are derived from classical sources, although the figures are dressed in stylish costumes of the 1820s. In 1836, a new set of blocks was carved for printing the figures in this scene. The new figures were dressed in updated *directoire* costumes. In 1849, the Zuber factory used the blocks, first carved more than a quarter century earlier as the background in *"Jardins Français,"* to produce a new paper called *"Jardins Espagnols."* To realize the geographical and cultural transformation, the factory simply eliminated the blocks for printing the figures, and artists painted in some figures in Spanish costumes, along with a few additional Hispanic details.

During the 1830s, Zuber issued two sporting scenes. One, a hunt shown against a background of Alsatian countryside, the area in which the Zuber factory is located, was called *Paysage à Chasse* (1831). A variety of famous races was depicted in *"Courses des Chevaux"* of 1837–8.

In 1832 the factory first printed an imitation of the eighteenth-century Chinese papers that featured flowering trees, birds, and insects. Called *"Décor Chinois" (color plate 35),* it remains in print to this day. It is unique among French scenic productions in directly using the format of the Chinese precedents and even reproducing their shallow spatial effect. (Compare with *color plates 15, 16.*) Its floral subject matter, however, is far from unique among the French papers. Large-scale displays of luxurious flowers were to become more popular from the 1830s through the 1860s. They succeeded landscape vistas as the major productions of the famous printers of scenics. The papers that featured flowers were among the most popular of the type of mid-century wallpaper decoration called *"décor"* rather than *"paysage."* (See pages 344, 353–359.)

At the Zuber factory during the 1840s and 1850s, gardens and flowers placed very close to the picture plane came to dominate the idyllic and exotic panoramas that had been the firm's most ambitious productions. These new *décors* included "Isola Bella" (1842–3), "Eldorado" (1848), and *"Les Zones Terrestre"* (1855). In "Isola Bella" and "Eldorado" large flowers and luxuriant plants rendered in muted, grayed pastel shades form a wide banding along most of the lower edge of the paper. This foreground of flowers dominates the paper almost totally; middle and background have faded in importance and are only vaguely suggested.

Just as the earliest scenic wallpapers arriving in America around 1800 had inspired advertisements verging on the rhapsodic, so Zuber's new style of grandiose floral decorations impressed writers of advertising copy. In 1854, Pratt, Hardenbergh and Company of New York offered "Isola Bella" for sale in New York, rejoicing:

> These Hanging Gardens bloom from every terraced hill-side; and to the delighted eye of the traveler, make all Italy seem like a Dream-land. . . . The walls of our palatial modern houses, begin to recall the scenes of those sunny lands as they are so faithfully pictured, on the satin Paper Hangings which the French artists have brought to perfection *(figure 9-26).*[16]

Although such *décor* papers as "Isola Bella" had become the focus of the efforts of famous French paper stainers, as late as 1861 the Zuber factory issued a new scenic. It was called *"Les Paysages Japonais"* or, at other times, *"Jardin Chinois"* by its manufacturers; apparently, the wallpaper trade never refined its appreciation for distinctions between Far Eastern cultures. As mentioned earlier, French landscape papers that show Oriental countryside seldom imitate the Chinese convention of flattening space *(color plates 18, 19),* nor does this Zuber example. Unlike the imported Chinese wallpapers of the eighteenth century in which distant scenes are brought forward and shown in layers above a foreground that fills the bottom portion of a wall, Zuber's Oriental view utilizes standard conventions of Western perspective. It presents only one large-scale non-repeating scene, which proceeds around all the walls of a room, creating an impression of continuous, panoramic space. Dufour and Desfossé et Karth also printed Oriental scenes organized in much the same way *(figure 9-21).*

The principal designers of these Zuber papers worked within formulas that conformed to and responded to the successful productions of the Zuber factory itself, as well as to the successes of its rivals. Many of the characteristics of the early Zuber scenics doubtless reflect the personal style of Antoine Pierre Mongin (1761–1827). He designed Zuber's first landscape paper in 1803 and stayed on to produce all of the factory's major landscapes during the next twenty years. He was probably responsible for the typical Zuber concentration of interest in a slightly distanced foreground, rather than in a foreground forced right up to the picture plane where figures would be rendered at large scale as in some Dufour papers. Mongin can also probably be credited with the rather relaxed, romanticized classicism in Zuber papers, which has been noted above. In the overall composition of a complete panorama, a fairly regular rhythm is apparent in many of the Zuber scenics he designed. Following one of his compositions horizontally all around a room reveals an alternation between densely detailed scenes of greater visual interest and relatively bland areas often made up of taller trees. It is a wavelike rhythm in which the high crests constitute the interludes, the pauses between the really interesting portions,

9-10

Figure 9-10 The firm of Joseph Dufour first produced this *"Paysage Indien"* about 1806. A full set included twenty lengths and was among the more moderately priced full-color subjects. American advertisements for "Indian Chase," "Views in India," "Views of Hindoustan," "Indian Hunt," or "A Tiger Hunt" could all have referred to this paper or to Zuber's *"L'Indoustan,"* designed by Mongin and printed for the first time in 1807. A set of a duplicate paper from the Putnam–Hanson house in Salem, Massachusetts, is now at the St. Louis Art Museum. Cooper-Hewitt picture collection

9-11

Figure 9-11 The printed French wallpapers were sometimes imitated in paintings made directly on plastered walls, as was this one, uncovered in the late 1970s in the parlor of a house in Amherst, Virginia. The painting is clearly derived from the imported paper shown opposite, and it is interesting to note the substitution of scenery, bearing a close resemblance to the mountains of Amherst County, for the exotic terrain of the printed version. Photograph courtesy of the Abby Aldrich Rockefeller Folk Art Collection, Williamsburg, Virginia

9-12

Figure 9-12 The monochromatic *"Les Vues d'Italie,"* first produced by Dufour et Leroy in 1822–3, was one of the most widely advertised scenics. Also called "The Bay of Naples," it includes Vesuvius. Similar views by several manufacturers have survived in this country in greater numbers than any others. Dufour's version, hung in Homewood Manor, the Ward–Boyleston house, in Princeton, Massachusetts, is shown in an early-twentieth-century photograph. Cooper-Hewitt picture collection

Figure 9-13 Four panels from *"Les Français en Égypte"* are pasted on a screen 7½ feet tall. The scenic was introduced about 1814 by Dufour. The inscription beneath the broken column is *"Le 20 Mars 1800, 10,000 Français Commandés par le Brave Kléber ont vaincu 80,000 Turcs dans les plaines d'Héliopolis."* Cooper-Hewitt Museum 1960-12-1; gift of F. Burrall Hoffman

9-13

where details are concentrated near the bottom of the paper. This compositional scheme not only worked well for Mongin, but was also used successfully by other French firms.

Mongin represents something of an exception in that he designed almost exclusively for Zuber. More usually, a designer of wallpaper scenes and *décors* worked for a number of rival wallpaper manufacturers in rapid succession, a fact that helps to explain why papers produced by several firms look so very much alike.[17] Indeed, stylistic analysis of a scenic paper of the 1820s or 1830s is less revealing of the probable manufacturer than is observation of a detail like the *ombré* effect in the ground coloring. If a paper has an *ombré* background, one can be fairly sure it is a Zuber, since that was a subtlety of little concern to other manufacturers.

After Mongin's death in 1827, Zuber employed several designers, including J. Michel Gue (1789–1843), who designed the Scottish scenic; Jean-Julien Deltil (1791–1863), who was responsible for the views of modern Greece, Brazil, and North America; as well as Eugène Ehrmann (dates unknown), who, in partnership first with Hermann Zipelius (dates unknown) and later with an artist recorded only as Schuler, created the floral *décors* that, in mid-century, supplanted the traditional scenic panoramas as the firm's most spectacular and ambitious wallpapers. Victor Poterlet, who worked for nearly every other major wallpaper factory of the mid-nineteenth century, designed the Zuber Oriental scene published in 1861.[18]

These designers produced scenic papers for Zuber that covered every major category of subject matter used by French manufacturers during the period: topographical views of countryside, exotic or domestic, with or without historical references; battle scenes; garden views and city scapes, especially as observed across or along bodies of water; classical scenes; views of hunting, shooting, and racing; and illustrations based on romantic literature. It could have been no accident that the Zuber factory covered all these categories. Its directors catered to popular tastes in literature and the arts in general, while conforming to current styles in the decorative arts. They must also have noted and responded to precisely what was selling well among scenic papers printed by their rivals.

DUFOUR

Chief among Zuber's rivals was the factory founded late in the eighteenth century by Joseph Dufour. Dufour entered the wallpaper business in Macon during the 1790s. With a brother, Pierre, Dufour opened his own manufactory. There, between

1804 and 1806, he printed his first scenic wallpaper, *"Les Voyages du Capitaine Cook,"* also called *"Les Sauvages de la mer du Pacifique."* Despite emphatic claims about its historical accuracy, this paper was based rather loosely on the celebrated events of James Cook's exploration of the Pacific, which brought the captain to his violent death there in 1779. (See *color plates 31, 32*.)

To design the paper, Dufour hired an artist from Lyon, Jean Gabriel Charvet, who was director of his own school of drawing in Annonay.[19] Well before the paper was ready for sale, Dufour began to promote it by publishing, in 1804 or 1805, a brochure describing it in great detail. The brochure specified that the paper would be made up of twenty lengths, each twenty inches wide. It identified the characters and incidents portrayed in every panel.

In his brochure, Dufour proclaimed that lofty educational motives had inspired him to undertake production of the paper:

> This decoration has for its object the idea of making the public acquainted with peoples and lands discovered by the latest voyagers, and of creating, by means of new comparisons, a community of taste and enjoyment between those who live in a state of civilization and those who are at the outset of the use of their native intelligence.[20]

So proud was Dufour of his fidelity to nature in the multi-colored botanical details depicting Pacific islands that he made a claim calculated to raise his wallpapers above the level of mere decoration. He appealed directly to mothers charged with the instruction of children at home:

> The mother of a family will give history and geography lessons to a lively little girl. The [several kinds of] vegetation can themselves serve as an introduction to the history of plants.[21]

By 1808 Dufour had moved his factory to 10 rue de Beauvau in Paris, into the former Abbaye de Saint Antoine, where, under the name "Joseph Dufour et Cie.," the firm enjoyed growing renown.[22] Dufour papers won prizes at all the French industrial expositions in which they were entered during the early years of the nineteenth century. In 1821, his daughter married Amable Leroy (d. 1880), whom Dufour took into the business, renaming it "Dufour et Leroy." Dufour died in 1827, and in 1836 Leroy turned the business over to the partnership of Lapeyre et Drouard.[23]

The scenic papers that Dufour's firm produced bear an interesting relationship to those made at the Zuber factory. A pattern of give and take, of responses and borrowings of ideas, is suggested by comparing the chronology of the introduction of new subjects into the offerings of the two factories. Zuber's topographical *"Vues de Suisse"* (1803–5) was originally marketed at about the same time Dufour was first selling his Captain Cook paper (1804–6). In 1806, Dufour again issued an exotic scenic, a paper called *"Paysage Indien"* or *"Vues de l'Inde" (figure 9-10).* The very next year, 1807, the Zuber factory countered with its *"L'Indoustan."*

Classical architecture was featured in Dufour's *"Portiques d'Athènes"* of 1808. In 1811, Zuber presented its classical landscape *"L'Arcadie."* Dufour's major offering of 1814, the scenic *"La Galerie Mythologique,"* by Xavier Mader, Dufour's most important designer, introduced a more archaeological classicism to the factory's wall-sized compositions. The paper is not really a scenic, but rather a figural *décor,* furnishing bold Empire-style ornaments for a room. It was the first of a series of classical subjects typical of the Dufour factory, to be surveyed in greater detail below.

Among the similarities between subject matter treated by the two firms, none is more striking than their views of Italy, the most popular subject of the time. At some time between 1818 and 1821, Zuber introduced Mongin's *"Vues d'Italie,"* a large, fully colored paper of twenty-five lengths that included a view of Vesuvius billowing smoke. By 1822, the Dufour firm had responded by bringing out its own *"Les Vues d'Italie" (figure 9-12),* showing Tivoli, Amalfi, and the Bay of Naples with Vesuvius in eruption. The Dufour paper was unusually large, including thirty-three lengths. Dufour undoubtedly hoped to undercut sales of Zuber's Italian scenes, since Dufour printed his version in monochrome and could therefore offer it at a lower price. Several copies of this paper with background coloring of a blue sky behind sepia-toned printing survive in museum collections.[24] A hurried look at one of these papers can nearly fool the eye into accepting Dufour's cheaper views of Italy for a fully colored scenic. Dufour et Leroy again illustrated Mount Vesuvius erupting in a later paper, *"Paysage Pittoresque"* (c. 1832?), a paper apparently appealing to the widely wandering tourist, since it incorporated glimpses of the coast of Normandy and of a French military encampment along with its Neapolitan vista. Because of the similarities of the Zuber and Dufour papers, verbal descriptions attempting to distinguish among the various papers featuring the same erupting volcano have led to numerous misattributions and widespread confusion.

Since the Zuber firm still exists in its isolated factory in Alsace, and its archives have been kept in relatively good order, it is possible to trace the history of most of its scenic papers, straightening out intervening misidentifications and misdatings. Unfortunately, however, the records and printing blocks of the Dufour firm were dispersed over time. Wood blocks first carved by the Parisian firm for printing its great scenic masterpieces were sold to a number of successors, including not only Lapeyre et Drouard (1836–41), but also Lapeyre (1842–50), Lapeyre et Kob (1851–9), Kob et Pick (1859–64), and finally Desfossé et Karth, who bought them in 1865.[25] Desfossé et Karth also bought the tools and wood blocks for scenics originated at other firms, including those of Xavier Mader, the same man who had designed many of Dufour's papers, including *"La Galerie Mythologique,"* mentioned above. The possibilities for misattributions and for inaccurate dating of Dufour papers have been multiplied by these circumstances. Nevertheless, the best French and German studies of the scenic credit Dufour and Dufour et Leroy with over twenty major sets.[26]

Many of these built on the success of Dufour's first formula. As in the Captain Cook paper, the formula combined historic adventures with exotic scenery. In 1814 Dufour issued *"Les Français en Égypte" (figure 9-13),* celebrating Napoleon's recent campaigns in Egypt. About 1826 the firm introduced *"Les Incas,"* fully colored, illustrating in twenty-five panels Pizarro's conquest of Peru in 1531. The paper improvised on an account by the historian Jean François Marmontel (1723–99). *"Campagnes des Armées Françaises en Italie"* (1829) featured recent military history in a setting romantic and picturesque, if less exotic than some of the earlier settings for Dufour scenics.

Closely related to this group, though lacking the historical dimension, were Dufour's landscape papers. One group of these featured exotic scenery: *"Paysage Indien"* of 1806 was followed by *"Paysage Turc"* in 1823 and by *"Les Rives du Bosphore."* This last paper is standardly dated 1829, but American advertisements of 1817 and 1826 offering papers called "River Bosphorus" and "Views of the Bosphorus" invite speculation that the paper has been assigned too late a date.[27] A second group of Dufour's landscape papers featured European views, especially views of cities set beside water. Among these, *"Portiques d'Athènes"* (1808), *"Monuments de Paris"* (c. 1814) *(color plate 34 and figure 9-14), "Les Vues d'Italie"* (c. 1822) *(figure 9-12),* and *"Paysage Pittoresque"* (c. 1832) have already been mentioned in other contexts.[28] Others in this group include *"Paysage Décor"* (date unknown), *"Vues de Lyons"* (date unknown) and *"Vues de Londres."* In the London views, major buildings, removed from their actual settings, are shown as if lined up along the Thames. Internal evidence suggests a date of about 1840 for this paper.[29]

Dufour's most important and distinctive essays within the genre of scenic wallpapers depict mythological subjects. Some of the papers in this group of wall-encompassing, non-repeating papers are not strictly scenics, but are Empire-style decorations blown up to a scale that covers the wall from chair rail to ceiling with classical ornament and large-scale figures. These include Mader's "*La Galerie Mythologique*" of 1814, mentioned above, an elaborate grisaille composition with touches of green and of blue, made up of six scenes illustrating the Vengeance of Ceres, Apollo and Phaeton, Venus and Diana, The Muses, The Judgment of Paris, and Time and the Seasons. It was accompanied by six wallpaper trophies planned to punctuate the composition.

Closely related to "*La Galerie Mythologique*" is the collection recognized in the nineteenth century as Dufour's masterpiece, by the company itself as well as by its admirers: "*Les Amours de Psyche*" or "*Cupid et Psyche*" *(figure 9-15).* First produced in 1816, after cartoons by Louis Lafitte and another designer, Merry–Joseph Blondel, it too is a grisaille composition, printed from 3,642 blocks on twenty-three panels of paper. Richly draped figures placed in the foreground dominate each episode. In most panels these figures appear in grandiose marble halls and pavilions like stage sets. Framed by these massive architectural elements of the foreground are glimpses of idyllic countryside beyond. The interiors are laden with the full panoply of the Empire-style renditions of classical architectural and decorative detail, including chairs, urns, and boldly geometric floor patterns, derived from recent archaeological finds. This set was frequently advertised for sale in America.[30]

Another grisaille paper in this classical group was called "*Les Fêtes Grec*" or "*Olympic Fêtes.*" Designed by Xavier Mader, it was advertised for sale in Washington, D.C., as early as 1819.[31] The figural groups in this paper are placed at a greater distance than in the two other grisaille classical compositions just described, and are shown amidst landscape. Thus, the paper is more accurately classified as "scenic," the scene in this case being the land of the gods.

Produced just before or just after "*Olympic Fêtes*" was another classical paper: Dufour's "*Télémaque*" *(figure 9-24).*[32] It became better known in America during the nineteenth century than any others of this classical group and was hung in many important houses. In this paper Xavier Mader used eighty-five colors to depict the voyage of the Greek hero to the island of Calypso. Sweeping vistas of Calypso's paradisiacal surroundings and of sparkling oceans provide the settings for a melodramatic representation of their romance and his escape.

Following the success of Telemachus, Dufour had Xavier Mader design "*Les Voyages d'Anthénor,*" another fully colored scenic, featuring dramatic episodes set at waterside. The paper depicts the wanderings of a Trojan prince who was spared during the siege of Troy and who later founded a city at Cyrene or Patavium.[33]

During this period a requirement for a successful Dufour scenic wallpaper seems to have been vast expanses of water. In the early 1820s, the Dufour firm again issued a multi-colored scenic wallpaper illustrating the ocean-going pilgrimage of another adventurer, though not in this case a classical hero, but rather a knight of the Middle Ages. *"Rénaud et Armide,"* or *"Jerusalem Délivrée,"* is supposed to have been the last set planned by Dufour himself.[34]

In 1823 or 1824 the company turned its attentions to the illustration of a more contemporary literary classic—the romantic tale of *"Paul et Virginie."* The book, written by Bernardin de Saint Pierre and published in 1789, ranked as a standard favorite throughout the nineteenth century. In 1869, eighty years after its publication, the American magazine *Appleton's* noted that "Long ago it became a classic in both languages, and there are few readers of the present generation who do not recall its perusal with unalloyed delight."[35] No doubt Dufour was counting on the public's sentimental attachment to the story to sell a wallpaper illustrating its high points, but the subject matter was also filled with the watery element that Dufour's designers were accustomed to treating with success: scenes of shipwreck and views of the island of Mauritius. The tale came ready-made with the seascapes that were so desirable for a successful scenic. Jean Broc (1780–1850) designed the monochromatic composition.

SCENIC WALLPAPERS BY OTHER MANUFACTURERS

During the first half of the nineteenth century, papers by Zuber and Dufour were the most popular of the French scenics imported by Americans. But papers from other French firms were also imported. The papers of François Robert have already been mentioned; in addition, the panoramic papers of five other French factories have been positively identified. Beyond these, there are a great many papers that have not yet been attributed to specific manufacturers.

The ambitious landscape productions of Jacquemart et Bénard probably ranked next in popularity to those of Zuber and Dufour during this period. Pierre Jacquemart (1739–1804) and Eugène Bénard (dates unknown) succeeded in 1791 to ownership of the renowned Réveillon factory. After the elder Jacquemart died in 1804, his son René continued the firm until 1840. The son remained in partnership with Bénard for a time, but by 1825, when only Jacquemart's name was appearing on the factory's publications, Bénard had apparently either left the business or died.

There are two multi-colored panoramas that have been found in American houses and that can be positively identified as Jacquemart et Bénard papers. Both are dominated by sporting themes: one is called *"La Chasse de Compiègne"* *(figure 9-20);* the other is *"Le Parc Français"* *(figures 9-17 through 9-19).* Eight examples of these papers, originally hung during the early- to mid-nineteenth century, survived into this century and were recorded by researchers; in fact, several examples still survive in place. *"La Chasse de Compiègne,"* first issued about 1815, was documented in five houses—one in Massachusetts, one in New Hampshire, one in Virginia, one in Kentucky, and one in Pennsylvania.[36] The scenic recently identified as *"Le Parc Français"* had been introduced in 1825 and was documented in two houses in Virginia and one in Massachusetts.[37]

Four examples of another paper sometimes attributed to Jacquemart et Bénard have been recorded in houses in New Hampshire, New Jersey, Massachusetts, and Vermont.[38] Printed in grisaille, the paper is called *"Les Quatre Saisons."* Like the scenics positively identified as Jacquemart et Bénard productions, it shows outdoor sports—shooting, iceskating, an outing in horse-drawn sleighs shaped like swans—as well as courting and haying scenes.

Another scenic paper is known to have been available in America because it is pictured in precise detail in a lithograph showing a Philadelphia wallpaper shop of 1849 *(figure 14-7).* This paper, *"Les Huguenots,"* was produced by the factory of Auguste Pignet, a lesser-known manufacturer from Saint-Génis-Laval near Lyons, by 1823. This paper represents the most elaborate exercise in the Gothic style among all the French scenic wallpapers *(figure 9-22).*

An ambitious mid-century wallpaper called "The Ages of Man," found in a Boston house, was made by yet another relatively obscure manufacturing firm, Campnas et Garat.[39] This firm was perhaps best known as the Parisian agent for Zuber.[40] "The Ages of Man" blows up to wall size a favorite theme of nineteenth-century prints and paintings, the life stages from infancy through maturity and old age to death. In this wallpaper version, the representatives of the various stages are given as large figures prominently placed in the foreground, dressed in medieval costumes.

Delicourt and Desfossé et Karth were the two most important Parisian wallpaper manufacturers of the mid-nineteenth century. Both specialized in the creation of elaborate Second Empire floral *décors* and panel decorations, which the American wallpaper trade chose to call "fresco papers." These will be discussed in greater detail in Chapter 15 *(color plates 73 through 77 and figures 15-15 through 15-19).* Nevertheless, some of their papers were scenes, although of a rather different style from earlier landscape views. Basically, in these mid-nineteenth-century scenics, the dados, columns, and friezes that had long been used to frame wallpaper scenes assumed a much greater importance, visually overwhelming the scenes they enclosed. This created the illusion that the framing elements in the wallpaper were placed right on the surface of the real wall. The scenes now looked like framed paintings rather than like vistas located in some imaginary space seen as if through a window. Both the floral *décors* of mid-century and the papers of the same period that can more properly be called "scenics," including "The Ages of Man" discussed above, Delicourt's "*La Grande Chasse,*" and the "*Rêve de Bonheur,*" produced by Desfossé et Karth, show the same tendency: visual interest is in the foreground; the distance to the apparent horizon is closed down; figures and major pictorial elements are presented at large scale, near at hand. Sometimes the elements in a *décor* seem to impinge upon the real space of a room.

Étienne Delicourt, born in 1806, had been associated with Dufour and then with the firm of Xavier Mader, before establishing his own wallpaper factory in Paris in 1838. Until 1859 he also acted, with Campnas et Garat, as Parisian agent for Zuber. Delicourt was repeatedly awarded medals for his firm's papers at French industrial expositions. At London's Crystal Palace Exhibition in 1851, the scenic masterpiece "*La Grande Chasse,*" designed by Dury (possibly Antoine, 1819–78), was enthusiastically awarded the only "Council Medal" awarded for wallpaper. In 1853 the firm won a bronze medal and received special mention at New York's Crystal Palace Exhibition "for general excellence of paper hangings."

Beginning in 1850, Delicourt papers were regularly offered for sale in New York by Emmerich and Vila, who listed themselves in the New York city directories as "Agents for Delicourts, Paris." In 1859, Frederich J. Emmerich began to list himself as New York agent for Delicourts, and he continued the same listing as late as 1884, despite the fact that Delicourt had turned the business over to Les Frères Hoock in 1860.

The elaborate productions of Desfossé et Karth also reached this country by way of regularly established agents during the nineteenth century.[41] In 1851, Jules Desfossé (1816–89) purchased the Mader firm. He operated it as a factory bearing his own name until 1863, when he formed a partnership with Hippolyte Karth (1842–1904). Karth on his own had succeeded the earlier firms of Dauptain, Brière and Clerc et Margeridon. He also brought to the new partnership some of the blocks for printing papers introduced by Dufour et Leroy.[42] In 1865, Desfossé et Karth bought additional blocks for printing Dufour scenics, and continued production of a number of important papers dating from earlier in the century. In fact, as late as 1931 the firm of Desfossé et Karth printed sets of the Dufour paper *"Cupid et Psyche."* [43]

Like the Delicourt papers, those of Desfossé were exhibited at New York's Crystal Palace in 1853, where Desfossé also received a bronze medal "for large and fine Tableau Decoration Paper." Examples of Desfossé's *"Rêve de Bonheur,"* designed by an artist recorded only as Durrant, have been found in houses in New York, Massachusetts, New Jersey, and Pennsylvania.[44] *"Rêve de Bonheur"* is very similar to the "Ages of Man" paper by Campnas et Garat. In both papers, figures are dressed in medieval costumes and shown at relatively large scale both near the foreground and throughout a dense middle ground closed off from any distant horizon by trees, rocks, and other objects.

Of the papers not yet attributed to specific manufacturers, one of the most distinctive illustrates the adventures of Don Quixote. Examples have been documented in houses in Bath, Maine, and Salem, Massachusetts; in addition, the town historian of Lynchburg, Virginia, recorded that a set of the paper was hung in a house on Main Street in that town in honor of a visit by then General Andrew Jackson in 1815.[45]

THE TRADE IN SCENIC WALLPAPERS

Despite earlier assertions to the contrary, recent scholarship has conclusively shown that French wallpapers were widely available in this country during the nineteenth century. Antiquarians fantasizing during the 1920s supposed that scenic papers found their way here from France as precious rarities for a distinguished few New England walls, "as a wedding gift or a birthday or anniversary present [brought] by some sea-captain specially charged with its transportation."[46] But the extensive records at the Zuber factory in Alsace make it clear that French panoramic papers were manufactured in quantity and regularly shipped all over the world, and especially to America, as wholesale commodities. Although detailed documentation of the production and trade arrangements of the other major French wallpaper manufacturers is lacking, the extensive records of the Zuber factory suggest a volume and regularity in shipping scenics to all the then-settled parts of America that are probably representative of the nature of the international business conducted by Dufour and perhaps by others of the early-nineteenth-century French factories.

Nineteenth-century account books which have survived in the offices of the old Zuber factory buildings in Alsace provide detailed information about the French scenic wallpaper trade with American customers. They record shipments of wallpaper to individual Americans as well as wholesale transactions with a great many

9-14

Figure 9-14 (above) At Friendfield, a South Carolina house in the Georgetown vicinity, a set of Dufour's "Monuments of Paris" *(color plate 34)* survived to be photographed early in this century. It was hung with an elaborate drapery swag frieze and with a floral border at the chair rail. Over the mantle, the picture is also wallpaper, a small-scale scene printed in France. In addition to this example, twelve sets of "Monuments of Paris" supposed to have been hung during the first half of the nineteenth century remained to be recorded in research files at the Cooper-Hewitt Museum. It also appears in a lithograph showing the window of a Philadelphia paper-hanging store of 1847 *(figure 10-1)*. Courtesy of the Charleston Museum

9-15

9-16

Figure 9-15 (opposite, bottom right) "The Reconciliation of Venus and Psyche" is illustrated in three panels from Dufour's great scenic set *"Les Amours de Psyche,"* which includes a total of twenty-six lengths. Designed by Merry-Joseph Blondel (1781–1853) and introduced by Dufour in 1816, it was kept in print as a product of Dufour et Leroy after Dufour took on a son-in-law as a partner in 1821. Later Desfossé et Karth bought the blocks; they were last used in 1931. Printed *en grisaille* on modern machine-made paper in 1923, this panel is 70 inches high. Cooper-Hewitt Museum, 1974-109-11; gift of Mr. and Mrs. Abraham Adler

Figure 9-16 (above) A photograph made early in this century at The Lindens, a house in Danvers, Massachusetts, that has since been moved to Washington, D.C., suggests the sad fate of many a scenic paper. Here the horns of the stag head have been thoughtfully placed so that the leaves on a palm tree on Calypso's island behind him will embellish his already impressive array of headgear. He appears to be unperturbed by Minerva's desperate gesture to save Telemachus by pushing him from a cliff. She does this to rescue him from the intrigue of Venus, which has made him fall in love with Calypso and thus divert him from the task of finding his father, Ulysses. See also *figure 9-24* for *"Télémaque."* Cooper-Hewitt picture collection

9-17

9-18

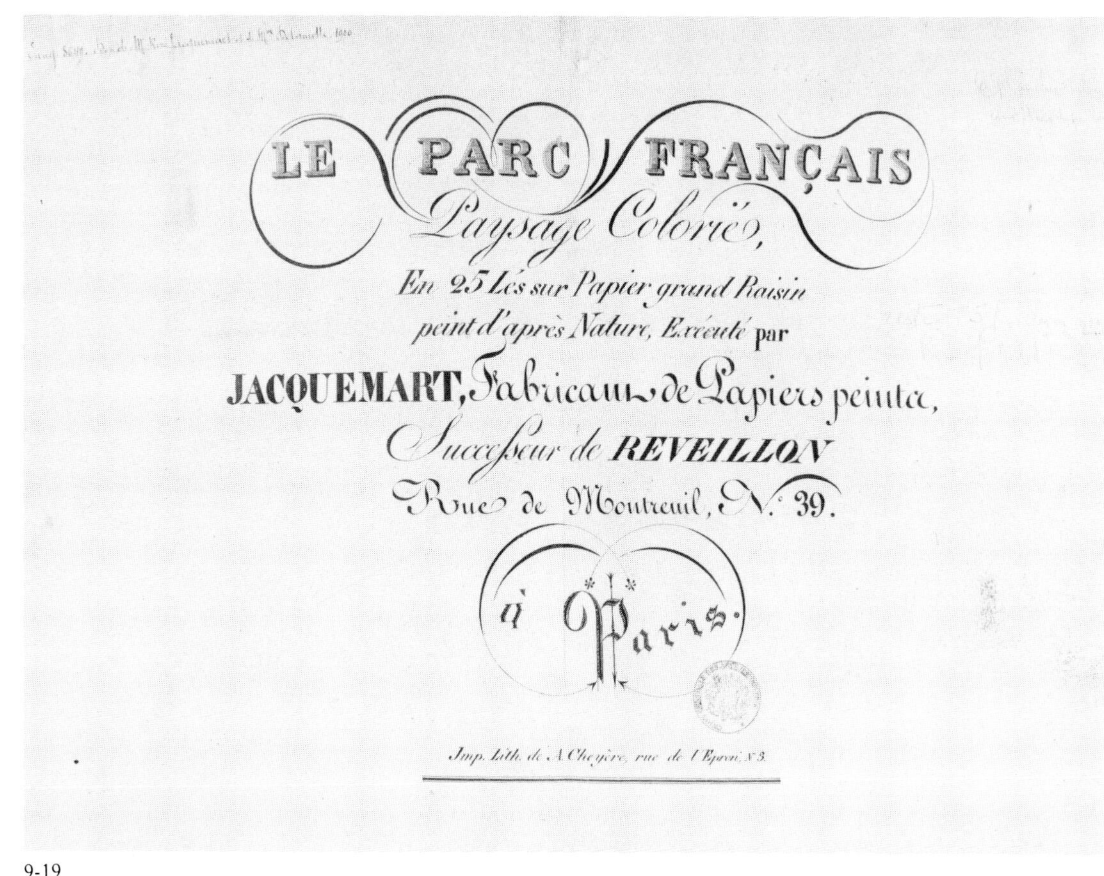

LE PARC FRANÇAIS
Paysage Colorié,
En 25 Lés sur Papier grand Raisin
peint d'après Nature, Exécuté par
JACQUEMART, Fabricant de Papiers peints,
Successeur de REVEILLON
Rue de Montreuil, N.° 39.
à Paris.
Imp. Lith. de A. Cheyère, rue de l'Eperon N.° 3.

9-19

Figure 9-17 (opposite, top) In the hall at Prestwould, the eighteenth-century home of the Skipwith family, near Clarksville, Virginia, Jacquemart's paper *"Le Parc Français"* still hangs. The Richmond firm of Francis Regnault put up this paper for Prestwould's owner, Humbert Skipwith, in 1831 and 1832. A duplicate of this paper hangs in the King Caesar house in Duxbury, Massachusetts. Photograph courtesy of the Association for the Preservation of Virginia Antiquities and the Prestwould Foundation

Figure 9-18 (opposite, bottom) This is a lithograph from an album issued by Jacquemart illustrating the paper shown in *figure 9-17.* Close comparison makes clear that the paper in the stair hall at Prestwould is indeed this Jacquemart product. Another set of this scenic was photographed in a Lynchburg, Virginia, house before it was removed and the house demolished. The paper in Lynchburg is supposed to have been hung by Dr. John C. Cabell in 1817 in honor of a visit by General Andrew Jackson. Courtesy of the Musée des Arts Décoratifs, Paris

Figure 9-19 (above) This is the title page of an album of lithographs that included the page illustrated in *figure 9-18.* Courtesy of the Musée des Arts Décoratifs, Paris

9-20

Figure 9-20 Regnault also hung this paper at Prestwould during 1831–2. It is a product of Jacquemart et Bénard called *"La Chasse de Compiègne,"* a hunting scene first printed in 1814. It survives with the floral borders and Empire-style dado originally hung with it. In the 1830s, an eighteenth-century paper with a leaf pattern was left on the wall under the new scenic, and its imprint can be seen where it has stained an impression through the "sky" above the trees. The dado pattern duplicates one illustrated as *color plate 41.* Photograph courtesy of the Association for the Preservation of Virginia Antiquities and the Prestwould Foundation

dealers and distributors. Study of these records suggests that it is safe to assume that examples of every major scenic wallpaper printed at the Zuber factory were readily available in several American cities almost as soon as they were introduced in France. These archives reinforce the testimony of advertisements in numerous American newspapers that by the 1820s, if not a little earlier, a customer who fancied a scenic for the front hall did not need to journey to the major commercial centers of the northeast, much less to France; nor was it necessary to have an agent abroad or a friend or relation who sailed the seas.

These documents have also disproved another myth of the 1920s, which held that "very few scenic papers found their way below the Mason and Dixon line" and that "they stayed mostly in or near the Northern seaport towns where they landed."[47] Not only do the Zuber accounts document sales to dealers in the Middle Atlantic states as well as the south, but examination of local newspapers for the period reveal a great many advertisements for scenics in all these regions. All this, coupled with the discovery of a number of old sets of the papers on walls in these areas, amends old misleading notions about scenics. The misinformation was published because wallpaper enthusiasts of the 1920s seem to have taken facts about the relatively small eighteenth-century trade in very expensive painted Chinese and English papers and erroneously applied those facts to the larger nineteenth-century commerce in the printed French panoramas.

A breakdown of the Zuber records reveals interesting details. The *Grandes Livres,* or master account books, for the years 1829 to 1834, record sales and shipment to ninety-nine individuals and firms, most of whom are known to have been manufacturers of, or dealers in, wallpaper. They were located up and down the American East Coast, and as far west and south as New Orleans. After 1834, most of Zuber's American trade was channeled through an agent or distributor in New York, Ernest Fiedler, and later through other primary agents. But before consolidating its American dealings, Zuber did business directly with forty New Yorkers, with thirteen Philadelphians, and with eleven Bostonians; with seven individuals and firms in Baltimore, with five in New Orleans, with three in Providence, as well as three in Washington, D.C. The *Grandes Livres* also record shipment to two dealers in each of six cities: Albany, Charleston, Montreal, New Bedford, Portland, and Richmond. In addition, there are accounts of dealings with a single buyer in each of five additional cities: Augusta, Norfolk, Portsmouth, Savannah, and Salem.

Some of the more detailed records preserved at the factory show that scenics were shipped to this country in wholesale quantities, and others list precisely what was included in individual shipments; among them were quantities of scenics.

On May 30, 1816, the factory recorded shipment to the Philadelphia firm, Geisse and Korkhaus, of three sets—designated "collections" on Zuber invoices—of *"Paysage l'Arcadie."* The monochromatic collections were shipped *"en grisaille"* as well as in two shades of brown: *"en chocolat"* and *"en bistre."* Each set cost 45 francs. On the same invoice appeared a collection of *"Vues de l'Indoustan"* at 60 francs and one of *"Vues de Suisse"* at 56 francs. *Paysages panoramiques* represented only six and a half percent of Zuber's total wallpaper sale to this Philadelphia firm during 1816. The complete account with Geisse and Korkhaus came to 3,933 francs in that one year.[48]

In the following year, 1817, for February 27, the Zuber records show shipment of larger quantities of scenics to another Philadelphia concern, Virchaux and Borrekins, wallpaper manufacturers. This included: "12 *Collections Vues de l'Indoustan, à nuages* [with clouds]" at 80 francs each. Copies of the same collection had been sent, without clouds, to Geisse and Korkhaus a year earlier, at only 60 francs. During 1817 the records show that Virchaux and Borrekins purchased twelve collections of the Indian views, plus "5 *Collections Vues de Suisse à nuages*" at 56 francs each, "2 *Collections Paysages l'Helvétie,*" a large, full-color panorama costing 120 francs for each collection, and "2 *Collections Paysages L'Arcadie en Nanquin*" (light beige). Each set of the smaller monochromatic *L'Arcadie* cost only 45 francs.

9-21 9-22

Figure 9-21 (above, left) A thoroughly Western version of a Chinese landscape, this French scenic paper of the early nineteenth century was photographed early in this century at the Lathrop house in Stockport, New York. (Contrast this with *figure 5-3* and with *color plates 18 and 19.*) Cooper-Hewitt picture collection

Figure 9-22 (above, right) *"Les Huguenots"* was made by Pignet et Cie., at St. Génis-Laval near Lyon, in the 1840s. It displays mid-nineteenth-century taste for the Gothic in one of its most elaborate manifestations. Cooper-Hewitt Museum, 1970-26-6; gift of Jones and Erwin

Virchaux and Borrekins' account with Zuber in 1817 amounted to 2,770 francs, fifty-six percent of which was for scenic panoramas. This is in sharp contrast to the other Philadelphia firm of Geisse and Korkhaus mentioned above, whose order for scenic papers the previous year had constituted only six and a half percent of the total amount.[49]

Scenics constituted a still larger percentage of the purchases of the only other American for whom detailed accounts of the early nineteenth century have been preserved at the Zuber factory. For December 21, 1822, the Zuber accounts record the sale of nine collections of scenics to Miles A. Burke of New York. At least one each of every major scenic the factory had in production up to that date was shipped to Burke: *Helvétie, Italie, Vues de Suisses, Vues de l'Indoustan, Arcadie (en nanquin et en gris), Jardins Français,* and *Petit Helvétie.* The cost of these panoramas was ninety-eight percent of Burke's total bill of 483 francs for 1822.

Two years later, Burke bought sets of only two panoramas—"*Vues de l'Indoustan à Nuages*" and "*Jardins Français*"—but he bought eight sets of each. Perhaps these two represented the best sellers from his earlier selection. These panoramas accounted for one hundred percent of his purchases from Zuber in 1824, which totaled 1,040 francs.[50]

The invoices also confirm that all kinds of Zuber wallpaper, not only scenics but also repeating patterns, borders, friezes, and dados, imitations of architectural elements and of statues, as well as imposing ornaments of every description, were being sent here. Among the listings, the high percentage of scenics seems logical. American craftsmen were able to print the simpler repeating patterns, but they were not up to making the magnificent block-printed masterpieces of panoramic sweep that were Zuber's specialty. If American manufacturers wanted to offer large-scale scenics for sale in their stock, they had to turn to France.

Considering the percentages of scenics listed on the few detailed invoices of shipments to Americans as rough indicators, one can calculate the quantities of scenics probably included in the other, undetailed, shipments that are entered in the *Grandes Livres* only as total sales. Even when scenics comprised only six and a half percent of one shipment in 1816, that figure represented five sets of scenics. Geisse and Korkhaus, the purchasers of those five scenics, were but one of the thirteen Philadelphia firms whose names appear in Zuber account books. Their account total for one year of 3,933 francs can be compared with the totals of other Philadelphia accounts, which reached sums as high as 10,547 francs over a four-year period.[51]

Zuber sold the largest and most expensive of the multi-colored scenic collections, "*Paysage de l'Helvétie,*" for the unusually high price of 120 francs, and the smaller, monochromatic *L'Arcadie* for as little as 45 francs. In between these extremes the factory charged about 70 francs for many other scenics. Seventy francs could, therefore, be considered a typical price for a scenic set. The very low figure of six and a half percent can probably be considered a minimum percentage of scenics in any American order of wallpapers from Zuber. Taking the 70,000 francs total for all accounts with Philadelphians recorded in the Zuber archives during the twenty-year period 1815–35, simple calculation yields a figure of seventy-five as a conservative estimate of the *minimum* number of Zuber scenics that must have reached Philadelphia from the Zuber factory alone during those years.

Applying what may well be a more representative percentage—fifty percent—to this same set of figures yields five hundred scenics that might have been included among the Philadelphia dealers' total purchases between 1815 and 1835. Application

9-23

Figure 9-23 (above) This sweeping view alternately called *"Les Bords de la Rivière"* and "Roman Ruins" would have filled a wall space just over 14 feet wide. Such block-printed French wallpapers brought into American homes of the early nineteenth century an excitingly close and realistic vista of the glamour and beauty of places and cultures far distant in time and space. A detail from another portion of this same scenic is shown in *color plate 37.*

Figure 9-24 (opposite) Andrew Jackson chose *"Télémaque"* or *"Les Paysages de Télémaque dans l'île de Calypso"* for the grand entrance stair hall of the Hermitage. The scenic wallpaper was designed between 1815 and 1820 for Dufour by Xavier Mader—2,027 blocks and 85 colors were required to execute Mader's interpretation of the adventures of the son of Ulysses ("Telemachus" in English) on the island of the nymph Calypso. Dufour's wallpaper was based on *Télémaque,* a prose work of 1699 by François Fénelon (1651–1715). In 1836 Jackson ordered sets of the paper for his house outside Nashville, Tennessee, from Robert Golder, a Philadelphia wallpaper dealer. Courtesy of the Ladies' Hermitage Association

Figure 9-25 (opposite) Zuber scenics displayed in the New York store of Sutphen and Breed were depicted in their advertisement of 1855 that appeared in volume 3 of the *American Portrait Gallery.* "Eldorado" designed by Ehrmann, Fuchs, and Zipelius in 1848 was shown over a paneled dado on the right and "Isola Bella" (figure 9-26) was shown over a wallpaper balustrade on the left. Courtesy of the Metropolitan Museum of Art, Elisha Whittelsey Fund, 1958

9-25

9-24

9-26

Figure 9-26 Eugène Ehrmann and Hermann Zipelius designed this "Isola Bella" *décor* in 1842 for Zuber. In 1854 an advertisement for Pratt, Hardenbergh and Company rhapsodized over "These Hanging Gardens of Isola Bella, rising in magical beauty from the glassy bosom of Lake Maggiore . . . to the delighted eye of the traveler, make all Italy seem like a Dream-land. The walls of our palatial modern houses, begin to recall the scenes of those sunny lands as they are so faithfully pictured, on the satin Paper Hangings which the French artists have brought to perfection." Zuber et Cie. still produced this scenic in 1975. Courtesy of Zuber et Cie., Rixheim, Alsace

of the same formulas—six and a half percent at 70 francs each, then fifty percent at 70 francs each—to the 52,000 francs total of Zuber wallpapers sent to Baltimore between 1829 and 1832 yields a minimum of 48 and a more probable figure of 370 for estimating scenics among those shipments. Two dealers in Charleston bought a total of 3,236 francs worth of wallpaper from Zuber between 1829 and 1832—perhaps as few as 3 or as many as 23 scenics might be estimated among these purchases. The five New Orleans purchasers mentioned in the Zuber archives bought 14,462 francs worth of wallpaper between 1829 and 1838—this total possibly could have included as few as 13 scenics or, more probably, as many as 103. Between 1829 and 1832 a single account in Norfolk, Virginia, totaled 8,631 francs—this could mean as few as 8 scenics or as many as 61.

Even greater quantities of wallpapers were regularly sent by Zuber to forty New Yorkers during the 1820s and 1830s. By 1838, when Ernest Fiedler had become the principal agent for Zuber wallpapers in the United States, the account with Fiedler for that one year totaled 363,809 francs. Using the formula, this could indicate as few as 338 sets of scenics or as many as 2,600 average-priced scenics sent to New York during 1838 alone.

Such numerical speculations make clear that the number of French scenic wallpapers imported during this twenty-five year period was impressive. The magnitude of the American trade in scenic wallpapers becomes even more impressive in light of the fact that American dealers who bought from Zuber standardly dealt in scenics made by other French firms as well. Between 1815 and 1826 detailed listings of scenic wallpapers were featured in the advertisements of the two principal wallpaper dealers in Washington, D.C., S. P. Franklin and Samuel Robinson. Franklin's name appears in the Zuber account books for 1828 and 1829 as a purchaser of 3,142 francs worth of wallpapers. But his advertisements in the Washington newspapers list for sale papers that seem more likely to have come from Dufour's factory. For instance, in February of 1823 Franklin announced that he had just received "a few Select views of the most beautiful colors, viz: Adventures of China, Ditto of Telemachus, Indian Hunts, Monuments of Paris, Views of Sicily."[52] While "Indian Hunts" and "Views of Sicily" could have been Zuber productions (though Dufour produced papers illustrating the same subjects), there is little doubt that the Telemachus paper and Monuments of Paris were Dufour's.

Advertisements of Franklin's rival, Samuel Robinson, are even more richly detailed. Again, they list papers that can be identified as the products of both Zuber and Dufour. On June 10, 1820, he announced in *The Daily National Intelligencer* that he had

> . . . just received an assortment of views, selected by a gentleman of Philadelphia, in Paris, of known taste in the above line of business—they are The views of Helvetia, in gay colors, affording a beautiful prospect of the country. The views of Hindostan, rich colors, &c Do. Do. Telemachus, representing an interesting and familiar tale of his adventures on the island of Calypso, in rich Colors. Also, the views of the Passage Nankin, in buff coloring.

While "views of Helvetia" probably came from the Zuber factory, and those of "Hindostan" could have been made by either of the firms, Telemachus was Dufour's exclusive product. In other advertisements published between 1819 and 1826, Robinson also specified among his landscapes "Spanish Views," "Views of Bayonne," "Views of the Bosphorus," and "Views of Italy," as well as "Views of Greek Fests," "Views of Greek Festivals," and "Views of Olympic Games"—these last three probably describing the same paper in Anglicized versions of Dufour's alternate titles *"Les Fêtes Grec"* or *"Olympic Fêtes."*

Hundreds of such detailed and informative advertisements for "Views" and "Long strip landscapes" were published in newspapers in almost every city of any size during the first half of the nineteenth century. Newspapers of the northeast abounded with them. The early-nineteenth-century popularity of scenic wallpaper in New England and other northern seaboard cities was recognized in early-twentieth-century books and magazine articles about old houses of these areas, as well as in all the 1920s magazine articles about wallpaper history. In addition, more recent research in the middle Atlantic states, the south, and the inland states of the old west, which were being settled during the early-nineteenth century, has uncovered quantities of advertisements and scenic wallpapers surviving from early-nineteenth-century installations.

One of the first of the advertisements from a place far removed from the seaport towns of New England appeared in a New Orleans newspaper in 1808:

> SAVAGES OF THE PACIFIC OCEAN. A beautiful Decoration in paper, lately painted in Paris, and received per ship Franklin from Bordeaux. The painting is composed from the discoveries made by Captain Cook, de la Perouse, and other travellers, it represents a shaded Landscape, done by Nos. and on 20 widths of paper, 20 inches wide on 90 high, it is for sale at L. Fournier's, Royal Street . . . where may be had a large assortment of fashionable papers which he has just received.

This advertisement reveals that the shipment of these French papers to New Orleans had been direct from Bordeaux, not by way of any northeastern port. In describing the paper as "painted," the New Orleans dealer used a word that continues to crop up when scenics are summarily described. In some cases it reflects an overly literal translation of the French *"papiers-peint,"* and in others it betrays the fact that most casual observers could not imagine a printing process that produced work so like a painting.

An advertisement including the Captain Cook paper was placed in local Lexington, Kentucky, newspapers by Downing and Grant, who described themselves as "Painters and Grocers." Along with the vegetables, they advertised in 1816:

> *French and American paper hangings* . . . Among them there are a few sets of the Monuments of Paris, Views of the City and Bay of Naples, with an elegant representation of Mount Vesuvius, Captain Cook's voyage in the Pacific Ocean, and a representation of his death by the Owyhee nation. A view of the Chase. Paul and Virginia, and some Views in India . . . They have also received an extensive assortment of GROCERIES . . .[53]

Records of the Captain Cook paper exist for houses in Salem, Massachusetts; Augusta, Maine; Hoosic Falls, New York; and Mecklenberg County, North Carolina.[54] Bearing in mind that these records only represent chance survival, and/or the reporting of old wallpapers to researchers who published the information, or recorded it in study files like that at the Cooper-Hewitt, it is probably no exaggeration to estimate that as many as a hundred or more examples of this Captain Cook paper might have been hung in America during the first half of the nineteenth century.

Views of Italy seem to have been the most popular subjects in scenic papers used in America. Downing and Grant's Kentucky advertisement had offered one of the several such views available in 1816, describing it as "Views of the City and Bay of Naples." In addition to seventeen examples of the Italian views printed by Dufour[55] and to several by Zuber, which can be fairly readily identified, at least four additional French scenic papers have been found in American houses that illustrate Italian cities and countryside. These include a paper called "Venetian Scenes," which survived to be recorded in the 1920s in ten houses spread across Massachusetts, Vermont, Pennsylvania, New Jersey, and Virginia.[56] Another Italianate view known as *"Les Bords de la Rivière,"* featuring classical ruins, has been found in a house in Dorchester, Massachusetts, as well as in Unionville, Ohio *(color plate 36).*[57] A duplicate paper sold in 1835 by a New York dealer to an innkeeper in Ohio was called "Suburbs of Rome" on the bill of sale.[58] Similar papers called "Views of Rome" and "Roman Ruins" have been recorded in houses in Colchester, Connecticut, Chelsea, Vermont, Manchester, New Hampshire, and Portsmouth, New Hampshire.[59] Other papers have been found in America, similar enough to create confusion, which do not necessarily depict Italian scenes, but scenes that could be either French or Italian —riverside settings adorned with classical structures, peopled by figures in early nineteenth century costumes, like one designated "Views of Town and Country."[60]

By 1819, a set of Dufour's *"Vues d'Italie"* as well as another paper showing the Bay of Naples, had been hung at Gay Mont, a house in Caroline County, Virginia. The paper had made such a strong impression on John H. Bernard of Gay Mont that when he visited Naples, it colored his impression of the actual scene. Bernard wrote about Naples in a letter of April 16, 1819, to his wife:

> I do not think that the bay itself is so beautiful as that of N. York with w^h it is frequently compared . . . A tongue of land and the ancient Castel dell'Ovo w^h is suited to it by a bridge, is the same w^h is nearly opposite the fireplace in the drawing room at home w^h really presents as correct a view of the whole scene as such a representation could be supposed to do.[61]

By 1850, at least one American company had attempted to manufacture a scenic wallpaper in the French manner. The Boston City Directory for that year carried an advertisement of the Boston and Chelsea Paper Company that offered "of their own manufacture, SCENERY PAPER, forming a continuous panorama, and embracing American Views of universal interest."

Despite this and other, limited, evidence that American firms did print scenic wallpapers during the mid-nineteenth century, the author has been unable to locate surviving examples. In *Antiques* Magazine for April 1961, Charles Woolsey Lyon advertised for sale and illustrated what he described as a hand-printed wallpaper panel. The paper, mounted on a three-leaf screen, depicts "Bowling Green, 1825." That illustration is the only one known to the author that shows what is possibly an American-printed scenic of the mid-nineteenth century.

In 1854, the writer of advertising copy for the New York wallpaper firm Pratt and Hardenbergh announced with the bravado of a P. T. Barnum that the firm "intended" to manufacture an American scenic wallpaper. In the announcement of this great event, the copywriter first assumed the voice of a reporter/critic to declare:

> There has hitherto been one thing to be desired in this department of commerce—*American Scenes* . . . We are rejoiced to learn that these accomplished young men [Pratt and Hardenbergh] are preparing to manufacture original styles of papers which will illustrate our own History and scenes.

He then appropriated the voice of a poet to bless his readers with five verses of "Lines inscribed to Pratt Hardenbergh and Co. On hearing that they had determined to manufacture Wall Paper illustrated with Scenes from American History and Landscape."[62]

Most of the surviving French scenic wallpapers have been found in rather grand houses. This has created the impression that their use was restricted to the homes of the very wealthy. Yet, by the mid-nineteenth century, they were apparently somewhat more commonplace. In the *New York Commercial Advertiser* for January 26, 1835, a two-story frame house was offered for sale on Prospect Street in Brooklyn, and "landscape wallpaper in three rooms" was noted. Though never cheap, scenic papers were enjoyed in the homes of aspiring merchants, businessmen, and professionals, not just in the mansions of the most elite. For instance, the National Park Service's restoration of Harper House, no grand mansion, at Harper's Ferry, West Virginia, uncovered fragments of a set of the Dufour firm's scenic *Les Fêtes Grecs*, which was probably hung by James Wagner, as part of remodeling, in 1832–3.[63]

Prices for sets of scenics varied. A bill of 1819 records the sale by J. W. Foster of Boston to Jacob Wendell of "1 set, Landscape Paper (Elysian Fields)" for 10 dollars. This perhaps was Zuber's *"L'Arcadie,"* one of that factory's least expensive small grisaille collections.[64]

Entries in the account book of Ephraim Gilman, general importer of Alexandria, Virginia, for July 3, 1819, assign values to five landscape subjects. These include "1 sett paper hangings, View of Antibes, Bordeaux, and Bayonne," priced at 35 dollars. An entry for two sets of "Environs of Rome" was followed by two different prices —27 dollars and 30 dollars, totaling 57 dollars for the two. A set of a "Bay of Naples with border and surbase" was valued at 40 dollars, and one of "Cupid and Psyche" at 40 dollars as well. Finally, "Five setts papers" were assigned the low value of 14 dollars each.[65] While prices ranging from 10 to 40 dollars for a set of wallpapers do not represent cheap wall decorations in 1819, such prices do not put them in a category possible only for the richest few. By 1819, the cheapest patterns in American-made wallpaper were being sold for 25 cents a roll, but advertised prices as high as $2.50 and even $5.00 appear in East Coast newspapers. At $2.50 a roll, twenty rolls—the number equivalent to an average-sized set of scenic wallpapers—would have cost $50.00. Numbers of prosperous Americans—not just the scions of old fortunes and those who had profited phenomenally from ventures in newly developing enterprises—were apparently prepared to pay such prices to decorate their walls.[66]

Scenic papers belonging to the rich and famous have been the most carefully preserved, as have been the documents of sale accompanying their wallpapers. During 1836, Andrew Jackson was twice billed for sets of Dufour's Telemachus paper that had been ordered for his Tennessee home, the Hermitage *(figure 9-24).* Among Jackson's papers survives a bill from George W. South of Philadelphia, who charged 40 dollars for each of the three sets sent to Jackson. Because the wallpaper was not delivered, later in 1836 Jackson ordered three more sets of Telemachus, this time from Robert Golder of Philadelphia. On Mr. Golder's bill the price per set was only 29 dollars.[67] One is left to speculate whether such variation in price for the same scenic paper reflects the fact that the examples of the papers first sent were embellished with particularly expensive clouds, or were printed on especially fine paper, or whether Mr. Golder was a political sympathizer who charged President Jackson very little mark-up.

A Mr. Van Gorder of Warren, Ohio, had had to pay substantially more than President Jackson for each of the two sets of a "color view" he had purchased in

1835. The bill of the New York dealer who sold the paper, Francis Pares, describes it as "Suberbs of Rome." The bill shows that he charged Mr. Van Gorder 60 dollars per set for the papers that had been ordered for his Inn and Coach Stop in Warren.[68]

Scenics were popular for such public places. The use of these papers in American hotels caught the eye of one traveler, Harriet Martineau, who published her *Retrospect of Western Travel* in 1838. In a chapter entitled "High Road Travelling" she commented:

> I observed that hotel parlors in various parts of the country were papered with the old-fashioned papers, I believe French, which represent a sort of panorama of a hunting party, a fleet, or some such diversified scene. I saw many such a hunting-party, the ladies in scarlet riding-habits. . . . At Schenectady, the bay of Naples, with its fishing-boats on the water and groups of lazzaroni on the shore adorned our parlor walls. It seems to be an irresistible temptation to idle visitors, English, Irish, and American, to put speeches into the mouths of the painted personages; and such hangings are usually seen deformed with scribblings. The effect is odd, and in wild places, of seeing American witticisms put into the mouths of Neapolitan fishermen, ancient English ladies of quality, or of tritons and dryads.[69]

This observant English visitor deemed French scenic wallpapers "old-fashioned" by the late 1830s. The comic-strip-like captions she described are perhaps indicative not only that there is a long-standing tradition of graffitti art in America, but also that travelers had become so tired of seeing the same grandiose scenes in place after place as to judge their redundancy in need of novel embellishment.

Certainly by 1853, the popularity of scenic papers was beginning to fade in America. A critic writing of wallpapers at the New York Crystal Palace Exhibition that year instructed his readers that, while elaborate papers were perhaps admissible in public halls or saloons, and sometimes in the entranceway of a private residence, they were not to be used in a private drawing room, library, or chamber:

> In such a situation the covering of the walls should have the same relation to the furniture, objects of art, and occupants of the apartment, that a background sustains in a good picture: not overpowering.

He went on to warn:

> Landscapes, therefore, whether large or small, groups of figures, imitative carving and panelling, *et id genus omne* should be carefully avoided in paper

> hangings. . . . Few things are more unpleasing to a cultivated eye than the bunches of gaudy flowers and foliage perspectivaly rendered in the intensest colors, and the landscapes, repeated with endless iteration on the walls.[70]

By the year 1880, the author of "*What Shall We Do With Our Walls?*" could jest: ". . . one can hardly estimate the courage it would take to own that one liked an old-fashioned landscape-paper in a hallway or in a dining room."[71]

In spite of the turn of fashion, many of these nineteenth-century scenic wallpapers have been kept in production to this day. Indeed, their survival in so many American houses attests to the special esteem they have enjoyed, even over the objections of tastemakers influenced by successive waves of design-reforming zeal generated by the English aesthetes and the moderns alike.

WHERE AND HOW SCENICS WERE HUNG

Over the time scenic wallpapers were popular, there were variations of fashion in their placement and embellishment. Almost always, elaborate scenic wallpapers were hung on the portion of a wall above the chair rail. Although a few were put up unaccompanied by additional wallpaper finishes—especially in rooms with elaborate cornices or other carved woodwork—it was much more usual to trim scenics with borders, friezes, or dados *(figure 9-20).* A balustrade of paper, pasted below the chair rail, was a favorite accompaniment for "views" in American rooms devoid of wainscoting *(figure 9-25).*

French manufacturers created imitations of architectural elements to frame scenic papers. Usually these included three distinct elements: (1) dados, imitating paneling and embellished with rosettes and medallions or decorated with the more complicated figural subjects favored by designers in the Empire style; (2) columns or pilasters with bases and capitals, sometimes conforming to, but often elaborating on, the classical orders; (3) friezes with related ornament. See *figures 9-4, 9-7* for examples of these elements.

Occasionally, a factory would show a view illustrating a medieval theme surrounded by appropriately Gothic architectural ornament *(figure 9-5).* Sometimes the finishes that manufacturers suggested for "views" simply looked like picture frames carved with acanthus, oak leaf, egg and dart, and other standard neoclassical elements *(figure 9-3).*

Such framing papers had been available in the United States since scenics were first introduced in the early nineteenth century. Yet Americans seem to have been restrained in their use of the full surrounds for scenic wallpapers until the middle of the century, judging from surviving examples. Up to this time, they were most often seduced by the elaborate dados *(figure 9-20).* The use of elaborate wide borders above the scenics, at cornice level, was also fairly common. Ornate friezes of drapery swags embellished with flocking, which a modern decorator would probably judge to be one elaboration too much for the already elaborate scenes, were frequently used *(figure 9-14).*

Before the 1840s, vertical elements like pseudo-columns were seldom allowed to interrupt the visual sweep of a French *paysage panoramique.* But, in the middle of the century, framing elements that featured strong verticals became very important parts of decorative schemes when scenic wallpapers were hung. The surrounds eventually usurped the position of scenics not only in visual interest on the walls but also as the focus of attention for the great French manufacturers' mid-century concentrations of design and promotion *(figures 15-15, 15-19).*

During this period, the use of frames around scenics was encouraged not only by the French who made them, but also by English critics who did not approve of these imitative, naturalistic pictures as wallpapers. The very theorists who argued that abstract flat patterns were the only ornaments proper for printed wall decorations could not resist the visual seductions of the French landscape papers. Christopher Dresser, reviewing the wallpapers displayed at London's International Exhibition of 1862, devoted paragraphs to decrying the French floral papers and criticizing the naturalism in wallpapers entered by the leading Parisian factories. Then he confessed that he "could not pass in silence" the large landscape shown by Monsieur Jules Desfossé because "it is so attractive." Dresser even condoned the use of this landscape paper if it was framed, and, like a work of art, marked off and visually divorced from the surface of the wall. In the following passage, Dresser assessed the ways in which this landscape wallpaper offended his principles, and rationalized:

> While this has merits of a high order, it must not be classed with surface decorations simply because it is produced by block-printing: indeed it is a work of pictorial art, and cannot be considered as ornament. Let this be applied to a wall, and we have a display so extravagant as to be offensive, and a treatment of the surface which would completely mar the beauty of all objects placed in the room thus papered; and the landscape, unless framed in some manner, would be very unsatisfactory.[72]

Forty years earlier, when scenics had enjoyed the appreciation of the stylish, their beauties were deemed acceptable with or without frames. They were best liked in halls. Scenic papers have survived most frequently in American entrance and stair halls, where they made grand first impressions. In 1825, S. P. Franklin of Washington advertised "a few Historical Views" as "Well calculated for Halls and Passages."[73] As late as 1857, W. H. Sackett of New York was still offering wallpaper "scenery for halls."[74] Scenics were also hung in other major rooms—parlors, sitting, drawing and dining rooms—and only occasionally in bedrooms *(figure 10-9).*

Surviving bills of sale, dating from 1831 and 1832, for wallpapers that Francis Regnault, Jr., of Richmond sold to Humbert Skipwith and hung at his magnificent eighteenth-century Virginia mansion, Prestwould, provide details documenting the period's preference for scenics in halls and major rooms. The bills also list pseudo-architectural elements that were used with the scenics, along with border papers for each. While, as the bills reveal, repeating patterns were used in almost every other room in the house, it is significant that scenics are the only papers succeeding generations of the family did not remove or paper over.

Regnault's bills specify *"Le Parc Français" (figures 9-17 through 9-19)* by Jacquemart et Bénard as the set used in the hall, that same firm's hunt scene as the dining room paper *(figure 9-20),* and Zuber's *"Jardins Français" (figure 9-9)* as the choice for the drawing room. Borders used with the scenics in these three rooms appear on the bill, and most of their borders survive in the house. "Capitals," "Bases," and "Columns" are also listed, and some examples of these too survive—over the mantel in the drawing room.

The papering of this house was perhaps unusually thorough. In addition to the landscape papers in the three major public spaces at Prestwould, the Skipwiths had papers and borders hung in the saloon, in chambers, in dressing rooms, in a lobby, and even in a closet. Eight fire screens were also covered with wallpapers and borders,[75] and although the bill does not document it, it might well have been about this time that the four-poster bed at Prestwould was embellished with wallpaper borders *(figures 8-3, 8-4).*

Figure 10-1 (next page) Fire boards were featured in the windows of John Ward's Philadelphia paper-hanging warehouse in June 1847. The accurate observations of the artist, W. H. Rease, as reproduced in a lithograph allow identification of Dufour's scenic paper "Monuments of Paris" *(color plate 34* and *figure 9-14)* in the front window behind a fire board print depicting a romantic couple. Around the corner to the right two more fire boards are displayed, one showing a vase of flowers, similar to that of *figure 8-1.* Courtesy of the Historical Society of Pennsylvania

PAPER HANGING.
J. WARD: PAPER WAREHOUSE
PAPER HANGING
WHOLESALE & RETAIL
CASH
PAID FOR
RAGS
PAPER HANGING
MARKET ST
OAK ST
JOHN WARD
Paper Hanging.
J. WARD
501
PAPER HANG
WAREHOUSE
June 1847
Market-street

10-1

10 FRENCH WALLPAPER ORNAMENTS OF THE EARLY AND MID-NINETEENTH CENTURY

Along with "scenery paper," architectural fakery was part of the elaborate stock in trade that American wallpaper dealers imported from France throughout the early and middle years of the nineteenth century. American shops could furnish their customers with paper columns, with coffered ceilings, with friezes imitating carved moldings, and with dados that looked like paneling or like balustrades. They supplied these not only for use with *papiers panoramiques,* but also to be hung with repeating patterns. They dealt in floral, figural, and geometric ceiling centers, friezes, and dados, as well as in such decorations embellished with scrolls and other flourishes. They sold to their patrons pseudo-statues and flamboyant displays of the paper stainer's craft that took the form of urns or of baskets overflowing with fruits and flowers. In addition, they continued to supply customers with fire board prints and with wallpaper-covered screens.

COLUMNS

"Columns" and "pillars" sometimes appeared in advertisements of early-nineteenth-century American wallpaper stores, denoting not products of the carver, carpenter, or wood turner, but of the French paper stainer or perhaps of his American imitators. As early as 1798 Thomas and Caldeleugh of Baltimore had offered wallpaper "pilasters" among their wares.[1] In 1816, Woodbridge and Putnam, Hartford manufacturers, included "pillars" in their wallpaper offerings.[2] The bill for wallpapers hung at Prestwould in Virginia, mentioned above (see page 229), details a sale of paper columns-by-the-yard. The Skipwiths bought from the Richmond dealer Regnault "7 y^{d} of Collums" at a cost of $1.16, and "10 caps and bases" for $2.00.

During the 1840s, the Boston firm of Josiah Bumstead used a billhead specifying that its "Gothic, Grecian, and Ionic Columns," along with "panel work," were intended "for Halls, Entries, and Vestibules."[3] As late as 1853, Bumstead's dated

10-2

10-3

Figure 10-2 (left) This wallpaper balustrade, 22 inches high, in shades of gray over a green ground, separated two papers in the stair hall in a Salem, Massachusetts, house. It was produced by Jacquemart et Bénard. A duplicate was used in the Martin Van Buren house in Kinderhook, New York. Cooper-Hewitt Museum, 1942-19-5; gift of Grace Lincoln Temple

Figure 10-3 (above) A border design on a grand scale, this 22½-inch-wide wallpaper of the early nineteenth century could have been used as either a frieze or a dado. It is block-printed in brown, yellow, and green. Cooper-Hewitt Museum, 1955-85-2; gift of Josephine Howell

handbill, which listed fine French imports as well as papers made in Bumstead's factory at Roxbury, included a category of wallpapers labeled "Columnar." On this billhead, J. F. Bumstead and Co. detailed that the "Gothic, Grecian, Ionic and other Columns" included in that category had been "executed in gold and plain, with caps and bases, friezes, mouldings, and cornices to match."[4]

CEILINGS

On the billhead used in the 1840s by Bumstead's firm appeared the suggestion "CEILINGS may be papered in plain or figured style with ornamented centre pieces and corners." The practice was hardly a new one, for the fact that papering a ceiling

10-4

Figure 10-4 This scheme for wall decoration in a London publication of 1810 provided one among many published prototypes for elaborate wallpaper border treatments. The print appeared in the architect C. A. Busby's book *A Collection of Designs for Modern Embellishments, Suitable to Parlors, Dining, and Drawing Rooms . . . ,* which was advertised for sale by a New York book seller of the 1820s. The original colored drawings for Busby's publication are preserved in the collections of the Cooper-Hewitt Museum. Courtesy of Avery Library, Columbia University

hid unsightly cracks made it very popular. During the 1820s, French "ornaments for ceilings" were repeatedly offered in the newspaper advertisements of Samuel Robinson of Washington.[5] In 1845, George Platt of New York featured "Parisien Decorations for Walls and Ceilings" in his newspaper notices.[6]

Surviving ornaments support the evidence of advertisements. The photograph of pseudo-coffering on a ceiling in a North Carolina bedroom of the early nineteenth century reproduced here as *figure 10-9* shows one of the most elaborate examples of wallpaper used to imitate high-style architectural detailing.[7] Figural compositions for ceilings could be equally if not more ambitiously stylish than the architectural. An example of the genre is illustrated in *color plate 47.* It depicts Dufour et Leroy's *"La Toilette de Vénus,"* found in a Virginia house. This finely detailed, vibrantly colored composition, inspired by Venetian ceiling paintings, required the mastery of so many printing blocks and incorporated so many different colors that it was cited in French publications of the nineteenth century as an outstanding specimen of block-printing in distemper colors.[8]

Most wallpaper decorations for ceilings made during the first half of the nineteenth century were smaller and less showy than these two examples. Ornamental rosettes, devoid of aspiration to fine art, like the decorative rounds shown in *figure 10-8,* were particularly popular. An example of the American use of similar rosettes survives in the hallway of a Maryland house, Swansbury, in Aberdeen (not illustrated). Surrounding the hook from which a lantern is suspended, the paper decoration includes scrolls and a wreath of realistic flowers on a blue ground. Decorating manuals of the nineteenth century pointed out an advantage of using such wallpaper decorations, rather than plaster rosettes, above a lantern: When smoke from the lantern's flame blackened the ornaments, they could simply be replaced with new pieces of wallpaper.[9]

Borders printed for use on walls were also used as edgings for ceilings. Corners were sometimes printed to match a border and continue its motifs, often with a decorative flourish, right around the sharp angles. The use of a patterned wallpaper stripe on a ceiling, with a border, was not the sort of scheme featured in the pattern books of the manufacturers of fine French wallpapers, but Americans did occasionally put such wallpapers on their ceilings. An example is shown in *figure 12-5.*

DADOS, FRIEZES, AND OTHER BORDERS

In addition to columns, to coffering, and to ceiling rosettes imitating plaster work, a great many more examples of architectural fakery in nineteenth-century wallpapers have survived as borders imitating dentil moldings, carved cornices, and simple moldings at chair-rail level. Wide friezes and dados were standard features of nineteenth-century wallpaper decoration.

Pseudo-balustrades, favored for use below the chair rail on walls covered with scenic wallpaper, have already been mentioned (see page 227). Two examples of wallpaper dados imitating balustrades are illustrated in *color plate 40 and figure 10-2.* Dados imitating carved wood paneling or wainscoting, akin to the eighteenth-century example shown as *figure 7-1,* were also printed in the nineteenth century and used with scenic papers and with repeating patterns. Within the panel format, the squares and rectangles were often adorned with elaborate, boldly rendered, brightly

10-5

Figure 10-5 Though showing signs of its age, this multi-colored paper pasted on wooden boards illustrates how fire board prints were mounted. It was used at Ringwood Manor in New Jersey, the country home of Peter Cooper and of his granddaughters who founded the Cooper-Hewitt Museum. The print, made by Jean Zuber et Compagnie in Alsace, was entitled *"Le Chien du Régiment"* in the Zuber album. The poignant scene shows officers pausing in the heat of battle to attend to the wounds of the regiment's pet. The board is 49½ inches wide. Cooper-Hewitt Museum, 1907-15-4; gift of Mrs. A. S. Hewitt

colored Empire motifs. Frequently these took the form of imitations of relief sculpture, depicting gods and goddesses, or more realistic depictions of putti at play.

A dado paper from a New Orleans house, now in the Cooper–Hewitt collection *(color plate 41),* epitomizes the extravagance of design, of vivid coloring, and of printing technique lavished on these wallpaper ornaments for the lower section of the wall. The original painted rendering used in the factory's production of this particular dado paper is preserved at the Bibliothèque Forney in Paris. On the design are written instructions for the printing sequence of the twelve colors incorporated in the dado's decoration.[10]

Additional documentation for the use of such elaborate dados in America is relatively plentiful. Among the more interesting examples are remnants of a mid-nineteenth-century dado made by the Zuber company that was found during the 1970s restoration of Belmont, a house in Nashville, Tennessee. The dado illustrates scenes from *Don Quixote,* seen through Gothic arches. It was used with Dufour's Telemachus panorama.[11]

In the 1840s, dado papers were still among the wallpapers for sale by the Bumstead firm in Boston. The billhead of that era, cited above, included on the list of standard offerings "PANEL PAPERS, to go around the room at the base."

These papers, designed to run along the bottoms of walls, were often matched by papers for the tops. Friezes reproducing carved elements borrowed from the vocabulary of classical architectural ornament were supplemented, and perhaps outnumbered, by friezes imitating drapery swags. These had appeared in America well before

1800 *(figures 4-6, 7-3, 7-4 and color plate 22),* but during the first forty years of the nineteenth century they grew ever more popular; they became more elaborate, more vividly colored, and much wider. In his 1831 advertisement in a Baltimore newspaper, Walter Crook described "Velvet-printed festoon borders, very rich and elegant." Crook was among Zuber's American customers whose names appear in the factory's account books during the 1820s and 1830s, and perhaps some of his festoons were the product of that factory.[12] The velvet—that is, flocking—featured in Crook's advertisement probably would have been intricately shaded during this period. Techniques for adding shadows to the folds of swags rendered in flocking, or to the petals of flocked roses on such papers, were currently being exploited by French manufacturers.

The use of drapery swags with scenics has already been noted, and illustrated in *figure 9-14.* In addition, swag borders were used with papers that disguised entire walls with imitations of draped fabric *(color plate 49).* Swags also topped walls hung with repeating landscape figures *(figures 12-4 through 12-6),* and they appeared above simple repeating patterns and stripes. The gaudy border illustrated in *color plate 38* was hung as an incongruous topping for a wall covered in a simple gray and white stripe, dotted with small green sprigs blossoming with red flowers.

Such friezes and dados were but elaborated versions of borders, the products of the wallpaper manufacturers that by the early years of the nineteenth century were judged essential to wallpaper decoration. John Claudius Loudon (1783–1843), an English writer whose publications on landscaping, building, and decorating were widely read in America, urged the use of borders with wallpapers. In his *Encyclopaedia of Cottage, Farm, and Villa Architecture and Furniture,* first published in 1833, Loudon emphasized:

Figure 10-6 (opposite, above) This page from Zuber's *Collection d'Esquisses* shows elaborate ornaments. Factory archives document sales of many of these same designs to New York and Philadelphia dealers in 1816, 1817, and 1822. Invoices in the archives include production numbers and descriptions that correspond to several shown here. The dealers bought varied selections in small quantities. Courtesy of the Metropolitan Museum of Art, Elisha Whittelsey Fund, 1957

Figure 10-7 (opposite, below) A page from Étienne Delicourt's album of prints illustrating available wallpapers shows ornaments similar to those of Zuber. A golden-yellow paper printed with the Delicourt pedestal base shown under the central figure is preserved in the Cooper-Hewitt collections. Cooper-Hewitt Museum library

10-6

Statues, Vase, Pilastres et Colonnes
Délicourt Fabricant de papiers peints, rue de Charenton 125bis à Paris
L'Automne
Vase Médicis
Le Printemps
L'Eté

10-7

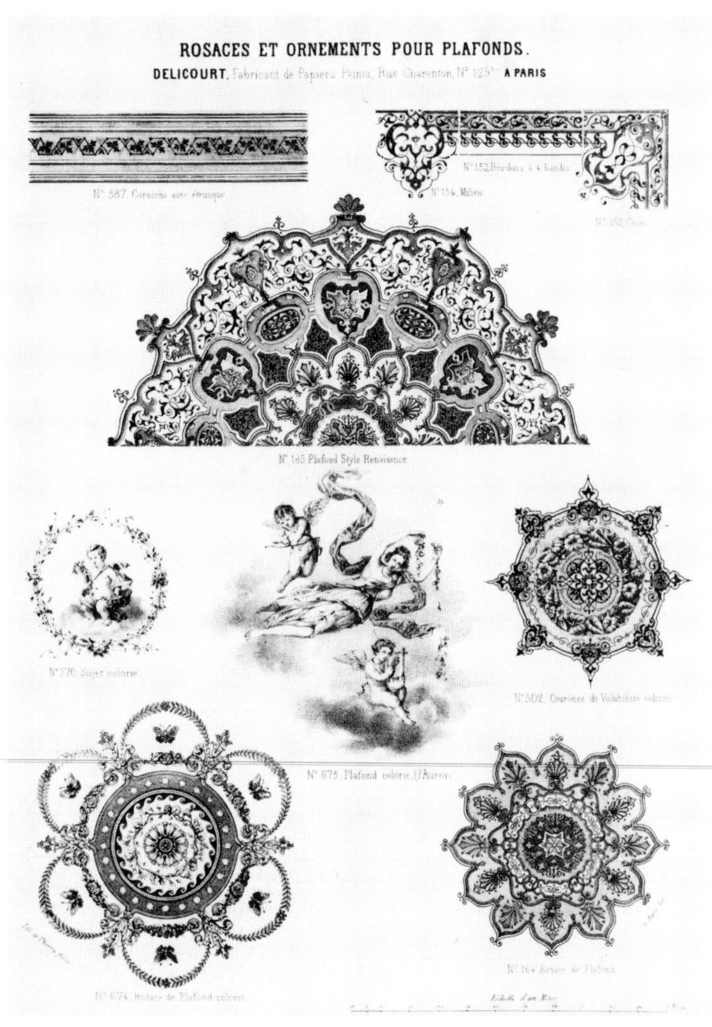

10-8

Figure 10-8 Étienne Delicourt included this page of ceiling decorations in an album illustrating his wallpapers. Although two of the medallions at the bottom of the page retain stylizations based on classical models and therefore relate to early-nineteenth-century ornament, other designs indicate that this was printed near the middle of the nineteenth century, especially the scrollwork combined with realistic flowers in the medallion in the center on the right, and the larger medallion shown as a half circle near the top. Cooper-Hewitt library

Figure 10-9 In sadly deteriorated condition when photographed by new, restoration-minded owners in 1941, parts of the wall and ceiling wallpaper are shown in a bedroom at Oak Lawn, an 1818 house in Mecklenburg County, North Carolina. The photograph provides fascinating documentation for the early use of elaborate wallpapers in an architecturally simple, though substantial, frame house in the inland south. One of two papered rooms found there in 1941, this one was decorated with Dufour's Captain Cook paper *(color plate 32),* hung on unplastered walls with elaborate borders and a tour-de-force of a ceiling paper: imitation coffering. Courtesy of Mrs. Wilson L. Stratton

10-9

> The side walls of a room equally ornamented in every part . . . by a rich paper would be intolerable were it not for the contrast produced by the plain ceiling and by the border with which the paper is finished under the cornice at top and above the base or surbase below.[13]

Such opinions influenced Loudon's most important American disciple, Andrew Jackson Downing. In Downing's book of 1850, *The Architecture of Country Houses,* his assumption that borders would be part of any paper-hanging scheme is apparent in such passages as:

> A *cornice* adds very considerably to the architectural character of any room, though it is seldom or never introduced in cheap cottages, except, perhaps, in a parlor or best rooms. When the walls are papered, its place is in a good degree supplied by the border, representing a cornice on the paper itself.[14]

In *The Paper-Hanger's Companion,* a mid-century English manual of instruction on the art of hanging wallpaper, James Arrowsmith waxed polemical in his encouragement of the use of border papers. The book was brought out in at least three American editions between 1852 and 1887. Arrowsmith, who described himself as a "workman of fifty years," devoted three pages to arguments for the utility and beauty of borders. One involved an analogy to classical architectural precedent:

> In advocating the invariable use of borders for the tops and bottoms of rooms, I venture to presume that my sentiments are borne out comparatively, on the acknowledged principle of correct architecture: for instance, how defective would a column appear without its capital?[15]

Other trade literature put forth more practical arguments for the use of borders, including the fact that borders covered any mistakes made in exactly fitting the cut ends of lengths of paper and that they helped to hold down and prevent peeling, which often began at the edges of papers.

Almost inevitably, when the word "wallpaper" appeared in a nineteenth-century advertisement, it was accompanied by the word "borders." Variations on a New York offering of "Plain and fancy patterns with borders for top and bottom," published in 1816, appeared all through the first half of the nineteenth century.[16]

The same pairing occurs on surviving bills that document sales of nineteenth-century wallpapers. Hundreds of such bills have been studied for this book, and among them, better than nineteen out of twenty include border papers.

Borders continued to be standardly embellished with especially bold colors and with materials that added glitter and texture, visually strengthening the impact of borders as edgings around fill papers. As had been true in the eighteenth century, a touch of gold was admired as an addition that would highlight a border design and reflect the flickering gleam of the flames used to light nineteenth-century rooms. "Superb gilt and flock borders" is typical of descriptions of borders published in advertisements of the 1820s.[17] "Velvet" or flocking, already mentioned as a specialty of French manufacturers during the 1820s and 1830s, was constantly appearing in listings of borders for sale in this country. On these borders, brightly colored realistic flowers, or ribbons, leaves, or other ornaments were often printed in distemper colors against "velvet" backgrounds. Their vivid hues stood out against the rich, dense flocking, which was usually dark. The eighteenth-century fondness for borders with patterning and coloring bolder than those on the repeating fill pattern continued into the nineteenth century.

Again, as in the eighteenth century, many borders were designed to match and to continue the motifs of repeating patterns. Sometimes pieces of a wall pattern were overprinted with a border planned to complete or close a design *(figure 12-30)*. Sometimes the border carried the line of a stripe up through a semi-circle and across a width of paper to join a second line in the stripe, forming an arch at the top. However, numbers of other borders shown with papers in sample books *(color plates 57 through 60)* and found on walls *(color plates 38, 62)* bear little apparent relationship, either in color or motif, to the papers with which they appear.

SCULPTURAL, FLORAL, AND OTHER ELABORATE ORNAMENTS

For the American hallway embellished with paper imitations of marble walls, and with architectural detailing that looked like carved stone, what could have been more desirable than sculpture that appeared to be carved marble or granite, or cast bronze? The paper stainer's versions of such works of high art found their way into many middle-class households during the nineteenth century. Paper and a little paste could bring the look of sculpture to the front hall, well within a modest budget. And wallpaper statues did not impinge on the floor space. Preposterous and pretentious as these papers may seem for home decoration, the combined evidence of advertise-

38

Color plate 38 (left) Swagged, feathered, tasseled, and bejeweled, this French drapery frieze is 21¼ inches wide. Block-printed with flocking, it was used with a simple wallpaper—gray stripes with green and red flower sprigs. From the early 1800s to the mid-1970s these papers hung in the parlor at Locust Thicket in Lynchburg, Virginia. Cooper-Hewitt Museum, 1973-10-1; gift of Mr. and Mrs. Hugh B. Jackson

Color plate 39 (immediately below) About the same width as that in *color plate 38,* this early-nineteenth-century French frieze draws on the vocabulary of classical architecture and relief sculpture for forms but on the paper stainer's fancy for colors. Cooper-Hewitt Museum, 1959-160-26A; gift of Josephine Howell

39

Color plate 40 (center right) A balustrade, installed with paste, was available for walls of the early 1800s. This French example is 17 inches tall. Such architectural fakery was often hung below a landscape paper. Cooper-Hewitt Museum, 1972-42-165; gift of Josephine Howell

40

Color plate 41 (bottom right) This dado, 22 inches high, was found in a New Orleans house. A duplicate in different colors still hangs at Prestwould *(figure 9-20).* The original painted design of 1825–30 is at the Bibliothèque Forney. Cooper-Hewitt Museum, 1935-213-1; gift of John Judkyn

41

Color plate 42 (immediate right) This diminutive column and its capital—combined, they stand just under 40 inches—are block-printed wallpapers. Jacquemart et Bénard's original design for the column, done about 1810, is at the Bibliothèque Forney, Paris. Cooper-Hewitt Museum, 1928-2-94

42

43

Color plate 43 (left, top) Neptune's consort Amphitrite is borne across the waves in her shell chariot drawn by dolphins, block-printed in golden yellows, gray, and white on a dark blue ground. Printed about 1805–10 in France as an overdoor or chimney picture, this example is 24 inches high, 37¾ inches wide. Cooper-Hewitt Museum, 1931-45-76; gift of *Antiques* Magazine

Color plate 44 (left, middle) Charming pictures like this one were block-printed by skilled French paper stainers during the nineteenth century for use over doors and in fireplace openings. This example of about 1830–40 is 23¾ inches high by 26½ inches wide. Cooper-Hewitt Museum, 1942-73-1; gift of Marian Hague

Color plate 45 (below) A French paper stainer in about 1825 copied a painting by Jean François de Troy (1680–1752) done in 1738. It depicts the Coronation of Esther, and is now in the Musée des Arts Décoratifs, Paris. The pattern is 30¾ inches by 39½ inches. Cooper-Hewitt Museum, 1931-45-3; gift of Eleanor and Sarah Hewitt

44

45

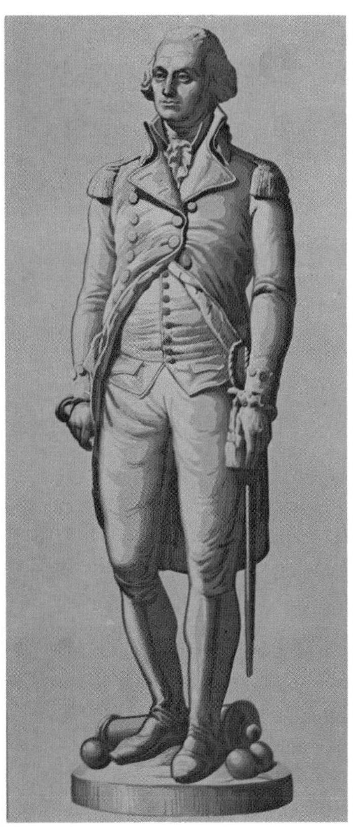

46

Color plate 46 (left) The Parisian firm of Jules Desfossé (1816–89) printed this imposing example of wallpaper statuary in 1856–7 for an eager American market. George Washington was one of *"Les Grands Hommes"* in a series of four revolutionary heroes, the others being Franklin, Lafayette, and Jefferson. The paper is 19 inches wide. Cooper-Hewitt Museum, 1949-78-1; gift of Dr. Gertrude Bilhuber in memory of her father, Ernest Bilhuber

Color plate 47 (below) This extravagant ceiling decoration was considered a masterpiece of block-printing by its maker, Joseph Dufour. He first printed examples of this design in Paris during the first quarter of the nineteenth century. It was often published and exhibited at industrial expositions of his era. This example survived until 1959 on the ceiling of a room in a frame house called Coxley, built prior to 1820 near Varina, Virginia. The diameter of this decoration is more than eight feet. Courtesy of the Valentine Museum, Richmond, Virginia

47

49

Color plate 48 (left) Elaborate satin draperies over a flocked ground are depicted in distemper colors on joined sheets of paper. This drapery paper was block-printed by the Parisian firm of Dufour about 1810–12. Elaborate Empire designs like this were advertised as "drapery panels" in contemporary American wallpaper advertisements. The panel is 92 inches tall. Cooper-Hewitt Museum, 1968-111-1; gift of Harvey Smith

Color plate 49 (above) This French wallpaper would have been described as a "drapery figure" in an American advertisement of the 1810–25 period. These block-printed draperies with lace trim and pearl tassels were hung with borders with black flocking. A narrow border outlined the walls, and a wide border on which a black scroll was printed against golden yellow provided a strong accent at the bottom of each panel. A corresponding "arch top" finished the paper at cornice level. The portion shown here is 25 inches wide. Cooper-Hewitt Museum, 1972-42-213; gift of Josephine Howell

48

Color plate 50 (below) A pattern formula typical of the early nineteenth century—two principal motifs alternating between stripes, over a ground spotted with small figures—was used for this French paper of 1813–14. (For variants, see *color plate 51 and figures 12-4 through 12-6, 12-12, 12-13.*) This paper commemorates the first important American naval victory of the War of 1812, in which the *Constitution* captured the British frigate *Guerrière.* The victorious American captain, Isaac Hull (1773–1843), is shown in a likeness that is a reversal of his portrait by Gilbert Stuart. Cooper-Hewitt Museum, 1955-4-1; gift of Jones and Erwin, Inc.

Color plate 51 (right) Another example of the pattern formula on the left, this French wallpaper of about 1810 had a swagged drapery frieze above and a border of daisies with an imitation of carved molding below. The sample is 20 inches wide. Cooper-Hewitt Museum, 1931-45-27; gift of Eleanor and Sarah Hewitt

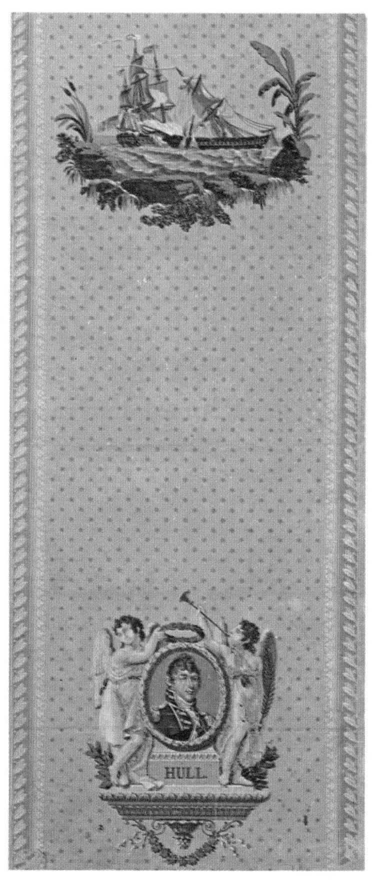

50

51

52

54

53

Color plate 52 (far left) Bright block-printed floral sprigs contrast with the gray of subordinate foliate patterning. This unused sample of an early-nineteenth-century French "sprig paper," printed in distempers on joined sheets 20 inches wide, was found in the attic of a Hudson River mansion built in 1803. Cooper-Hewitt Museum, 1974-81-2

Color plate 53 (bottom) Anthony Chardon's Philadelphia firm block-printed this stripe between 1814 and 1825. It was found in the Schuyler house in Saratoga, New York. The stripes with wheat are less than 1 inch wide. Cooper-Hewitt Museum, 1959-118-1; gift of the National Park Service

Color plate 54 (top, right) This large-scale pattern imitates a brocaded silk with velvet patterning. Diagonal lines simulate twill weave in the red medallions along the edges. Flocking imitates velvet in the large motifs; oil-based pigments in details give the sheen of silk. French, the sample of 1815–25 is 24 inches wide. Cooper-Hewitt Museum, 1931-45-54; gift of Sarah and Eleanor Hewitt

Color plate 55 (opposite) Another French wallpaper of about 1810–20 also imitates textile patterning, even recreating woven effects with the diagonal brown and yellow striations that form the dominant motifs. Cooper-Hewitt Museum, 1928-2-110; gift of Sarah and Eleanor Hewitt

55

56

ments, surviving examples, and the Zuber archives makes clear that such papers were widely used in early-nineteenth-century America.

"Ancient Statues" had been advertised among the wallpaper offerings of a Boston dealer in 1800, and similar offerings continued to appear in newspapers from every American city into the 1850s.[18] Samuel Robinson included an unusually specific description of wallpaper statues in an advertisement of 1821 in a Washington newspaper: "Bronze Figures of the Seasons."[19] These may well have been the wallpaper statues made by Zuber after designs by Évariste Fragonard (1780–1850).

The archives of the Zuber factory record the sale of Fragonard's "Seasons" to American dealers. Those records document that the New Yorkers and Philadelphians whose scenic purchases during 1816, 1817, and 1822 have been described (pages 214–20) also bought wallpaper statues.[20] The 1817 invoice listing the wallpapers Zuber shipped to Virchaux and Borrekins of Philadelphia includes twenty-four wallpaper statues, a total of six sets, each made up of four statues—one for each season. Three sets were printed in "marble"—gray with gold—and three in "bronze." The statues looked very much like the statue illustrated in a print published by the factory, shown as *figure 10-6.* The statue illustrated here (bottom, center) is "Old Age" from the same Fragonard's "Four Ages of Man." The numbers that appear on the invoice, 1067–70, duplicate numbers given under the statue illustrated here, where it is specified that they are the numbers for Fragonard's "Four Seasons." Another Zuber invoice of papers sent to Philadelphia in 1816, to Geisse and Korkhaus, shows shipment of eight of the statues representing the seasons. One set was printed in gray and gold, and one imitated white and bronze marble. In addition, four pedestals bearing the number 1071, another of the numbers that appear in the illustration reproduced in *figure 10-6,* were included on the order of Geisse and Korkhaus. Perhaps they were duplicates of the pedestal supporting "Old Age" in this illustration.

Zuber's records also document the sale of other kinds of elaborate wallpaper ornaments, ornaments depicting grand sculptural stands for vases laden with fruit and flowers, as well as baskets filled with towering arrangements of naturalistically rendered blossoms. Some of these ornaments are shown in the four corners of this

Color plate 56 "Rainbowing" *("irisé")* produced the color-blending apparent in this pattern, probably American, dating from 1835–50. Very simply printed by a single roller on a machine over ungrounded paper that is now a light tan color, found in Guilford, Connecticut. The sample is 20 inches wide. Cooper-Hewitt Museum, 1938-62-13; gift of Grace Lincoln Temple

same illustration, *figure 10-6.* Zuber's invoices list the production numbers of the very ornaments identified by number in this illustration. The invoices include descriptions that also match these pictures.[21]

In 1844, a writer for an English decorating journal reported, "About a quarter of a century ago, papers to represent sculpture and relief had a sudden run of brief duration." He stated that in England, "Sculptural designs are generally abandoned; it was indeed absurd to suppose that panels of paper-hangings would ever do more than caricature the effect of a recess containing a statue."[22]

While the English may have abandoned these *trompe-l'oeil* papers by 1844, they apparently enjoyed a longer and stronger popularity in America well into the 1860s. In 1846, Howell and Brothers of Philadelphia advertised that they had in stock "Architectural Designs for Halls and Entries, Consisting of Columns with Ornamental Pannelling, Statues and Pedestals, Niches, &C." In the 1840s and 1850s a St. Louis dealer featured "Statuary and Poetic Figures" in his wallpaper advertisements.[23] Among listings on the 1840s billhead, which has been cited so often above, the Bumstead firm included: "STATUES of various subjects, with or without niches."

A group of ornaments reminiscent of eighteenth-century decorative schemes continued to be sold during the first half of the nineteenth century. These were wallpaper "trophies," imitating relief-carved and plaster architectural decorations. One example of their nineteenth-century use in this country is documented by papers surviving from the Philadelphia house of Joseph Bonaparte. In an advertisement in a Washington newspaper of 1819, Samuel Robinson listed what were probably similar designs. He included "trophies" along with "Rich friezes" and "Chimney screens."[24] Earlier trophies are illustrated in *figures 3-10 and 3-11.*

Whether tricking the eye into believing a room was finished with elaborate architectural carving or plasterwork, adorned with marble statues, or filled with grandiose floral displays, French wallpaper manufacturers delighted in their deceptions. One last example from an American newspaper suggests that anything one might place on a wall was fair game for imitation in wallpaper. In 1800, the listings in a Boston advertisement for wallpaper concluded with: "a few sheets of false library."[25]

FIRE BOARD PRINTS

Offerings of "fireboard prints" and "chimney pieces" appeared constantly in American wallpaper advertisements of the first half of the nineteenth century. The French paper stainers used their refined block-printing techniques to produce small scenes and self-contained decorations for use over doorways as well as over and in front of fireplaces. American references to the eighteenth-century introduction of their products have been cited in Chapter 8. During the nineteenth century they became a much more important item in the American wallpaper trade.

"Figured pieces for fire boards," "patterns for chimney boards," and "fancy chimney pieces" are among the phrases used to advertise these popular decorations. Sometimes the scenes on the prints produced during the early nineteenth century were very much like *paysages panoramiques,* reduced and condensed. This is hardly surprising, since they were made in the same factories by the same designers and craftsmen who were making the scenic wallpapers. Perhaps the "landscape and figured pieces for fire boards" in a New York ad of 1810 were of this type.[26]

Among all the ornaments named in this chapter, these brought to the broadest group of American consumers examples of the skills of French block printers of wallpaper. The frequency with which ornaments for fire boards were advertised and the prices at which they were offered suggest that they reached relatively humble hearths in America, some of them in places far removed from the Eastern seaboard. By 1845, Richardson's Paper-Hanging Warehouse in Albany, New York, was selling "Fireboard Prints from 25¢ upwards."[27] But it was not only such inexpensive examples that were advertised both in the East and in the hinterlands. In St. Louis, G. B. Michael advertised in 1847 that his chimney screens were the finest, imported "direct from Paris."[28] Representative prices of fire board prints seem to fall in a range from $1.00 to about $3.50 during the 1820s to 1840s. Some indication that these were the going prices is provided on bills of sale that have survived. One, documenting that two fire board prints were bought by a customer in North Bloomfield, Ohio, from Charles Grant of Boston in 1821, records that they cost $1.50 and $1.00 respectively.[29] In 1835, Andrew Jackson paid $1.00 for a "Chimney Screen Vase of Flowers," $3.50 for another chimney screen noted to be "green velvet," and $2.00 each for two "landscape" chimney screens.[30]

By about 1800, American craftsmen had begun to make their own chimney board prints. A charming fire board showing Boston harbor in block-printed distemper colors, framed by scrollwork, is preserved at Sturbridge Village. Perhaps it is the work of an American paper stainer of the mid-nineteenth century. Not only the paper stainers, but also the printers who made illustrations for books and lithographs of the type described as "suitable for framing" seem to have taken advantage of the popularity of fire boards as a surface on which to paste their products. In 1836, a Philadelphian, advertised "lithographic and common fireboard figures very low."[31]

11 FRENCH WALLPAPERS AND THE DEVELOPMENT OF THE AMERICAN CRAFT

The elaborate ornaments surveyed in Chapter 10 and the scenic wallpapers that were the subject of Chapter 9 were the wallpapers that secured for the French industry its long-lasting dominance of the American wallpaper market. Throughout the first seventy years of the nineteenth century, some importation from England continued, and a few wallpapers were shipped to America by German manufacturers.[1] At the same time, the American wallpaper industry was mushrooming. However, "French" was apparently the word to catch the eye of the American wallpaper consumer well past the middle of the nineteenth century. It is the word printed in caps, underlined, or heading the list in advertisement after advertisement not only of those who were strictly importers and dealers, but also of Americans who made their own wallpapers at the same time they were selling French imports.

J. Milbert, a French traveler in America between 1815 and 1823, published an account of his sojurn. In it, he proudly noted the superiority of the wallpapers made by his fellow countrymen:

> America produces a large number of papers, England tries to manufacture some, but most of the business is done by the French, whose inexhaustible imagination, talent for matching colors, and ability to produce a charming effect, defy all imitation.[2]

His statement summarizes the popular consensus about wallpaper that prevailed from the early nineteenth century right through to the 1860s. While the author of a book of 1811 describing the industries of Philadelphia, including wallpaper, might assert, "We no longer depend upon Europe for excellent and handsome paper hangings," papers sent from France still delighted the stylish.[3] The magnitude of their consumption of French wallpapers continued to attract the attention of national legislators formulating tariff laws, which many thought should protect American industries.[4]

During the War of 1812, a thirty-two and a half percent duty was established for imported paper hangings, in line with generally high import duties reflecting the government's need for money. Although that war was followed by a period of relatively high protective tariffs, in April of 1816 the duty on paper hangings was reduced slightly—to thirty percent.

Protective sentiments urging high tariffs as well as nationalistic impulses to "buy American" seemed to except French wallpapers during this period. The most politically sensitive and carefully observed house in the country was hung with stylish French patterns in 1818. During that year, a visitor to the "President's House" reported on the redecoration in progress for James Monroe: "The walls of some of the rooms which have been finished are covered with very rich French paper studded with gilt flowers."[5]

By 1819, post-war disruptions in the economy and what was decried as unchecked importation led to general financial distress in America. A convention of "Friends of National Industry" deemed paper hangings in need of tariff protection when they included them on a list of manufactures "recommended to particular attention" of the legislators.[6] The *Digest of Accounts of Manufacturing Establishments* for 1820 noted that of two paper-staining manufactories in Providence, Rhode Island, "the business of one . . . is dull, from the markets being overstocked with French papers, and the other from the same cause is not in operation." In addition, the *Digest* reported that for wallpaper manufacturing in Philadelphia, "Business [was] reduced one third within two years in consequence of large French importations."[7] In 1821, much higher duties were imposed on most imports, but the thirty percent rate was maintained for paper hangings. American paper-hanging manufacturers were doubtless dissatisfied.

The legislators did impose a higher duty in 1824—forty percent. It appears to have been effective. Senate records for 1830 reported reduced importation. But by 1831, the merchants who sold the imports had joined forces to oppose the high duty. In a "Memorial of Sundry Merchants of New York, praying that the duties on paper-hangings be reduced," they explained to the Senate:

> Since the passing of the said [40%] tariff act 8 years ago, it is now clearly established that our manufactories have so far improved in the fabrication of paper hangings that the interests of manufacturers no longer require high protective duties.[8]

The list of signers of the memorial included a number of the names of Zuber's customers recorded in the *Grand Livre* for this period.

Their memorial had little apparent effect. In 1842, when rates were lowered to a general level of about twenty percent, the tariff rate on paper hangings and on paper for screens and fireboards was reduced only slightly, to thirty-five percent.

But in 1846 a duty of only twenty percent was established for paper hangings, a rate lower than that for other kinds of paper. By this time, Democratic control of legislation favoring Southern interests and American manufacturers who wanted to trade in foreign markets is credited with the lowering of most other tariffs to about thirty percent. The even lower rate on paper hangings suggests that by mid-century the American industry was strong enough to withstand the competition of French imports. By now, for the most part, only the most extravagant of French productions held much appeal for Americans, as good-quality papers of a more ordinary sort were being made in this country. And even the appeal of the most expensive French wallpapers was beginning to decline. By 1857, Edwin T. Freedley, writing of paper hangings in his book *Philadelphia and its Manufacturers,* noted with what appears to be some accuracy that "now the importation of foreign papers is an unimportant item, said to be not more than five per cent of the whole amount consumed, and confined to French goods of the first quality."[9]

While elaborate French wallpapers held the attention of the most stylish American decorators during the first sixty years of the nineteenth century, wallpapers made in America commanded an ever-growing market of consumers more interested in economical decorating. Between 1800 and 1840, the years just preceding the substitution of large-scale factory production for hand-printing, numbers of little American shops were established. They were able to make imitations of many of the simpler French patterns and were beginning to work out styles of their own.

In his *Report . . . on American Manufacturers* of 1810, Secretary of the Treasury Albert Gallatin included the manufacture of paper among those "branches [which] are firmly established" in the United States. He further reported: "The manufacture of hanging paper and of playing cards are also extensive."[10] Tench Coxe, an assistant in the Treasury, and from an early date in the 1780s a spokesman for industrialization as essential to a balanced national economy, compiled a digest of the census of manufactures. Congress published Coxe's *Digest of the Census* in 1813.[11] In that publication, he listed only four paper stainers in America, a count far short of the number of manufacturers known from advertisements and city directory listings of the period. He credited those four with a combined annual production of 148,000 pieces of paper hangings. Another early Federalist made note of the state of the American wallpaper business in a published study of manufacturing. Timothy Dwight, the New England intellectual who was president of Yale between 1795 and 1816, recapitulated Gallatin's report and added some statistics in his *Travels in New England.* He reported that 63,000 rolls of hanging paper were manufactured in Massachusetts paper mills in 1810. These were valued at $33,500. Such early studies of American manufacturing give scanty information about the making of wallpaper, but they do confirm that the industry was large enough to command the notice of those concerned with the encouragement of American manufacturing interests.

In 1833, Louis McLane, Secretary of the Treasury, submitted to Congress an extensive report on manufactures in the United States. Although it too failed to provide a full count of the paper-hanging manufactories known to have been active at that date, it did include specific information about seven Massachusetts manufactories. An account of one of them, the manufactory established in 1830 by W & W&C Laflin at Lee, Massachusetts, is particularly detailed. It lists materials used in the factory, their sources, and costs. The Laflins had established paper mills at Lee five years before going into the wallpaper business. At the time McLane compiled his report, the firm annually produced 132,000 pieces of paper hangings worth $19,800. Their products were sold in New York. In New Bedford, a paper-hanging manufactory which is not named, but which was doubtless that of John Perkins, who advertised extensively about this time, annually produced 70,000 rolls worth $14,000, sold in Massachusetts. The five Boston manufactories on which McLane published statistics each made an annual average of $20,000 worth of paper hangings, which were sold in Massachusetts and in other New England states.[12]

Taking these federal reports and the publications that resulted, plus information in city directories, advertisements, and local industrial histories, we can estimate that between 1800 and 1845 at least forty-five New York establishments were making paper hangings, as well as twenty in Boston, fifteen in Philadelphia, and fifteen in other cities. See Appendix B for a listing.

By 1810, there is evidence that craftsmen in major American centers of wallpaper making had developed styles that were recognized to be "local styles." In 1810, Zecheriah Mills of Hartford offered for sale: "Newest Philadelphia, New York, and Boston patterns, together with those in the Hartford stile."[13]

It is probable that none of the early-nineteenth-century establishments for making wallpaper would have attained the size of anything the modern world classifies as a factory. The numbers of people employed in even the largest was probably relatively small—under fifty. For the early 1830s, the McLane report provides some information about the numbers of workers in the seven Massachusetts manufactories listed. In addition, it gives their salaries. The Laflin manufactory in Lee employed seventeen men at seventy-five cents a day and six women at thirty-one cents a day. In New Bedford, the paper-hanging manufactory paid a dollar a day to four men, and fifty cents a day to twelve women. According to the report, the five Boston manufactories each employed fewer hands: an average of three men each, at dollar-a-day wages, one boy at thirty-seven cents, and two women at fifty cents a day. Between 1800 and 1830 the average wage of a carpenter or skilled mechanical laborer was

$1.10 a day. Between 1830 and 1848 mechanics earned $1.50 a day. Since unskilled labor after 1820 was paid 75¢ to $1.25 a day, the figures for workers in the Massachusetts wallpaper factories suggest they were considered semi-skilled labor.[14]

The underpaid women of the Massachusetts wallpaper manufactories whose salaries were listed by Louis McLane were not the only females in the trade whose work is a matter of record. City directories and newspaper advertisements of the period preserve the names of women who carried on the businesses of their wallpaper-manufacturing husbands and kinsmen. The assumption of such roles by wives, and even by unmarried daughters, was relatively common in nearly all the American crafts. And there was good precedent for women of the early nineteenth century assuming responsibility for paper-staining manufactories. In 1788, Ann Dickinson of Philadelphia had advertised that she was continuing the business of her husband, Joseph Dickinson, specifying "Paper Hanging Manufactory . . . continued by his widow at the same place."[15] In 1800, Margaret Hovey of Boston was advertising wallpapers, apparently continuing at least some of the business of the manufactory Joseph Hovey had established in Boston in 1786 and operated until his death in 1794.[16] In the Boston city directory for 1807, the name Margaret Edes replaced that of William Edes as "paper stainer" on Cambridge Street. The New York City directory of 1810 did not even bother to give the name of the woman whom it listed only as "Colles, Widow of John" at the paper-hanging manufactory on Pearl Street, which John Colles had operated since 1794. Since 1789, Colles had been listed in the New York directories as a paper stainer. In the 1810 directory, the name Stephen Bates was also given at Colles's Pearl Street address, suggesting that he was in fact assuming the role of successor to Colles in some arrangement with the widow.

Beyond the information related in Chapter 2 about the specialties of block-carving and block-printing, little is known about the internal organization of American paper-hanging manufactories of this period, or about the workers within them. Craft techniques of the eighteenth century were taught to apprentices who carried them on into the nineteenth.

By 1844 there were enough paper stainers in Philadelphia to warrant organization within the trade. A notice in *The Public Ledger* for January 31 read:

> Paper Stainers, attend: A Meeting of the Journeymen Paper Stainers will be held at the House of the Widow Crilly . . . this EVENING . . . at 7 o'clock. Punctual attendance is earnestly requested by order of the Committee.

We can only speculate what important issue might have provoked such a degree of organization: possibly it was the growing threat of mechanization within their craft. In any case, the simple fact that the advertisement addresses itself to "Journeymen Paper Stainers" suggests craft organization much like that in other trades, where master craftsmen or entrepreneurs owned the business and directed the daily work of these hired craftsmen whose proficiency had earned promotion from the status of apprentice to that of journeyman. These skilled workmen would have been the ones whose livelihood was most threatened by newly developed wallpaper printing machines like the one the Philadelphia firm of Howells and Company imported in 1844 (see page 308), the same year this meeting of journeymen paper stainers was called.[17] Some of the papers made by Americans of the early nineteenth century will be included in Chapter 12.

11-1

Figure 11-1 This is Anthony Chardon as portrayed about 1800 by Saint-Mémin (1770–1852). Chardon arrived in Baltimore in 1789, one of the three "gentlemen" mentioned on page 110 who came from Nantz. During the early 1790s he was associated with Burrill and Edward Carnes and a fellow Frenchman called LeCollay in a Philadelphia manufactory. In 1794 Anthony Chardon and Co. bought out Carnes. The same year, Chardon sold wallpaper to George Washington. In 1806 W. S. Austin was named as Chardon's partner in one reference. In 1826 he offered at auction his "stock in trade," acknowledging "liberal encouragement during the last 40 years." Courtesy of the National Portrait Gallery; the illustration is taken from Elias Dexter, *The Saint-Mémin Collection of Portraits* (New York, 1862), no. 275.

12 REPEATING PATTERNS OF THE EARLY NINETEENTH CENTURY

American craftsmen made only a few attempts to copy the elaborate French ornaments—the friezes, dados, statues, and ceilings that provided flamboyant touches for papered walls. However, they standardly copied and adapted French repeating patterns. Because it is instructive to follow the sequence from French prototypes to American adaptations in illustrations arranged in an order that visually emphasizes the connections, French repeating patterns were not presented with the more elaborate French products but, instead, are being surveyed here. Their late inclusion in this book should not be misunderstood to imply that they were exported to America at a later date.

DRAPERY PAPERS

Among repeating patterns, some of the most extravagant French paper hangings of the early nineteenth century that were imported and then copied were designed to look like curtains and elaborately draped textile hangings. Draping a room with textiles arranged in imitation of the walls of elaborate tents was in fashion in France during the period of the First Empire. These drapings were inspired by the campaign tents of Napoleon and his generals. The tents and fabric-draped walls had classical precedents. A famous room hung in imitation of a tent survives at one of Napoleon's residences, Malmaison. Draped with yards of red brocaded fabric, the room was a stylish antecedent for this wallpaper style, within which French paper stainers of the early nineteenth century produced myriad variations.

In the wallpapers, sometimes a full length of cloth was represented extending from the top of a wall to the bottom *(color plate 48).* Such wallpapers will be called hereafter "drapery panels." In some "drapery panels" a continuous vertical length of cloth was shown as a gathered panel, stretched from rods top and bottom, and cinched near the center with an ornamented knob, as in *color plate 48.* In other "drapery panels," lengths of cloth took the form of curtains, hanging in realistically shaded folds that fell unbroken down the length of the panel, or were shown with supplementary vertical lengths that were caught halfway down in wreaths or ornamental devices around which they were elaborately knotted *(figure 12-1).* The height of a full "drapery panel" could not be altered. While the panels might be hung in repeat around a room, no elements within individual panels were repeated vertically. While examples of such large-scale "drapery panels" were hung in American houses, they do not seem to have been copied and adapted by American paper stainers. Their manufacture required large-scale printing blocks. The smaller-scale imitations of draped patterns that were made up of vertically repeating elements *(color plate 49, figures 12-2, 12-3, 13-8)* were easier to produce and were copied or adapted by American craftsmen.

This second group of wallpapers imitating draped textiles with vertically repeating elements will be called "drapery figures." The patterns often depicted wall hangings made up of layer upon layer of swags. Some looked like the modern "Venetian Curtains". Much more frequently, however, layers of overlapping festoons made from pieces of fabric that appeared to hang horizontally across the wall were shown on these papers. And there were other variations in the wallpaper depictions of drapings. Some showed walls hung with lace, tulle, and fringe, or ornamented with brass bosses, pearl swags, feathers, and jewels. *Color plate 49* only suggests the elaboration of some of these repeating "drapery figures."

Evidence of the use of both kinds of "drapery papers" survives from American houses and in the advertisements of American vendors.[1] In these advertisements, the word "drapery" could have been used to describe both the vertical panels and the swagged patterns. A vague distinction, however, is indicated in the verbiage of the early-nineteenth-century offerings of such papers, and those nineteenth-century terms have been appropriated for the description above. "Drapery panels" and "draperies" were probably words that stood for the vertical panels *(color plate 48, figure 12-1),* and "drapery figures" probably described the papers made up of repeating elements *(color plate 49, figures 12-2, 12-3, 13-8).*

During the first decade of the century, John Perkins advertised "draperies" among the papers he imported to New Bedford, Massachusetts, as well as among those he made. In 1812, Perkins lured customers with descriptions of "New Drapery Patterns, some very large, with lace, fringe, curtain, and arch top borders to match." Through the decade, the New Bedford paper stainer continued to advertise "Large and small curtain and drapery patterns with Curtain and Lace figured borders to match."[2]

An 1808 bill for papers from Thomas and Caldeleugh of Baltimore survives in Chillicothe, Ohio. The papers hung there in a house called Adena. The bill itemizes:

18 ps New Drapery Paper No. 154	$18
3½ ps Borders, Do	10.50
7½ ps Do Narrow	5.62
6 ps Common Paper	4.50

Papers that seem to correspond to this bill survived at Adena to be photographed about 1900. Although the photograph is now too faded to reproduce well, it shows layers of draped swags and tassels covering the walls from floor to ceiling, interrupted only briefly with border above and below the wooden chair rail, along the top edge of the baseboard, and with the same border, which appears to have been less than two inches wide, outlining a fireplace (*figure 12-2)*.[3]

Fragments of another swagged drapery paper were uncovered in the 1970s during the course of restoring the eighteenth-century John Wheeler house in Murfreesboro, North Carolina. The pattern would probably have been called a "drapery figure" when it was installed early in the nineteenth century. It is made up of layers of swags. Each swag hangs from knobs at the two edges of the paper's width and each swag cradles a bouquet of flowers. Two bouquets alternate vertically: one features a large rose, the other three smaller flowers tucked between the folds of the drapery.[4]

Examples of the use of "drapery panels" in American houses have survived only rarely. One of the most striking of such papers, preserved since its nineteenth-century installation in this country, is displayed at Deerfield Village. The paper was originally hung in the Williams house in Augusta, Maine *(figure 12-1)*. It virtually defines in visual terms the French phrase used to describe such papers, "*Rideau à Bayadère,*" meaning curtains made from a striped fabric. This was the phrase Josiah Bumstead used to describe some of the French wallpapers he offered for sale in 1821. Bumstead mentioned "a few sets of "*Rideau à Bayadère*" as "a new and most splendid paper for drawing rooms."[5] The Boston wallpaper dealer probably copied his title for the newly arrived wallpaper from the French manufacturer's working designation on an invoice or crate, or possibly from a title in a printed catalogue.

SATIN PAPERS

"Satin" is a word that frequently appeared in advertisements for drapery papers. Within the context of the early-nineteenth-century wallpaper trade, that meant "having a shiney, polished finish." Beginning in the late eighteenth century, satin papers were widely advertised, and they were offered with increasing frequency through the first half of the nineteenth century. In 1800, Moses Grant included in a Boston advertisement for paper hangings "Striped and small figures, on satin grounds," and in the same year, Josiah Bumstead advertised his own "Sattin, Frosted, and Common Borders" as well as "English Sattin Paper Hangings."[6]

Every kind of pattern seems to have been printed on satin grounds. Some of the plain colored wallpapers were also described in the early nineteenth century as satin, in phrases like "plain satined Papers for rooms."[7] On satin grounds, figures, stripes, and plaids appeared. Among the hundreds of early-nineteenth-century advertisements that could be used to emphasize this point, one of 1826 from Baltimore that boasted of the Parisian origin of "splendid paper hangings . . . Velvet, Satin, and Half Satin papers with Velvet Borders to match" serves to suggest the fascination with contrasting slick and fuzzy surfaces in wallpapers during this era.[8]

Hardly a wallpaper advertisement of the 1820s and 1830s omitted the word "satin." In an advertisement of 1825, James H. Foster found it the key adjective for describing nearly everything in the twelve "cases and bales" of newly arrived French paper hangings, which included: "Damask, Satin, Satin Stripes, with rich cloth borders . . . satin, marble, and common papers."[9] His use of the word indicates that the shiny finish was felt to distinguish fine paper hangings from the run-of-the-mill or common papers. "Satin Paper" was one of the attractive features noted in a house at 77 Willow Street, Brooklyn, offered for sale in a New York newspaper of 1835.[10]

A DISTINCTIVE PATTERN TYPE

While French drapery patterns inspired some American imitation, a much larger group of French patterns was arriving in the early part of the nineteenth century that inspired large numbers of copies and adaptations. Many of the patterns within this group can be classified as Empire or neoclassical, but many of them have none of the distinctive characteristics of those styles. Because this group of patterns incorporated motifs from several decorative styles, it is more helpful to distinguish them by categorizing them as a pattern type made up of elements positioned in conformity to a set formula. Examples of these patterns are shown in *color plates 50 and 51* and in *figures 12-4 through 12-13, and 12-15.*

Complete versions of this pattern type include three elements that were always placed in a set relationship within the width and length of a roll of wallpaper. These included: (1) a dominant motif (or motifs), often a bouquet, figure, or self-contained little scene placed in the center of a width of wallpaper; (2) a spotted background made up of a subordinate motif, usually very small and simple and not much darker or far removed in hue from the ground color (dots, florets, leaves, stars, for example); and (3) vertical stripes along the edge or edges of the paper's width. The stripes were usually formed of small-scale geometric shapes—dots, dashes, ovals, circles—printed in stronger colors than the motifs of the background spotting and stacked in close-spaced vertical lines so that they formed definite stripes. The stripes often incorporated leaves, simple flowers, and other motifs.

Often, the dominant motif was matched by an alternate motif of equal importance, a close variation on a given theme, and these two were alternated vertically right down the center of the paper's width. In many other examples, a larger dominant motif was alternated with one slightly smaller—a subordinate element spacing the repeats of the more important motif. Occasionally, in more expensive papers, a variety of these dominant motifs were introduced, perhaps six or eight. The vertical spacing between the most important motifs in these papers varied from example to example. At one extreme, only three or four inches might be left between the large central elements, and six repeats might occur between the chair rail and the ceiling. At another extreme, which really represents a variant from the formula, only one rather large motif might be printed between the stripes within a wall-length of paper.

When the large central elements were eliminated, such papers became stripes with spotted backgrounds. But when the spotting was dropped from the scheme, the effect, though simplified, was quite similar to that achieved when all three elements were used.

"LANDSCAPE FIGURES"

Figures 12-8 and 12-9 illustrate an American wallpaper closely related to the pattern type described above. No background of spotted motifs is included in this paper, but the vertical stripes still serve as framing elements for the dominant motifs, which would have been called "landscape figures" at the time they were printed. Between 1810 and 1820, in the Boston newspapers, a number of sellers of wallpaper advertised "small landscapes" and "landscape figures." The terms are confusingly similar to one of the period's names for scenics—"landscape views"—but "landscape figures" almost certainly meant papers that were close variants on the one illustrated in *figures 12-8 and 12-9* rather than the wall-sized non-repeating French scenes. "Landscape figures" and small vignettes with human figures were also worked into patterns organized within frameworks other than that of the stripe. They appeared in simple drop repeats and amidst framing elements that encircled them with flowers and foliage like that illustrated in *figures 12-26, 13-6.*

Two illustrations in Hartford, Connecticut, newspapers of 1814 and 1815 provide good visual evidence of what was meant by "landscape figures" during this period. The newspaper cuts are illustrated in *figures 12-7 and 12-11.* In the text under the illustration with its prominent label—"New Style of Paper Hangings. Landscape Figures."—Mills and Danforth noted that they had

> . . . at great expense, executed the Manufacture of the above style of Room-Papers, in a superior manner, and in a variety of figures; obviating former objections to too much sameness in the room, when on.[11]

A fisherman on a bank under an awkwardly stunted tree was the subject both of the "Landscape Figure" shown in this 1814 advertisement and the wallpaper printed by the Boston firm of Moses Grant, which is illustrated as *figures 12-8, 12-9.* An advertisement Moses Grant had published a year earlier in Boston uses phrases similar to those in the Hartford notice. Grant had announced in 1813 that he had

> . . . just completed additional figures of the present fashionable Hangings for Rooms, obviating the objection of too much sameness, by introduction of a variety of views.[12]

The samples of Grant's wallpaper show the variety he mentioned: within the striped framework there is not only the above-mentioned fisherman, but there are also five other motifs—or "variety of views"—in an earnest attempt to "obviate the objection of too much sameness."

The similarity of subject matter and pattern suggests that Hartford paper stainers like Mills and Danforth probably saw examples of these very patterns printed by Grant. This assumption is substantiated by Hartford advertisements that offer Boston-made wallpapers for sale. Mills and Danforth themselves offered a "handsome selection from the Philadelphia and Boston Manufactories" in 1813.[13] The Hartford paper stainers could have based their "landscape figure" of a fisherman under a gnarled tree on Boston patterns, or they, as well as the Bostonians, could have gone directly to French wallpapers for motifs. A fisherman and his lady, again under a stunted tree, appeared on the French overdoor or fire board paper illustrated as *color plate 27.* It suggests the kinds of motif that appeared on French imports and were adapted for inclusion in such repeating patterns.

SMALL-SCALE REPEATING PATTERNS

Small-scale motifs, including the simplest suggestions of flowers or leaves rendered as conventionalized, very basic shapes and printed in solid, flat colors, appeared on hundreds of striped, diapered, squared, circular, and ogival configurations on wallpapers made right through the late eighteenth century and the first half of the nineteenth century. Because they were so often hung by houseowners over a long period of time, when fragments of these simple patterns are found in old houses it is difficult to determine their dates of manufacture.

Old Sturbridge Village preserves a rare document that pins down the dates for some of these patterns. It is a sample book that bears the mark of Janes and Bolles. That Hartford firm was in business only briefly, between 1822 and 1828. Janes and Bolles placed advertisements in local newspapers that illustrated simple striped patterns related to examples in the pattern book now at Sturbridge. The advertisement and a selection of patterns from the sample book are illustrated in *figures 12-13 through 12-20.*

Simple foliate, floral, and geometric forms, printed in solid flat colors, with no shading, make up the repeating patterns and stripes in the sample book. Like a modern wallpaper sample book, patterns are shown in numbers of alternate colorings. Soft blues, mustard yellows, pale yellows, chalky reds and pinks, and greens are featured. Motifs are combined in various ways, indicating the use of the same wood blocks to print a variety of patterns.

The Janes and Bolles sample book served as the basis for identifying fragments of wallpaper found during the 1973 restoration of an early-nineteenth-century house in Natchez, Mississippi. In a second-floor room of the Hugh Coyle house, bits of patterning survived *(figure 12-19)* that fit like jigsaw puzzle pieces into a Hartford-made pattern included in the Janes and Bolles sample book. Close comparison of the samples indicates that they were printed from the very same wood blocks. Finding Hartford-made wallpapers of the 1820s in a house on the Mississippi substantiates the evidence of advertisements that there was extensive domestic trade in the products of East Coast wallpaper makers during the early years of the nineteenth century.

Sprigs of flowers or foliage and many other small-scale motifs are always evident in the wallpaper trade. The motifs spotted or powdered in evenly spaced rows, drop-repeated to create optical diamond grids all over a wall, might illustrate any-

Color plate 57 The vocabulary of neoclassical ornament was embellished with color-blending *(irisé)* in this pattern, which is shown with a flocked floral border just as it was bound in a volume of designs produced in 1828–9 by the Zuber company in Alsace. The volume is one of the long series surviving in the archives of the still-operating factory in Alsace. A fragment of this particular Zuber pattern was found in a bedroom at Fort Hill, the house of John C. Calhoun in Clemson, South Carolina. Built in 1803, the house was enlarged in 1825, when Calhoun became vice president; the paper duplicating this pattern was probably hung shortly after that date. Courtesy of Zuber et Cie., Rixheim, Alsace

57

58

Color plate 58 (left) Another Zuber pattern with a flocked floral border is shown as it was mounted in Zuber's volume for 1832–3, which has been preserved in the company archives. Two duplicates are preserved at the Society for the Preservation of New England Antiquities. One was found in the Daniel Busley house in Newburyport, Massachusetts, in color-blended shades of green, gold, and white. The other, in green and gray with no color-blending, was found in the William Blake house in Camden, Maine. Courtesy of Zuber et Cie., Rixheim, Alsace.

Color plate 59 (opposite) Bound into the Zuber volume for 1829–30, this is a geometric pattern that depends on the color-blending technique for much of its visual impact. It is shown with a flocked floral border of a type the factory apparently deemed appropriate to embellish the varied range of pattern types incorporating color-blending. Courtesy of Zuber et Cie., Rixheim, Alsace

59

60

Color plate 60 (left) The Zuber volume for 1832–3 included this example of subtle and delicate color-blending for shading leaves and flowers. The factory featured this technique, used in a similar way for *"Décor Chinois"* *(color plate 35).* Here again a flocked floral border is Zuber's suggested finish for an *irisé* paper. Courtesy of Zuber et Cie., Rixheim, Alsace

Color plate 61 (opposite, left) A single repeat of a French *ombré* paper of 1815–30 is shown mounted between borders. Here, the ground rather than the pattern has the color-blending. It is 18½ inches wide. Cooper-Hewitt Museum, 1931-45-55; gift of Eleanor and Sarah Hewitt

Color plate 62 (below) This strong geometry was the earliest wall decoration in the stair hall of the Jonas Wood house, built in 1804 at 314 Washington Street in New York. Each diamond is a little over 6 inches long. The border used with it is just under 3 inches wide. Duplicates of the diamond pattern have survived in several places: the Harrison Gray Otis house, Boston; on a small bandbox at the New-York Historical Society; as the lining inside a picture frame of about 1815 bearing the label of Isaac Platt of New York; and lining a trunk found in an eighteenth-century house in East Hampton, Long Island. Cooper-Hewitt Museum, 1973-8-1A; gift of the City of New York, Housing and Development Administration

61

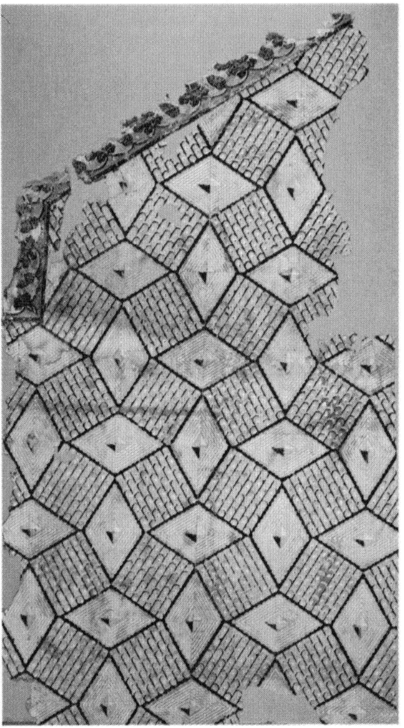

62

AND ROFF PAPER
PUTNAM
HANGING &
MANUFAC
BOX
BAND
CLAYTON
AND ROFF PAPER
PUTNAM
HANGING &

Color plate 63 (opposite) A selection of bandboxes at the Cooper-Hewitt Museum suggests the variety of papers used to cover them. Their decorations here include a manufacturer's advertisements, commemorations of current events, classical mythological scenes derived from French Empire ornaments, and naive pictures of animals. Many patterns were printed especially for bandboxes; others were wallpaper patterns.

Color plate 64 (right, top) George Putnam and Amos B. Roff advertised on their bandboxes, which must have attracted attention as they were carried about like shopping bags. Because they announced both the formation and dissolution of their partnership in Hartford newspapers, this can be dated between 1821 and 1824. It is 12¼ inches tall. Cooper-Hewitt Museum, 1913-45-10; gift of Mrs. Frederick F. Thompson

Color plate 65 (right, bottom) Block-printing in distempers on joined sheets of paper was here used to make decorative paper specifically for a bandbox 12½ inches tall. This scene commemorates a new pump wagon introduced by The Eagle Engine Company #13. The New York scene was used by a Philadelphia bandbox maker who affixed his label under the lid of this box: Henry Barnes, active 1829–1844. Cooper-Hewitt Museum, 1913-17-15; gift of Eleanor and Sarah Hewitt

64

65

66

Color plate 66 (left) Wallpaper with a strong geometric repeating pattern was used to cover this cardboard bandbox of the mid-nineteenth century. Fifteen inches tall, missing its lid, this box has two wooden bracing strips across the bottom. Other bandboxes were made entirely of wood, then covered with wallpaper. Cooper-Hewitt Museum, 1968-44-6; gift of Harvey Smith

WAGNER & Mc GUIGAN.
EUGENE ROUSSELS.
GEDDES PRINTER
S.W. STOCKTON & Co
STOCKTON & Co.
JOHN C FARR.
WATCHES
JOHN C FARR
JEWELRY
PERFUMERY E. ROUSSEL. FANCY ARTICALS
HOWELL & BROTHERS.
116
PAPER HANGs
JOHN C. FARR,
WATCHES,
JEWELRY, &c.
WHOLESALE AND RETAIL,
NO. 112 CHESTNUT STREET,
PHILADELPHIA.
EUGENE ROUSSEL,
MANUFACTURER AND IMPORTER OF
PERFUMERY,
NO. 114 CHESTNUT STREET.
PHILADELPHIA.
Fancy Soaps of every description,
COLOGNE WATER, HAIR OILS, &c.
WHOLESALE AND RETAIL.
Gold and Silver Medals awarded by the Institutes of New York, Philadelphia and Boston.
BRANCH STORE,
No. 159 BROADWAY, NEW YORK.
HOWELL & BROTHERS,
NO. 116 CHESTNUT STREET.
PHILADELPHIA,
MANUFACTURERS OF EVERY DESCRIPTION OF
PAPER HANGINGS,
Suitable for Parlors, Entries, Dining Rooms, &c. which they offer Wholesale and Retail at the lowest rates
A new article of wide heavy bodied Window Curtain paper, admirably suited to the country trade.
BRANCH STORES.
No. 137 Broadway, New York, and No. 217 Market St. Baltimore.
Where may be found all their variety of Patterns of Paper Hangings.
H. & B. have been awarded five Silver Medals for their superior manufactured goods, by the Cities of Boston, New York, and Philadelphia.

67

thing from a dragonfly to a puppy dog to a little cottage. Most often, flowers were evenly spaced over a wallpaper's surface, as in *color plate 52.* Flower sprigs, like the ones in that example, were among the most popular motifs, but bouquets, circlets, vases, and pots of flowers were also incorporated in patterns put together much as this example is.

Floral sprigs found their way into middle-class houses as well as into rooms decorated by the very rich during the early years of the nineteenth century. In more elegant dwellings they were sometimes relegated to unimportant rooms—bedrooms and servants' quarters—while in simpler houses they were used in parlors. The fact that the pattern illustrated in *color plate 52* was found in an elegant Hudson River mansion, while a similar floral sprig wallpaper is depicted in a painting of a relatively modest room shown in *figure 12-21,* suggests one aspect of the range of appeal of such patterns.

TEXTILE IMITATIONS AND RAINBOW PAPERS

During the 1820s and 1830s, the vocabulary of neoclassical and of floral and foliate ornament, which French paper stainers had been using since the eighteenth century, was stretched and embellished in thousands of ingenious ways to create an apparently inexhaustible outpouring of variants. Many of these are preserved in French archives.[14] Earlier in the century, that ornamental vocabulary had often been used with conventionalized motifs, appearing as flat two-dimensional ornaments on a flat surface, or it had been stylized to look like the shallow relief carving and plaster work of architectural trimmings. But by the 1830s, more often that classical vocabulary of ornament was overlaid with more three-dimensional naturalistic representations of flowers and foliage, elaborately shaded so that the blossoms seemed to await plucking.

In addition, many of the patterns of the 1820s and 1830s were printed as eye-

Color plate 67 About 1849, Howell and Brothers of Philadelphia used their wood blocks and distemper colors to print a posterlike trade card 22½ inches wide. The firm was founded in Albany, New York, in 1813 by John B. Howell, who had come to America from London in 1793. He moved to Philadelphia about 1817. After 1835, his sons carried on as Howell and Brothers. They introduced new English machines for printing wallpaper to America in the 1840s. Praised by taste makers like A. J. Downing, their papers were exhibited at London's Crystal Palace of 1851, New York's Crystal Palace of 1853, the Paris Exposition Universelle of 1867, and at Philadelphia's Centennial Exhibition of 1876. By this time Howell was the biggest wallpaper manufacturer in the United States. Courtesy of the Library Company of Philadelphia

fooling imitations of textiles. This applied not only to the drapery papers, but also to patterns that made a wall look as if a flat piece of cloth had been stretched across it. It is also true of simulations of woven fabric, where tiny lines represented thread patterns—twills, satin weaves, and brocades *(color plates 54 and 55).*

A newly developed printing technique distinguished many of the patterns of this period from similar patterns made during other periods. About 1820, the Zuber factory in Alsace began to produce papers featuring color-blended effects. At the Zuber factory shaded coloring produced by special techniques was called *irisé.* Some trade literature used the word *ombré* for what Zuber called *irisé,* and, less frequently, the phrase "*fondu* style" was used in referring to papers that featured color shading. In the American wallpaper trade, they were called "rainbow papers."[15] Samuel Robinson of Washington, D.C., used this term in his 1826 advertisement for "Rainbow Papers, all colors."[16] The kinds of papers he was selling have survived in quantities from houses in all the then-settled parts of this country. Similar color-shaded effects were popular in textiles and carpets, especially in ingrain carpets (pile-less woven carpets usually with patterned surfaces rather like double-woven coverlets), during the 1820s and 1830s.

Though the popularity of rainbow, or *irisé,* papers had peaked nearly a quarter century earlier, as late as 1853 they were still admired by the important English critic Matthew Digby Wyatt, who mentioned them in his *Industrial Arts of the Nineteenth Century,* a book published in 1853. Wyatt described "rainbowing" as "the production of a shaded or blended ground, consisting of graduated tints of a particular color," and called rainbowing "one of the most effective processes" in the production of paper hangings. These color-blending techniques were used for the ground coloring of repeating patterns made by many manufacturers, and Wyatt noted that in 1825 an English firm, "Messrs. Clarke and Henderson of Bond Street," had introduced machines using cylinders to apply rainbow grounds, which he deemed an improvement over the French handwork processes.[17]

Not only did the Zuber factory use color-blending techniques to create shaded grounds for their scenics and repeating patterns, but, in addition, they perfected and exploited color-blending techniques for the patterning applied with wood blocks. Zuber patented his processes for printing in the rainbow colors.

An Englishman, Andrew Ure, in his *Dictionary of Arts, Manufactures and Mines,* first published in 1839, described the production of what he called the "*fondu* or rainbow style of paper-hangings." Ure seems to have known Zuber's technique, or closely similar techniques. He described "rainbowing" as:

> produced by an assortment of oblong narrow tin pans, fixed in a frame, close side to side, each being about one inch wide, two inches deep, and eight inches long; the colors of the prismatic spectrum, red, orange, yellow, green, &c. are put in a liquid state, successively in these pans; so that when the oblong brush . . . is dipped into them across the whole of the parallel row at once, it comes out impressed with the different colors at successive points . . . of its length, and [the brush] is then drawn by the paper stainer over the face of the woolen drum head, or sieve of the swimming tub, upon which it leaves a corresponding series of stripes in colors, graduating into one another like those of the prismatic spectrum. By applying his block to the . . . [surface of the swimming tub] . . . the workman takes up the color in rainbow hues and transfers these to the paper.[18]

Zuber used brushes bearing graduated multi-colored hues to spread the colors in circular, chevron, and other geometric configurations on the swimming tub, so that even more complex color-blended effects could be picked up by the printing blocks and stamped in patterns on the papers.

At the Zuber factory, the *irisé* technique was featured in a few startling geometric patterns that are related to some Op art of the 1960s *(color plate 59).* The technique was also used to create realistically blended tones for naturalistic designs, adding richness of detail for leaves and for the petals of flowers *(color plate 60).* Most frequently Zuber used *irisé* effects in stylized floral and foliate patterns derived from textile prototypes *(plates 57, 58).* In many of these papers, the colorful effects dominate the patterning, and the increasing and diminishing strength of the colors creates wavelike rhythms that are the most arresting features of the papers when they are hung on large expanses of walls.

Hundreds of *irisé* papers are preserved in dated sample books at the Zuber factory. Rainbow papers that can now be positively recognized as Zuber products introduced in a given year in the 1820s and 1830s have been found in houses in towns as widespread as Clemson, South Carolina; Alexandria, Virginia; Branden, Vermont; Newburyport, Massachusetts; and Camden, Maine.[19] Such papers were probably imported directly from Alsace by some of the ninety-nine American dealers with whom the Zuber firm did business during the early nineteenth century. Complete photographic records of the Zuber patterns are now housed at the Museé des Arts Décoratifs in Paris and could be of great assistance in identifying other Zuber patterns as they turn up in American houses. In addition to the *irisé* patterns printed by Zuber, rainbow papers by other makers have survived in American houses *(color*

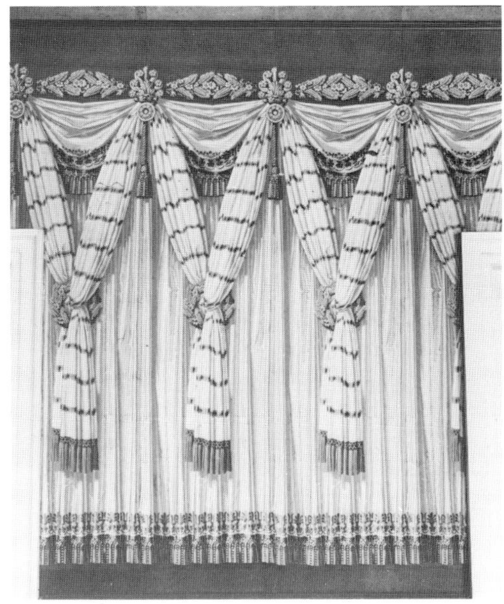

12-2

12-1

12-3

Figure 12-1 (above left) This wallpaper drapery, imitating panels of heavy green-striped white silk fabric with gold fringes and ornamentation, was block-printed in France about 1815. The paper was originally hung in the Ruel Williams house in Augusta, Maine, and is now displayed, as shown here, in the Stebbins house at Historic Deerfield, Massachusetts. The presence of a stamped English excise mark on the paper is puzzling. The pattern duplicates a paper known to be French that survives in the Carlhian collection in France. Courtesy of Historic Deerfield, Deerfield, Massachusetts

Figure 12-2 (above right) Creating the illusion that the drawing room walls were draped in endless swags and tassels, this wallpaper and its narrow border were hung at Adena, in Chillicothe, Ohio, shortly after their purchase in October of 1808 from a firm of Baltimore paper stainers. They probably were printed by Thomas and Caldeleugh, whose bill to Thomas Worthington for these papers survives in the Adena State Memorial. They imitate elaborate French drapery patterns. The photograph was made about 1900, before the papers were removed. Cooper-Hewitt picture collection

Figure 12-3 (left) This fragment of a drapery figure was almost certainly printed in America to imitate French prototypes. It was found at the Hartwell Tavern, in Lincoln, Massachusetts, built in 1732 by Ephraim Hartwell, and enlarged in 1830. The paper, which dates from the early nineteenth century, was probably used in the parlor. The fragment, about 10 inches wide, is crudely block-printed in brown on paper with no layer of ground color. The paper stock is now tan. Courtesy of the National Park Service, Concord, Massachusetts

plates 56, 57, figures 12-27, 12-28). They do not show the technical brilliance of the Zuber papers, but they attest to the popularity of the style.

PATTERNS PREFERRED FOR SPECIFIC ROOMS

While contemporary documents suggest where scenic papers, wallpaper statuary, and some of the more elaborate wallpaper ornaments were generally deemed appropriate, they were less specific about the multitude of repeating patterns. It is difficult to know just which patterns were thought stylish for which rooms. However, some generalization seems safe and fairly obvious. Throughout the eighteenth and nineteenth centuries, the larger-scale, more elaborate, and more formal papers were preferred for halls, sitting and dining rooms, while small-scale patterns, especially repeating floral ones, were preferred for more intimate rooms like bedrooms.

During the first half of the nineteenth century, the wording of advertisements indicates that there were prevailing tastes dictating the suitability of certain styles or motifs for certain rooms. A Baltimore advertiser of 1826 offered, for example, paper hangings from Paris "suitable for drawing rooms, parlors, chambers, stores, and passages."[20] However, he gave no details indicating which patterns belonged in which places, nor did many contemporary advertisers.

As was true for the eighteenth century, more evidence has survived about papers thought to be appropriate for halls than for any other rooms. "Entry papers" and "hall papers" lead the list, compiled from advertisements of this period, of patterns designated for particular rooms. Landscape views, statues, columns, and panel papers of the types described in Chapters 9 and 10 were the favorite hall papers, but "large figures" were also described as "suitable for halls, entries, &c" in advertisements of the first quarter of the century.[21]

Patterns found in a New York house of the early nineteenth century provide a typical and interesting sequence of wallpaper choices for use in the hall of a modest dwelling. Most of the series of papers removed layer by layer from the stair hall of the Jonas Wood house, built in 1804 on Washington Street in Manhattan, is shown in *color plate 62 and figures 12-31 through 12-33.* These papers date from 1804 through the time in the mid-nineteenth century when the installation of wooden paneling sealed them all in, to be uncovered when the house was moved to a new site and restored during 1972 and 1973.

The little house was built by a grocer named Jonas Wood, who occupied it until 1807. The strongly delineated diamond pattern of *color plate 62* probably decorated

12-4

12-5

Figure 12-4 (left) In this French wallpaper of 1810–15, a third motif has been introduced to the formulaic scheme illustrated in *color plates 50 and 51.* Here, all the striping has been printed along the right edge of the paper width, but on the wall the striped effect would have been like that of *figures 12-5 and 12-6.* Predominantly green, the sample is 20 inches wide. It was preserved by the Kilham family in Beverly, Massachusetts. Cooper-Hewitt Museum, 1955-86-1; gift of Teresa Kilham

Figure 12-5 (above) Though in a sad state of decay when photographed in 1941, these wallpaper remnants document the use—in inland North Carolina—of French wallpaper patterns very similar to those of *figure 12-6* hung about the same time in Canada. The elaborate combination of borders and patterns both on walls and ceiling is particularly interesting. These papers were pasted directly on unfinished boards in a bedroom at Oak Lawn, built in 1818 in Mecklenburg County, North Carolina. Courtesy of Mrs. Wilson L. Stratton

12-6

Figure 12-6 (above) The same basic framework for the patterning as seen in *figures 12-4 and 12-5* is apparent in this early-nineteenth-century French wallpaper, incorporating pink flowers with green foliage and gray ornaments on a gray ground. With its matching borders, it was hung in this room, the parlor of the French–Robertson house, shortly after the house was built about 1820 at Mille Roches on the St. Lawrence River in Canada. The house is now part of Upper Canada Village in Ontario. Courtesy of the St. Lawrence Parks Commission

Figure 12-7 (right) In Hartford's *Connecticut Mirror* for May 9, 1814, Zecheriah Mills and Edward Danforth ran this woodcut over an announcement that "at great expense" they had "executed the above style of Room Papers, in a superior manner, and in a great variety of figures, obviating former objection to too much sameness on the walls." These figures were based on French prototypes like *color plate 27.* Motifs similar to the fisherman were used in papers like those of *figures 12-4 through 12-6* and *12-8, 12-9, 12-12.* "Landscape figures" were printed between stripes as well as at the centers of connecting circles and other geometric patterned frameworks. (See wallpaper on bandbox illustrated in *figures 13-5, 13-6, 13-8.)* Such figures were also simply stamped at regular intervals over plain grounds. Courtesy of the Connecticut Historical Society, Hartford

12-7

12-8

12-9

12-10

Figures 12-8, 12-9, 12-10 (this page) The mark of Moses Grant, Jr., was stamped on the reverse of this wallpaper, removed in 1968 from the Ephraim Wood house in Concord, Massachusetts. Grant's father was making wallpaper in Boston in 1789. The junior Grant probably printed these "Landscape Figures" around 1813. In addition to these four motifs—which include a fisherman like that in *figure 12-7*—two others were found in the same room printed in brown, tan, red, and black on a gray-blue ground. Courtesy of the Concord Museum, Concord, Massachusetts

Figure 12-11 (opposite, top left) On October 26, 1816, two years after Mills and Danforth illustrated a landscape figure in a Hartford newspaper, James R. Woodbridge and George Putnam, who had bought out the firm, ran another landscape figure in Hartford's *American Mercury.* The vignette is similar to the cows and pigs on Moses Grant's papers, and to the girl and sheep in *figure 12-12.* All were drawing on French sources for the motifs. Courtesy of the Connecticut Historical Society, Hartford

Figure 12-12 (opposite, top right) This French-style pattern was probably printed by Janes and Bolles (active in Hartford 1822–8), since it was found with the album including one marked paper of that firm and is numbered on the reverse "861 3/6" in the same script found on the Janes and Bolles sample. It is printed in blue and white on a pale blue ground. Courtesy of Old Sturbridge Village

12-11

12-12

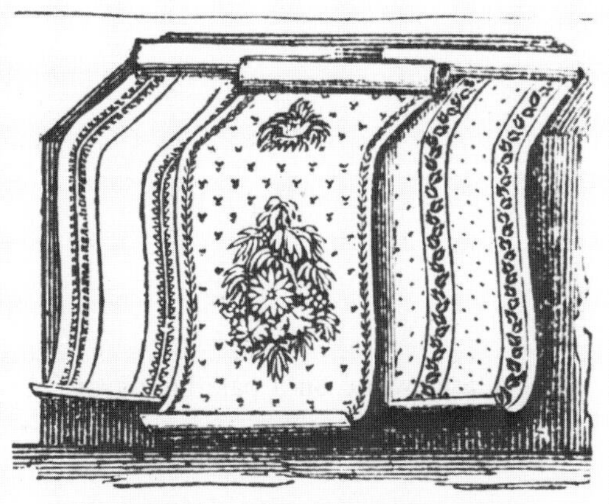

12-13

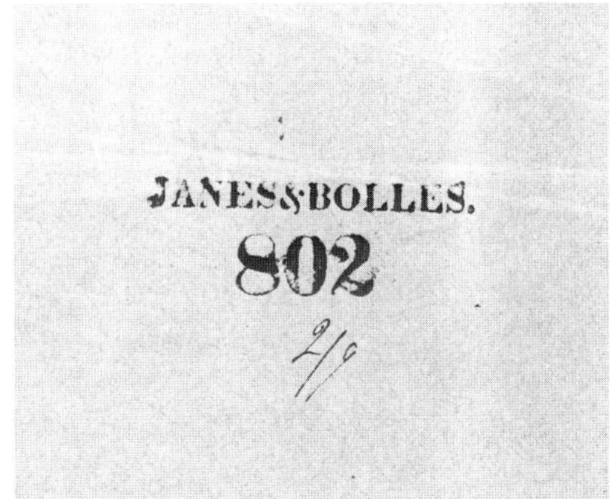

12-14

Figure 12-13 (bottom left) An advertisement placed by Adrian Janes and Edwin Bolles in Hartford's *American Mercury* for June 1, 1824, shows a French-influenced pattern, similar in format to another Janes and Bolles pattern *(figure 12-12).* Courtesy of the Connecticut Historical Society, Hartford

Figure 12-14 (bottom right) This mark of Janes and Bolles, stamped on the reverse of *figure 12-15,* was bound with 29 other samples, all priced in the same script—"4/." Courtesy of Old Sturbridge Village

12-15

12-16

12-17

12-18

Figures 12-15 through 12-18 These are four of the twenty-nine wallpaper samples included in the Janes and Bolles sample book. All the patterns are block-printed in distemper colors on joined sheets, marked on the reverse in script with a pattern number and a price that depended on the number of colors. The base price of 2/ (probably shillings) appeared on every pattern printed in one color on an ordinary ground color. A golden-yellow ground color was more expensive: One color on golden yellow cost 2/3. Three colors on golden-yellow were the most expensive: 4/.

12-15 (top left) bears the mark shown in *figure 12-14.* This pattern appears twice in the sample book: here in blue and white on pale blue; elsewhere in white and green on gray.

12-16 (top right) Here printed in white on mustard yellow, this pattern also appeared in blue on white.

12-17 (bottom left) This appeared in five combinations of colors: here in blue, black, and white on blue; elsewhere in green, black, and white on blue; in green, black, and white on gray-blue; in green, black, and white on yellow; and in green, black, and yellow on yellow. In yet another sample, the chain link of flowers was eliminated and replaced by tiny configurations of dots.

12-18 (bottom right) This pattern appeared in red on white, red on yellow, blue on white, and white on yellow.

Courtesy of Old Sturbridge Village

Figure 12-19 (top) Fragments of wallpaper were found pasted directly on the boards in a second-floor room of the Hugh Coyle house, built in 1793–4 in Natchez, Mississippi. Though faded, the color of the patterning is clearly blue, white, and black on a light tan paper stock without any coating of ground color. These fragments duplicate a pattern included in the Janes and Bolles sample book, and appear to have been printed from the same wood blocks. The board shown here is 8 inches tall. Cooper-Hewitt Museum, 1973-13-1; gift of the Natchez Historical Society through Thomas B. Buckles, Jr.

Figure 12-20 (bottom) In the Janes and Bolles sample book at Sturbridge Village, the pattern duplicating that found in Natchez was shown twice: the sample illustrated is pink, red, and white on yellow; another is green, white, and black on pink. Courtesy of Old Sturbridge Village

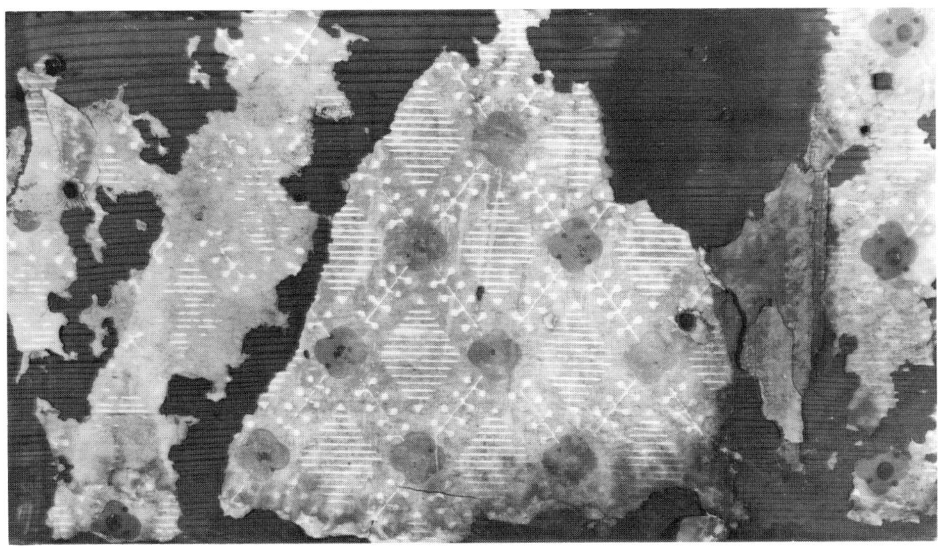

12-19

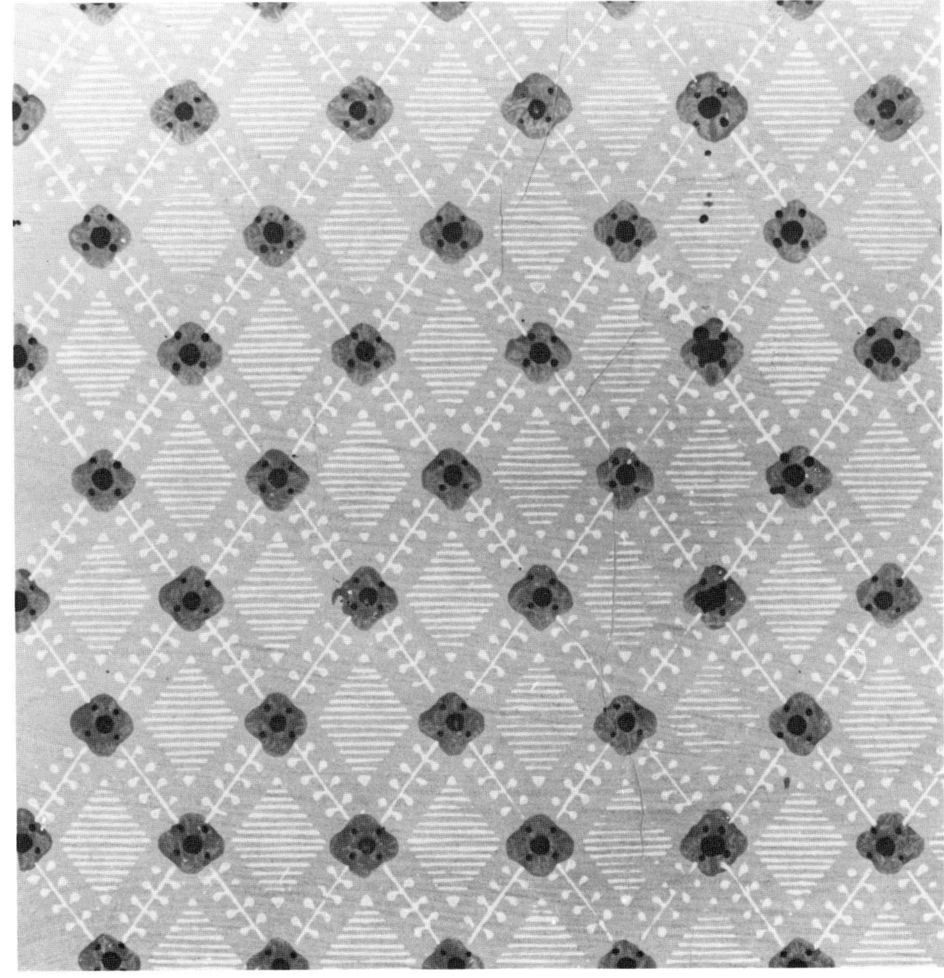

12-20

12-21

Figure 12-21 (top) Green and red floral sprigs on a yellow ground covering the walls in the Talcott family's New York State home were meticulously detailed by Deborah Goldsmith, the artist who painted this portrait about 1832. Courtesy of the Abby Aldrich Rockefeller Folk Art Collection, Williamsburg, Virginia

Figure 12-22 This sample, block-printed in one color, bears a label: "MANUFACTURED/ AND SOLD BY / OLIVER SAGE/ New Haven, Conn." Sage's working dates have not been found, and this could have been printed at any time during the late eighteenth or early nineteenth century. Courtesy of the Henry Francis du Pont Winterthur Museum

12-22

12-23

12-24

Figure 12-23 In 1842 Sarah Louise Spencer, aged 1 year, 1 month, stood on a boldly patterned carpet before a papered wall for her portrait by Henry Walton. Walton's painting documents the persistence of neoclassical taste in patterning on the wall and in the tablecloth. The medallions framing the child's head are adaptations of classical anthemion and palmette motifs, motifs used on French wallpapers and American imitations of French wallpapers all through the early nineteenth century. Courtesy of the Abby Aldrich Rockefeller Folk Art Collection, Williamsburg, Virginia

Figure 12-24 (top right) Large circular motifs like this, akin to the pattern in the painting of *figure 12-23,* but derived from vernacular rather than classical sources, were popular during the second quarter of the nineteenth century. This one is machine-printed (see Chapter 14) in blue and green on glazed white paper. The sample is 16½ inches wide. Cooper-Hewitt Museum, 1939-58-1, gift of Mrs. Carola R. Green

12-25

Figure 12-27 (opposite, left) A color-blended ground of blue shading into green and then to yellow forms the background for an overscaled medallion pattern printed in shades of gray and white. Each octagonal circle of acanthus leaves is nearly 20 inches wide. Perhaps French, the paper was used in the stair hall of the Horace Loomis house built in Burlington, Vermont, in 1800. Cooper-Hewitt Museum, 1941-52-4; gift of Mrs. William A. Hutcheson

Figure 12-25 (above) At Rose Hill, a house built in 1802 in Yanceyville, North Carolina, French wallpapers and floral borders with black-flocked grounds survive in the parlor wing added to the house early in the nineteenth century. According to family tradition, these papers were brought back to North Carolina by Bradford Brown and his bride, Mary Lumpkin Glenn Brown, after their wedding trip of 1816 to England and Scotland. Within a framework of neoclassical foliate patterning printed in a tone just darker than the creamy white ground color there are naturalistically rendered bouquets of fruit and flowers in bright multi-colors with strong accents of red. Courtesy of Mrs. J. Williamson Brown

Figure 12-26 (right) Motifs like the girl in the swing, derived from French rococo sources, were simplified by American paper stainers and incorporated in patterns of the second quarter of the nineteenth century. Thousands of motifs could have been substituted within this framework of flowers and birds, and it in turn might have been made up of any one of a number of alternate designs, like conventionalized foliate motifs derived from classical sources. The dark ground—here black—was relatively unusual, although papers with such ground colors were occasionally advertised. The pattern—an 18-inch-long repeat—was block-printed in pink, green, and white. The same pattern printed on a light ground survives on a bandbox in the Shelburne Museum. Cooper-Hewitt Museum, 1970-26-17; gift of Jones and Erwin, Inc.

12-26

12-27

12-28

12-29

Figure 12-28 (top right) In a crude use of color-blending on a paper probably made in America, one printing block or roller produced a two-color patterned stripe on ungrounded paper. The sample is 20 inches wide; the colors pink to white on tan paper. Cooper-Hewitt Museum, 1970-26-7; gift of Jones and Erwin, Inc.

Figure 12-29 (bottom right) This paper, from the Mary Waters Homestead, Stafford Street, Dodge, Massachusetts, is block-printed in buff and in shades of bright green on a cream-colored satin ground. It is a Zuber pattern introduced in 1832–3. Cooper-Hewitt Museum, 1940-74-1; gift of Mrs. Rollin Stickle

12-30

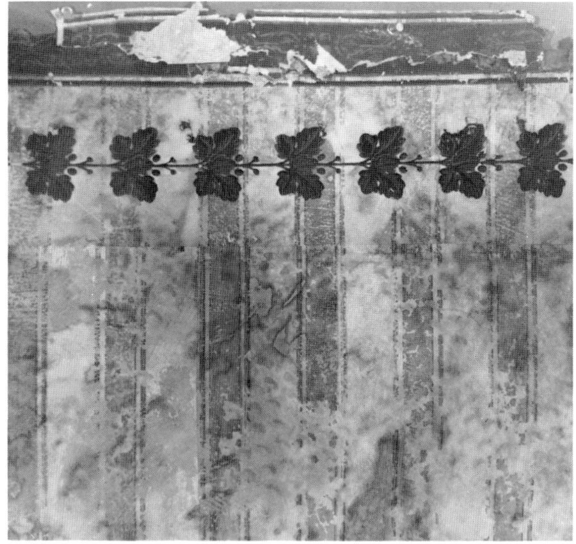

12-31

12-32

12-33

12-34

12-35

Figure 12-30 (opposite, top left) Found in the Horace Loomis house, built in 1800 in Burlington, Vermont, this was finished with two borders: a simple flocked edging and a variation of the wall pattern, made by overprinting the green and white stripes, each 4½ inches wide, with leaves in two shades of green and flowers in orange, green, and brown flocking. Cooper-Hewitt Museum, gift of Mrs. William A. Hutcheson

Figure 12-31 (opposite, top right) These stripes, in shades of yellow edged with brown, were pasted on top of the diamond pattern in the stair hall of the Jonas Wood house *(color plate 62).* Staining from the earlier pattern is apparent here. With its black and green flocked border, it appears to be French of 1805–15. The light area between each stripe is 1½ inches wide. Cooper-Hewitt Museum, 1973-8-2A, B; gift of the City of New York, Housing and Development Administration

Figure 12-32 (opposite, bottom left) This was the fourth wallpaper in the hallway of the Jonas Wood house. Its imitations of tiny cut stones—each only 3⅛ inches long—were printed in two shades of gray and white over a lighter gray ground, about 1810–30. Cooper-Hewitt Museum, 1973-8-4; gift of the City of New York, Housing and Development Administration

Figure 12-33 (opposite, bottom right) This imitation of stones laid in bond (1845–60) was the last wallpaper hung in the Jonas Wood hallway, and the third imitation of stone. Between this and *figure 12-32* two other patterns were hung, including a duplicate of *color plate 68.* Each "stone" shown here is 18½ inches long, machine-printed in gray, white, black, and turquoise on a gray ground. Cooper-Hewitt Museum, 1973-8-7; gift of the City of New York, Housing and Development Administration

Figure 12-34 (this page, top) In the 1830s or 1840s, the hallway of the Bliss–Keep house in Longmeadow, Massachusetts, was papered with this marble imitation block-printed in gray, blue, and white distempers over light gray on machine-made paper. It is probably French. Cooper-Hewitt Museum, 1974-61-1; gift of Robert B. McTaggart

Figure 12-35 (this page, bottom) Blocks of ornament, not unlike those in stonework papers, were often chosen for halls with heavy traffic. A matching block could be cut from extra pieces and pasted over any damage with little trace. This paper was block-printed about 1830–50 in green and gray on pale gray. It is 18½ inches wide. Cooper-Hewitt Museum, 1951-3-1; gift of Mrs. Charles P. Morgan

the stair hall throughout the grocer–owner's occupancy. The paper was found pasted directly on plastered walls and appears to be original to the house.

It is not clear which of three renters and owners who lived in the house between 1808 and 1815—they included a broker and a merchant–grocer—papered over the nearly dizzying diamond pattern, but the second paper on the wall seems to fall within that period. Whoever was responsible turned from the earlier, optically strenuous, geometry to a restrained and sober stripe, trimmed with a tastefully modest bordering of flocked leaves. This paper and its border are shown in *figure 12-31,* in which staining from the earlier diamond pattern can be seen penetrating the stripe pasted over it.

Between 1816 and 1834 a merchant, a widow, yet another grocer, and the owner of a china store were among the occupants of the house. Only fragments of one pattern installed sometime during that eighteen-year period survive. The rest of that paper must have been removed, although the two earlier layers remained in place, when one of the early nineteenth-century occupants chose to paper the same stair hall in a pattern imitating a stone wall. Each gray stone was about the size of a very small brick. That paper, shown in *figure 12-32,* was probably hung sometime prior to 1835.[22] It is typical of wallpapers then recommended for halls, and represents a taste that was to dominate the wallpaper choices of successive decorators of the hallway in the Jonas Wood house.

"Stone Patterns," or "ashlar patterns," like this and like two more wallpapers found pasted over it, were so popular for hallways that perhaps they are the papers the early-nineteenth-century dealers had in mind when referring to "entry papers" and "hall papers." In the *Encyclopaedia of Cottage, Farm and Villa Architecture and Furniture* of 1833, John Claudius Loudon advised:

> One of the best plain papers for the entrance lobby and the staircases of cottages, is one simply marked with lines in imitation of hewn stone; because, when any part of this paper is damaged, a piece, of the size of one of the stones can be renewed, without having the appearance of a patch.[23]

The early nineteenth-century decorators of the Jonas Wood house might have been following such advice when the gray stone imitation of *figure 12-32* was hung. Tiny fragments of a second pattern imitating stone work were found pasted two layers above the little stones of *figure 12-32.* These fragments fit precisely into the mid-nineteenth-century stone-with-scrollwork pattern shown as *color plate 68.* In the Wood house duplicate of the pattern in *color plate 68,* and in a still later and larger stone pattern of the 1850s or 1860s, illustrated in *figure 12-33,* which was pasted directly above, the nineteenth-century preference for hallway wallpaper that imitated stones was demonstrated by successive occupants of the house on Washington Street. The pattern shown in *figure 12-33* was the last layer of wallpaper pasted on the walls before the wooden paneling was installed.[24]

Dozens of variations on the fake stone theme are preserved in wallpapers of the Cooper–Hewitt collections. Many were found on the walls of halls and entries. One example from a house called Teviotdale in Liwithgow, New York, survived on walls of both hallway and drawing room. It bears the mark "L. Steel/New/234/Factory/Albany." Lemuel Steel was manufacturing wallpaper in Albany between 1822 and 1848.[25] And in many halls the division of the pattern into easily replaced blocks—whether square or rectangular—met Loudon's requirement that the pattern be easily replaced by cutting out new blocks and fitting them between the sharply defined divisions in the patterning even when the design did not imitate the surfaces of stones. Some of these papers were made up of what looked like small framed pictures, miniature scenes—they might well have been called "landscape figures"—stacked up like cut stones laid in a wall.

Personal whim, cost, and the availability of various kinds of patterns often appear to have dictated the selection of patterns for the other rooms of a house; rules of taste were no more binding then than they are today, although they certainly exerted an influence on choice. Formal, high-style, and expensive papers were more frequently found in the more public rooms of houses, while floral patterns and less expensive papers were used in bedrooms. Conversely, quite simple papers do sometimes survive in early-nineteenth-century parlors and hallways, despite recommendations of wallpaper sellers or writers of books on house-decorating and home economics.

Prevailing tastes in wallpaper were not often the subject of writings other than those published for the decorating trade and its customers during the early nineteenth century. It is therefore interesting to find comments on wallpaper fashions in the correspondence and other writings of individuals not in the trade. In 1836, Margaretta Brown of Frankfort, Kentucky, wrote from New York to her daughter-in-law who was building a house in Kentucky, and for whom the older Mrs. Brown was shopping. She said that she was sending two lots of paper and bordering for lesser rooms and added that she had sent from the shop of Pares and Faye two other papers

> intended for your drawing and dining rooms—it is not the fashion now to furnish the rooms opening into each other exactly alike; but there is some difference in the paper and carpets.[26]

13 BANDBOXES

As early as the seventeenth century, utilitarian lightweight wooden or pasteboard containers were made in England to store linen neckbands and lace bands worn by gentlemen. In 1755, Dr. Samuel Johnson defined "bandbox" in his famous dictionary as "A slight box used for bands and other things of light weight."[1] The colorful profusion of American bandboxes shown in *color plate 63* only begins to suggest the range and variety of boxes covered with wallpaper and with papers made by the paper stainers especially for covering such boxes. These bandboxes enjoyed a peculiarly American popularity during the second quarter of the nineteenth century.

The wallpaper trade in this country took these lowly "slight boxes," enlarged them, embellished them with papers that were block-printed in distemper and varnish colors, and transformed their function. By the period of their greatest popularity —between 1820 and 1845—"bandbox" had come to have a slightly different meaning in America: a large pasteboard or lightweight wooden box, oval or round in plan, usually twelve to fourteen inches tall, and covered with colorful decorated paper. The boxes served as the great catch-all for America's overflowing paraphernalia of travel, for closet storage at home, and probably for the occasional storage of the collar bands from which their name derived.

Many examples of bandboxes were the size of large modern hatboxes, and some were indeed used for carrying and storing hats, as is made evident by the labels of milliners pasted and printed on some of these boxes. While to modern eyes they may all look like hatboxes, their role as receptacles for hats was but one among many more common uses to which they were put. A printed handbill dated "Albany, July 28, 1835," now in the Cooper–Hewitt Museum, provides but one among many bits of evidence from the period that bandboxes were used to carry almost anything. The handbill advertised a five-dollar reward for the return of a "BAND BOX . . . LOST between the Canal Bridge at Port Schuyler and the Patroon's Bridge." The contents of the lost box were described as "Ladies' and Children's Wearing Apparel."[2]

If bandboxes were used by travelers during the second quarter of the nineteenth century much as we use shopping bags today, they were also used by manufacturers and sellers of wallpaper just as stores today use colorful, eye-catching paper bags to advertise themselves and their wares. Hartford paper stainers of the 1830s emblazoned their names across their own products to bring them to the public's attention. Putnam and Roff's colorful eagle, which appears on the box in *color plate 64,* grasps in his beak a banderole bearing the partners' names and the words "Paper Hanging: Band Box Manufac, Hartford Con."

Manufacturers of wallpaper did not only produce bandboxes as a sideline. By the late eighteenth century, a few American craftsmen were devoting their efforts exclusively to the making of bandboxes. One "Widow M'Queen" was listed in the New York City directory of 1791 as "ban-box maker." While other kinds of decoration might have adorned earlier examples, by the 1790s wallpapers were certainly being used to cover such boxes. Evidence of this is given in an advertisement of 1795 for "cheap paper hangings" including "some low priced flowered paper suitable for covering trunks and bandboxes."[3]

As early as 1800, Silvain Bijotat began to list his "Bandbox Manufactory" in the New York City directory. In 1824, following his death, an inventory of his possessions was made. It includes a listing, with evaluations, of the bandboxes left on hand when he died. This inventory furnishes precise documentation for information implied in other sources; the information that bandboxes were cheap, and that "bandbox papers" as well as wallpapers were used in their decoration. The compilers of Bijotat's inventory made separate entries for "paper hanging," "band top paper," and "sides" as well as for "lining paper" and "Box board."[4] Heading the list on the inventory are "2 doz nests of band boxes @ 7 / [shillings, for a total value of] $21." These "nests" were sets of boxes in graduated sizes, which could be shipped packed one inside another, like Chinese boxes. Although it is not specified exactly how many boxes were included in each nest, there must have been at least three. Next on the list, ten additional "nests" received valuations of 5 shillings 6 pence each for a total of $6.87. At about $1.12 and 70¢ for each nest, these must have represented handsome items. Even though these sums are not large, they do seem grand in comparison with other prices appearing in the list: some of the single boxes were assigned average values of 2¢ and 1½¢. Such low prices help to account for the wide circulation of bandboxes during the 1820s.

The makers of bandboxes frequently pasted printed labels bearing their names and addresses inside the box lids. Comparing the information on the labels with newspaper advertisements and city directory listings, it is possible to pin down the dates during which a bandbox maker was working at the particular address given on a label. It is also possible to establish an approximate date for any wallpaper that originally covered the box. Since only one American wallpaper sample book for which dates can be documented has survived, and since there are pitifully few illustrations that can be accurately dated showing wallpaper patterns, bandboxes have proved to be particularly important sources of dates for patterns printed in this country during the second quarter of the nineteenth century.

Manufacturers frequently lined their boxes with dated newspapers. Although these can provide additional clues about the date of a covering paper and about the area in which it was used, one cannot assume that any dated newspaper found in a bandbox automatically establishes a date for either the box or its patterned covering. Because the bandbox maker could have used old newspapers, or because the box could have been relined long after it was first made, with then-current newspapers, or because the wallpaper itself may have been an aging scrap when it was pasted on, caution is needed in trying to establish just when the box and its paper were made.[5]

As bandboxes became more and more popular, paper stainers began to print scenes on papers specifically designed to fit the proportions and shapes of the boxes. These papers were neither designed nor sold for use on walls. The designs were printed horizontally along a length of paper, as borders were printed, but each motif, though repeated, stood as a discrete scene, of the proper size to ornament the side or the top of a box; it was not visually connected by any elements of patterning to the next motif printed on the papers.

The New York business directory of 1840 included three lists of closely related tradesmen and craftsmen: "Band Box Paper Manufacturers" appeared as a separate, if small, category in addition to "Band Box Manufacturers" and "Paper Hanging Warehouses." Ten names appeared under the heading "Band Box Manufacturers." In 1820, only five names had appeared on a comparable list published in the directory for that year. The proliferation of these craftsmen during the intervening twenty years gives some indication of the growing popularity of their product. By 1840, there were twenty-five names under the heading "Paper Hanging Warehouses" in the New York directory. On that list ten names were marked with asterisks indicating firms that actually manufactured paper hangings. Of those ten manufacturers, two were also listed on the shorter list of four "Band Box Paper Manufacturers." And of the four "Band Box Paper Manufacturers," three had duplicate listings as Band Box Manufacturers.

The four New York manufacturers of bandbox papers whose names appeared in the directory in 1840 were Thomas Day, Jr., Archibald Harwood, J. H. Hazen, and George Peuscher. These men may well have been the principal sources of the colorful papers that appeared on the products of the much more numerous group of craftsmen who manufactured the boxes. Among them, these four may have been responsible for the scenes featuring New York landmarks that survive on numbers of bandboxes: Castle Garden—a concert hall on the lower tip of Manhattan, Holt's Hotel, The Merchant's Exchange, The Deaf and Dumb Asylum, and the lighthouse at nearby Sandy Hook.

Printers of bandbox papers apparently provided the same patterns to box makers in several cities. For example, one paper, commemorating the New York Fire Department's introduction, in 1830, of its new fire engine, # 13, was used on bandboxes labeled by Philadelphia and New Bedford bandbox makers. Henry Barnes of Philadelphia used the paper to cover his labeled bandbox illustrated in *color plate 65,* and a duplicate paper was used by Joseph Tillinghast, a New Bedford wallpaper importer and bandbox maker, to cover the box bearing his label that is now in the Shelburne Museum.[6]

A bandbox scene that has survived in large numbers depicts a chapel and what appear to be academic buildings, as yet unidentified *(figure 13-4).* At Winterthur, a bandbox decorated with this paper bears the label of Henry Cushing and Co., bandbox makers of Providence, while an example of the same pattern, differently colored, survives on a bandbox in the Boston Museum of Fine Arts bearing a label of Joseph Tillinghast of New Bedford.[7] Many other examples indicate that in some cases there were common sources of supply for duplicate examples of these papers while in others one paper stainer apparently copied another's pattern.

American landmarks of contemporary interest appeared on a number of bandbox papers made during the 1830s and 1840s. One on which the design incorporates the words "The Grand Canal" shows the Erie Canal, completed in 1825. Another shows the port of Buffalo. The new capital at Washington was depicted on another popular bandbox paper.

Coaching scenes, views of the new locomotive engines, of sailing ships, and of paddle-wheel steamboats were probably deemed especially appropriate for these boxes used by travelers on the very conveyances they depicted. One printer gave his customers a "Peep at the Moon," shown as a pastoral landscape on a disk floating in a star-filled sky above another landscape on earth.

The subjects of some bandbox papers commemorated events that had recently captured public interest, like the balloon ascent of an Englishman, Richard Clayton, who in 1835 traveled 350 miles through the air from Cincinnati to Munroe County, Virginia.[8] Bandboxes also played a part in political campaigns. In 1840, William Henry Harrison, presidential candidate of the Whig Party, was the butt of a political joke illustrated on bandboxes. During the campaign, his opponents the Democrats contended that he was such a yokel that

> Upon condition of his receiving a pension of $2000, and a barrel of cider, General Harrison would no doubt consent to withdraw his presidential pretensions and spend his days in a log cabin on the banks of the Ohio.[9]

The bandbox illustration shows the general greeting an old veteran in front of a cabin, with a barrel of cider on hand, and the Ohio River in the background. (Harrison won the election, but died only a month after his inauguration.) Andrew Jackson, called "Old Hickory," and General Zachary Taylor, known as "Old Rough and Ready," were also depicted on bandbox papers.

Animals, exotic and familiar, ranging from bushy-tailed squirrels to giraffes, camels, and boa constrictors, were featured on one numerous group of bandboxes, as were fantastic animals drawn from mythological sources and many kinds of birds, including patriotic eagles.

Scenes from mythology were also popular for bandbox papers. In addition, wallpaper patterns depicting gods, goddesses, and putti, like those illustrated in *figure 12-4,* were appropriated for use on bandboxes. Since these and other motifs from the group of wallpapers illustrated in *color plates 50, 51 and figures 12-4 through 12-12* were often depicted as self-contained decorative elements of a size appropriate for these boxes, many of this type were used.

These were not the only kinds of wallpaper patterns frequently pasted on bandboxes. Hannah Davis of Jaffrey, New Hampshire, used repeating wallpaper patterns on most of the wooden boxes she produced between 1825 and 1855. Because their wooden construction was sturdier than that of pasteboard examples, bandboxes bearing her labels have survived in some numbers. For examples of repeating wallpaper used on bandboxes, see *figure 13-6 and color plate 66.*

Wallpaper borders were often used to trim bandboxes covered both with repeating wallpaper patterns and with bandbox papers. Sometimes wide swag borders of the type shown on walls at cornice level in *figures 9-14 and 12-5* were used to cover an

Figure 13-1 (opposite, left) This advertisement appeared in *The Ladies Magazine* for December 1829, testifying to the importance of boxes within the early-nineteenth-century wallpaper trade. The advertiser, J. P. Hurlbert, was listed as a seller of paper hangings and maker of bandboxes in the Boston directory of 1826. In 1829 and 1832 his name appears in the Zuber company records as a customer. Another man named Hurlbert—S. M.—who was probably related, was also in the bandbox business in Boston from 1833–46, when he went into partnership with S. H. Gregory. Courtesy of the New-York Historical Society

Figure 13-2 (opposite, top right) This little box of about 1835 was of a size appropriate for storing collar bands, from which the larger boxes derived their name. It is 3¾ inches tall, covered with a border paper—possibly French or perhaps made in New Bedford—block-printed in green, white, pink, yellow, and brown on a gray ground. Cooper-Hewitt Museum, 1968-71-1; gift of Elizabeth Dennison

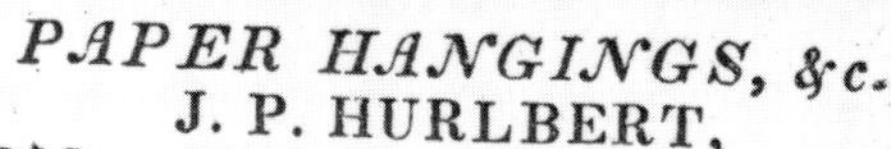

J. P. HURLBERT,

No. 222, *Washington-Street,* 2*d door North of Summer-street, Boston.*

INFORMS his friends and the public that he continues to Import the latest styles and fashions of French Paper Hangings and constantly on hand a large assortment of American do. likewise imports fancy boxes of all patterns and sizes, fancy papers, &c.

Also, Manufactures Paste board boxes for packing all kinds of goods and for stores, and made to order of every dimension.

Also, All sizes Paste board Bonnet, Hat and Packing Boxes, Paste board Bonnet, do. Cane, &c.

The above named articles will be sold wholesale and retail on the most favorable terms, and sent to any part of the city free of expense.

13-1

13-2

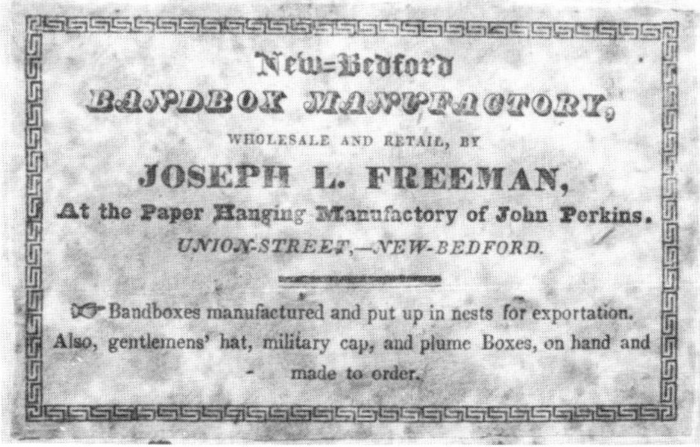

New-Bedford
BANDBOX MANUFACTORY,
WHOLESALE AND RETAIL, BY
JOSEPH L. FREEMAN,
At the Paper Hanging Manufactory of John Perkins.
UNION-STREET,—NEW-BEDFORD.

☞ Bandboxes manufactured and put up in nests for exportation. Also, gentlemens' hat, military cap, and plume Boxes, on hand and made to order.

13-3

Figure 13-3 (immediately above) This label, pasted on the underside of the lid, identifies the box as the product of Joseph Freeman, who, during 1834, advertised in New Bedford newspapers "Bandboxes by the 100 dozen and made to order" (*New Bedford Gazette,* Nov. 3, 1834). Freeman's business was of relatively brief duration, since he died in 1841. But John Perkins, at whose wallpaper factory the label announces that Freeman worked, made wallpaper in New Bedford from 1809 to 1859. Perkins's name appears in the Zuber records as a customer in 1829, 1830, and 1831. In 1853, in partnership with a Mr. Smith, John Perkins exhibited his own block-printed papers at New York's Crystal Palace Exhibition, receiving an honorable mention.

13-4

13-5

13-6

13-7

Figure 13-4 (top left) The scene here printed in pink, white, and green on a blue ground appears in a variety of colors on other bandboxes bearing several manufacturers' labels. The scene showing what is apparently an academic institution has not been specifically identified. This box is 12½ inches tall. Cooper-Hewitt Museum, 1913-45-6; gift of Mrs. Frederick F. Thompson

Figure 13-5 (opposite, top right) A repeating wallpaper pattern was used to cover this bandbox of the 1830s. At the center, a naive American interpretation of an Empire-style motif is block-printed in green and white with pink accents on a blue ground. Compare the white foliate patterned banding here with those of *figures 13-6 and 13-8.* In each, curved bandings outline circles and ovals framing various motifs. The bandbox shown here is 13 1/3 inches tall. Cooper-Hewitt Museum, 1913-45-14; gift of Mrs. Frederick F. Thompson

Figure 13-6 (opposite, bottom left) The vignette on this bandbox was crudely block-printed, probably by an American, in pink, green, and white on the blue field of what was obviously a repeating wallpaper, not a special bandbox design. The box is lined with a dated newspaper of 1840, which provides a clue to the wallpaper's age. The box is 13 inches tall. Cooper-Hewitt Museum, 1913-12-8; gift of Mrs. James O. Green

Figure 13-7 (opposite, bottom right) These lively squirrels, of apparently monstrous proportions amidst diminutive trees, are block-printed on an American bandbox paper of 1825–40 in red, green, white, and brown on a yellow ground. The box is 12 inches tall. Cooper-Hewitt Museum, 1913-9-1; gift of Alexander W. Drake

Figure 13-8 (this page, top) The outlines of a French-inspired "drapery figure" (see page 258) have been used by an American paper stainer as the framework for these vignettes featuring a gentleman who appears to be Napoleon. The repeating pattern is crudely printed in pink and green on a yellow ground. Hannah Davis (1784–1863), a bandbox maker of Jaffrey, New Hampshire, active from the 1820s into the 1840s, used this wallpaper to decorate the wooden box that bears her label. It is lined with a newspaper dated 1833. Cooper-Hewitt Museum, 1918-19-3; gift of Eleanor and Sarah Hewitt

Figure 13-9 (this page, bottom) A coach drawn by spirited horses rearing before an inn seems especially appropriate as a motif decorating boxes so closely associated with the era of stagecoach travel. This American design is duplicated in many museum collections. This one, 12½ inches tall, is block-printed in red and green on a yellow ground. Cooper-Hewitt Museum, 1913-17-15; gift of Eleanor and Sarah Hewitt

13-8

13-9

entire bandbox. In twelve- to fourteen-inch widths, swag borders were the perfect size for bandboxes. Swag as well as other kinds of borders in narrower widths, like that shown in *figure 13-2,* were also taken straight from the stock of paper decorations intended for walls and applied to these boxes.

Tucked away as they usually were on closet shelves, bandboxes preserved numbers of wallpaper and border samples in unusually good condition and, with them, rare bits of documentation permitting dating of many of these patterns.

Although much has been written of their popularity among women who worked in the New England textile factories of the early nineteenth century, little has been published about the use of bandboxes further south.[10] Yet they certainly were sold and used in the south. This is made evident by advertisements like one that appeared in the *Daily Mercantile Advertiser* in Richmond, Virginia, for February 25, 1822. A "Mrs. White" advertised the receipt of "bonnets from New York" and the fact that she had for sale "BAN BOXES by the quantity."

By 1854, when Elizabeth Leslie published *Miss Leslie's New Receipts for Cooking,* the moment of glory for the colorful bandbox apparently had passed. Miss Leslie, whose cookbook included advice and guidance on activities beyond the confines of the kitchen, wrote a brief section entitled "Travelling Boxes." In it she noted:

> As bandboxes are no longer visible among the travelling articles of *ladies,* the normal way of carrying bonnets, caps, muslins, & c. is in small square wooden boxes, covered with black canvas or leather . . .[11]

After that cutting observation, what stylish traveler would have dared to appear with bandbox in hand? Apparently very few, if we may judge from the dwindling numbers of advertisements for bandboxes, and the shrinking lists of names given as "bandbox makers" in city directories during the late 1840s. The fact that the category for "bandbox makers" had been dropped from the New York City directory by 1850 is but one more indication that as commodities in the wallpaper trade and as items of regular use by most Americans they were fast disappearing.

Although stylish travelers of the late nineteenth century left their old bandboxes behind in attics, the name remained in currency. Connotations from the days of their glory survived in the still-used line describing the smartly dressed—looking as if they have "just stepped out of a bandbox."

14 MAKING WALLPAPER BY MACHINE

During the 1840s industrialization began to transform the business of making wallpaper in America. For seventy years, the relatively simple process of hand-printing wallpaper, closely related to textile block printing, had been practiced in workshops scattered among American cities. These workshops were soon to be replaced by larger factories using machines that required capital investment far surpassing anything that had been needed when a few printing tables, sets of carved blocks, and barrels of color could launch a paper stainer in business.

An appetite for pattern on floors, upholstery, and window hangings as well as on walls was fed and stimulated by manufacturers of carpets, furniture, and textiles just as it was by the wallpaper makers. Machine-made papers became so cheap that all but the very poorest could afford wallpapers. The wallpaper industry's success in satisfying the popular appetite for mural patterning was reflected in statistics of soaring production.

THE PAPER STOCK

The invention of machines for making endless paper was the technological innovation that made possible the industrialization of wallpaper making. The first such machine was the Fourdrinier, which used a continuous moving belt of wire screen to form the paper. It was patented in England in 1799, and after further experimentation, an improved version was patented in 1807. In 1809 an Englishman patented another machine for making endless paper; this one was developed by John Dickinson, and instead of a continuous wire belt, it used a cylinder to form the paper. Both machines were widely used and adapted in the paper-making industry internationally.

These English machines were soon to transform American paper mills. In 1817, Thomas Gilpin produced the first machine-made paper in America, at his mill on the Brandywine Creek in Delaware. Gilpin patented his machine in 1816. He probably derived it from the cylinder machine patented by Dickinson. The Gilpin mill was producing machine-made paper for sale by 1820. In 1822, another American, John Ames of Springfield, Massachusetts, patented improvements on the Gilpin machine. In 1827 the first Fourdrinier machine was imported to the United States.[1] So, by 1830, the paper mills of America were feeling the influence of both of the important English paper-making machines of the early nineteenth century, and some quantity of paper was being made by machine in this country.

However, determining just how long it was before this series of inventions had a real effect on the American wallpaper factories is difficult, because no wallpaper factory records documenting the earliest use of machine-made paper have been found. In France, in the Zuber archives, there is evidence that the factory began using continuous paper by 1820. In England, tax laws requiring the marking of each of the individual sheets in a roll of wallpaper prohibited the use of the newly developed endless paper. Not until after 1830 could endless, machine-made paper be used legally in English wallpaper factories.[2]

The wallpaper industry's need for long strips of paper would seem to point to it as a logical early user of endless paper. However, handmade paper was heavier and stronger, and it was still available well into the middle years of the nineteenth century. In America, many paper mills continued to make paper by hand until about 1870. In 1829, of sixty paper mills in Massachusetts, only six were using machinery.[3] It is all but impossible to establish a cutoff date for the use of joined sheets for American-made wallpaper. There was doubtless a transition period of mixed use of handmade and machine-made paper in the wallpaper industry between 1820 and 1840—and perhaps later. Two advertisements of the 1830s suggest that during that decade endless paper was being introduced in America. A New York handbill of 1835–6 described "paper-hangings and borders manufactured in this City" as being "Of one Entire Sheet . . ."[4] A New Bedford advertisement of 1834 offered French wallpaper that included an example described as "a superior article without joints."[5]

In their perpetual effort to reduce production costs, paper makers substituted cheaper materials for the cotton and linen rags traditionally used in paper making. Owners of paper mills had been forced to advertise constantly their willingness to buy rags, which were in short supply in America. When the need to gather rags was eliminated, and when cheaper commodities went into the pulp for the mills' production of stock, the price of wallpaper dropped. Its reduced price was among the important reasons wallpapers became so ubiquitous in houses of the late nineteenth century. A writer for *The American Builder and Journal of Art* reported in 1869:

> The lowest grade of wallpaper is made entirely of straw; the next higher grade of straw and wool mixed; the next from Manilla hemp; and the best qualities, of cotton or linen rags.[6]

In addition, the introduction of wood pulp was important for the whole paper-making industry. Practical processes using wood to make paper were first worked out in England early in the 1850s and introduced to America in 1855.[7] By the 1880s, the bulk of commercial hanging paper stock had been greatly cheapened by the introduction of wood pulp, straw, and other less expensive ingredients. The brittle, highly acid papers of the late nineteenth century, familiar in cheaply made books in which the leaves are browned by the acids of the wood pulp, are easily differentiated from rag pulp stock in wallpapers as well. Smoothly finished, the wood-based papers, like all machine-made wallpapers, lack the imprint of water marks and of the woven screen visible in handmade papers, and the edges are cut rather than deckled (rough-edged, since they are formed in a frame, or deckle; see page 31). And, most obviously, there are no joints linking individual sheets into rolls.

Paper-making machines and the printing machines for wallpaper dictated new standardization in roll sizes. At the wholesale level, wallpaper printers bought paper from the mills in "rolls about 1,200 yards long, and from 20 to 35 inches wide," according to Edwin T. Freedley's report of 1858 on the manufacture of wallpaper in Philadelphia.[8] A British publication of 1856 reported the standard sizes for European wallpapers as sold at retail:

English wallpaper:	12 yards long by 21 inches wide
French wallpaper:	9 yards long by 18 inches wide[9]

An American trade magazine of 1869 gave the same dimensions for European papers, and reported that American wallpapers were 8 yards long and 18 inches wide.[10] Those standardized sizes prevailed into the twentieth century, although by the turn of the century the "American double roll" sixteen yards long and eighteen inches wide had become the normal selling unit.

In addition to these nineteenth-century changes in the ways paper was made, a means of coloring and texturing paper in the pulp was introduced in 1878, when James S. Munroe patented a process for making what he called "ingrain" paper.[11] Munroe's patent describes a process for using mixed cotton and woolen rags, dyed before pulping, which produced a thick paper with textured coloring ingrained in the paper itself. Marketed as "Munroe's patent ingrain paper," it required no application of a coating of ground color. Papers that made similar use of ingrained coloring and textured surfaces formed of small particles introduced into the pulp were called "oatmeal papers" in the trade. Sometimes the texturing looked rather like the surface of modern chip board, slightly refined. These "oatmeal" and "ingrain" papers were popular well into the early twentieth century in a range of murky colors, and were often hung as plain, unpatterned wall coverings.

It was not only the new processes for paper making that were to have great impact on the wallpaper trade; so were the numerous new ways of making wall coverings with high-relief embossed patterning. Although some, like the Japanese leather papers (see page 441), were made of paper, not all were, strictly speaking, paper products. Rather, they were wall coverings made up of various compositions, the ingredients of which were concocted to simulate embossed leather wall coverings. They became stock items in the American wallpaper trade of the late nineteenth century. One of the greatest commercial successes among these was "Lincrusta," a thick, linoleum-like wall covering developed in England in 1877 and, beginning early in the 1880s, manufactured in Stamford, Connecticut, under English patents (see page 442).

GROUNDS AND STRIPES

The simplest processes in paper staining—grounding and striping—were the first to be mechanized. Forerunners of machines like that shown in *figure 14-2,* devices using cylindrical brushes or rollers on machines set up to spread even coatings of solid coloring, were probably replacing workmen with their hand-held brushes by about 1800.

Machine-laid grounds can often be distinguished from those laid by hand. Machines applied the color with an even, steady motion, and the directional turnings of long brushes and rollers is sometimes apparent on the finished paper as regular vertical streakings. Brush strokes made by hand-held grounding brushes will, in contrast, appear in random lines and curves, following the directions in which the workman moved his hand.

What seems to be a way to use mechanized brushes for striping, not unlike those used for laying grounds, was described by Matthew Digby Wyatt in his book *Industrial Arts of the Nineteenth Century at the Great Exhibition 1851.*

> In 1829 the first machine for producing an endless line by means of stencil plates and a revolving brush was introduced into Paris; through the house of Dauptain . . . The introduction of this system produced a strike amongst the workmen, . . . the Government were obliged to send soldiers to protect the establishment. Dauptain became frightened and gave it up. The same fear pervaded the other Parisian manufacturers and it was abandonned til 1831 when the Messrs. Zuber established a machine with great success fabricating by means of it pieces in one length of thirty feet long.[12]

Similar combinations of stencils and revolving brushes also probably were used to apply patterning as well as lines to wallpapers. By running those same brushes over stencils with areas of patterning cut out, colored shapes could be quickly if crudely added to long sheets of paper.

Other mechanical devices also used brushes to apply striping. Stripes could be brushed on to a paper's surface using mechanisms that moved rolls of paper under a line-up of stationary brushes, fixed at appropriate intervals to produce stripes of the desired width, and each fed with the required color. Conversely, mechanisms that moved a series of brushes over stationary lengths of paper could also be used.

Yet another mechanical contrivance designed to speed up the application of stripes was used at the Zuber factory during the nineteenth century, and survives there. It used long metal troughs, V-shaped when viewed on end, to carry the colors. Two long plates of metal formed the V, the two sides of which did not quite meet at the bottom of the long trough, so that a narrow slit ran along its length. Inside, the trough was divided by metal plates into compartments corresponding to the desired widths of stripes. By feeding different colors into each division of the compartmentalized trough, colors were spread directly through the narrow slit at the bottom to form stripes on the paper. Similar devices were probably developed by other paper stainers in Europe and America.

WALLPAPER PRINTING MACHINES

The basic and dramatic transformation of the paper industry by the Fourdrinier, Dickinson, and other machines was a necessary prerequisite for the mechanization of wallpaper manufacture. Beginning in the 1820s, manufacturers added to the grounding and striping devices an ever-increasing number of machines for speeding up wallpaper production. They designed these to make more efficient use of the abundant, cheaper paper. As they had traditionally copied textile pattern designs, wallpaper manufacturers attempted to imitate the new textile-printing machines of the early nineteenth century.

English, French, and American patent records show attempted adaptation of calico-printing machines to wallpaper production. These machines, which used engraved copper cylinders for printing, were not commercially successful for wallpaper, probably because the thick distemper colors preferred for wallpapers could not be carried in engraved grooves as successfully as the thinner pigment bodies used for textile printing. In addition, paper, especially grounded paper, did not receive the

14-1

Figure 14-1 "*Taille douce*"—engraving—is the French name for the technique developed at the Zuber factory for using engraved metal surfaces to impress images on wallpaper. All the floral and foliate forms on this paper of the 1820s, preserved in factory archives, are built up of thin, sharp lines, printed in a shiny, thin-bodied pigment. Courtesy of Zuber et Cie.

coloring matter carried in grooves below the surface of a cylinder as readily as did the more pliant, absorbent textiles. And while patterning might look the same at a distance, the surfaces of the papers printed from engraved cylinders looked very different from those already in style. All the coloring on them was confined to narrow lines. To build up a form that appeared to be solidly colored required imprinting dozens or hundreds of parallel or cross-hatched lines. The printed surface of the paper had the character of engraved prints more familiar in book illustrations than in wallpaper.

The Zuber factory in Alsace did begin to use engraved copper rollers for wallpaper printing in 1827. Using thin-bodied varnish colors, the factory produced engraved papers that are easily distinguished from the bulk of printing done on machines developed in England and America during subsequent decades *(figure 14-1)*. Although Zuber continued to print papers from engraved copper well into the 1830s, these papers were never leading articles of commerce for the factory. Nor did similar experiments with engraving techniques undertaken by other manufacturers enjoy significant commercial success. The English and the Americans tried, like the French, to adapt textile roller printing machines to wallpaper. The United States patent records preserve a series of six charming illustrations of primitive machines for printing wallpapers using engraved wooden rollers, patented by Peter Force in 1822 *(figure 14-3)*.[13]

Experiments of 1839 undertaken by an English firm, Potters of Darwin in Lancashire, led to the development of the most commercially successful printing machines for wallpaper. By 1841 the Lancashire firm had developed a steam-powered machine with an efficient system for feeding color to cylinders that printed from raised, rather than engraved, surfaces. This invention successfully applied the old principle of printing from the raised surfaces of wood blocks to the curved surface of a cylinder on a printing machine.

The standard wallpaper-printing cylinder as adopted in wallpaper manufactories had a wooden core with a raised printing surface formed by tapping into that core strips of brass, which formed raised outlines of shapes—little walls (like the cloisons in enamel work) into which felt was tightly stuffed to carry the colors for the solid areas of patterning. Lines and dots were printed by appropriately shaped brass pieces *(figure 14-5)*.

Such cylinders were used on machines very much like some still used today for large-volume wallpaper printing. At the center of one of these machines is a large revolving drum, or giant cylinder, upon which the blank paper rides while it engages in sequence a series of smaller cylinders, each of which has a raised surface for printing one color of the pattern. Each printing cylinder is coated with its individual coloring by a roller-fed belt from a trough that holds the appropriate color *(figure 14-6).*

With the printing machine developed by Potters in 1841, the era of machine printing of English wallpaper had begun. Although there had been earlier experiments with engraved cylinder printing and some mechanical improvements in devices to speed up block-printing, significant mechanization did not take place and indeed was not possible until this invention.[14]

Nearly a decade passed before the Zuber factory in 1849 installed a steam-powered English machine, made in Manchester. That machine produced a great deal more wallpaper much more quickly than did the hand-powered machines used by the Leroy firm in Paris. Leroy had begun printing in one or two colors with raised-surface cylinders in 1843. But 1849 marks the beginning of full-scale machine production in France.[15]

It is not possible to state with authority precisely when Americans first used machines to print wallpaper. Three divergent accounts of the introduction of machine printing to the American wallpaper industry were published in volumes on industrial history written during the nineteenth century, and in wallpaper trade journals of the late nineteenth and early twentieth centuries. Since neither factory nor patent office records for the crucial years have been found to answer this question, it is difficult to assess which account is correct.

One account posits 1844 as the year when the machine printing of wallpaper was introduced to the United States. In a generally well-researched history of wallpaper industry written in 1895, the president of the National Wallpaper Company, Henry Burn, points to 1844, stating: "In that year, the first machine for printing wallpaper was imported from England and introduced into the Howell factory [in Philadelphia]; it printed only a single color."[16] This 1844 date is the one accepted by Nancy McClelland in her book of 1924, *Historic Wall-Papers.* In that book, which has remained ever since its publication the most extensive compilation of information

Figure 14-2 (opposite, top) *Scientific American* (December 31, 1881) showed this machine for applying ground color. A series of rollers and brushes spread the liquid color from the trough onto the paper, carried on the drum. After the color was smoothed by the brushes, the paper was mechanically festooned on wooden rods to dry. Photograph courtesy of the Science and Technology Research Center, New York Public Library

14-2

Figure 14-3 (right) Peter Force registered this drawing on August 22, 1822, as one of six accompanying his application for a patent for grounding wallpapers and for printing patterns from engraved wooden rollers. Courtesy of the United States Patent Office

14-3

A GLIMPSE OF THE ART ROOM

RECEPTION AND SALE DEPARTMENT

PRINTING TWELVE COLORS AT ONE IMPRESSION ON THE CHROMATIC PRESS

REELING UP

EMBOSSING

DECORATIVE ART IN THE HOMES OF THE PEOPLE—ITS PROGRESS ILLUSTRATED IN THE MANUFACTURE OF WALL-PAPER.

14-4

Figure 14-4 Frank Leslie's Illustrated Newspaper carried this page on May 27, 1882. Demonstrations of manufacturing processes proved popular in this period. At the Philadelphia Centennial celebrations paper hangings were actually printed in machinery hall (on Waldron Company machines built for Howell and Brothers) in what was described as a "beautiful operation . . . by experienced hands." Cooper-Hewitt Museum, from the Kubler collection

14-5

14-6

Figure 14-5 (opposite, bottom right) This cylinder or roller for printing wallpaper was used by the F. E. James Company during the mid-nineteenth century. Cylinders like this, each adding details in different colors, would have been used on machines like the one pictured at the bottom left in *figure 14-4.* On the wooden core of the roller, 19¾ inches long, the raised printing surface has been built up by hammering in strips of metal (usually brass), which form little walls standing out about a fourth of an inch; they appear here as the dark outlines around solid shapes. Those shapes are filled with felt, which carries the colors. The circumference of the cylinder is 16 inches. That measurement dictates the repeat length of the printed pattern. Cooper-Hewitt Museum, 1943-54-1; gift of Myrta Mason

Figure 14-6 Four machines for printing wallpaper are shown in the factory of Christy, Shepherd and Garrett in New York as illustrated in the July 24, 1880, issue of *Scientific American.* The largest machine, second from the left, was equipped to produce patterns in twelve colors, printed from twelve cylinders or rollers ranged around the giant central drum on which the paper was carried. The workman's left hand rests on the color trough, while with his right hand he adjusts the belt that feeds color to a printing cylinder. Photograph courtesy of the Science and Technology Research Center, New York Public Library, Astor, Lenox, and Tilden Foundations

about wallpapers used in America before 1840, the author used the introduction of machine printing as the cutoff point for her study.

J. Leander Bishop in his hefty book about most of the major industries of nineteenth-century America, *A History of American Manufactures,* puts the date for the introduction of machine printing of wallpaper at 1843. He cites machines "for printing two colors" that had been developed in 1843 by the J. R. Bigelow Paper Hangings Manufactory in New England as marking the advent of machine printing to the American wallpaper industry.[17] In *The Great Industries of the United States,* edited by Horace Greeley and published in 1872, this same account appears.[18]

Yet a third version of the story places the advent of machine printing at 1835. This date survives in another group of trade histories. The trade magazine *The Decorator and Furnisher,* in an unsigned 1890 article entitled "Wallpaper," relates that in 1835 the Boston firm of Josiah Bumstead introduced "a machine to print wallpaper in one color which, though crude, was a vast improvement to the hand process for rapid work."[19] The same date and "a Boston firm" are mentioned in a generally garbled promotional booklet by William R. Bradshaw, published by a firm of decorators and wallpaper manufacturers, The Joseph P. McHugh Company of New York, in 1891. Bradshaw's booklet is grandly titled *Wallpaper: Its History, Use and Decorative Importance.*[20] What is probably a third instance of this same account was published by an anonymous writer in *The Wallpaper News and Interior Decorator* in August of 1911. The article, "Early Wallpaper Machinery," relates that this author:

> obtained from William Campbell and Co., the Old New York Manufacturers, the following date which they had preserved as authentic: In 1835 a machine was invented in America to print wallpaper in one color . . . It was a machine but was operated by hand. . . . The first wallpaper ever made in New York by machinery was in 1839.

Probably, there is truth in all versions of the first-wallpaper-machine-in-America story. The American-made "machines" of the 1830s seem to have been closely related to the hand-cranked affair illustrated in *figure 14-3.* What distinguished the machine imported by Howell in the 1840s from these earlier contrivances was steam power. Efficient use of steam power catapulted wallpaper machine printing to commercial significance and then quickly to dominance in American factories. Therefore, the 1840s stand as the decade marking the beginning of commercially significant machine printing of wallpaper in America as well as in France and England.

This review of technological innovation in the wallpaper industry during the 1830s and 1840s leads to the conclusion that old papers bearing the impression of the raised-surface cylinders used on the wallpaper printing machines *(color plate 68)* cannot safely be dated before the 1840s. The little metal outlines with felt filling left a distinct impression—an outline of thicker coloring around edges of solid shapes; and when this impression is combined with traces in the colored areas of the directional movement of the cylinders, these indicate a date of no earlier than 1841 for English papers, 1843 for French, and, very probably, the mid-1840s for American wallpapers. In the earliest of these papers, the number of colors in patterns was limited to only one or two. Only gradually was the number of colors increased to figures as high as twenty, which could be printed by wallpaper patterning machines of the 1880s.

Although a series of refinements in wallpaper production followed the initial improvements, the development of endless paper and the introduction of steam-powered machinery using raised-surface cylinders stand as the two most crucial technological improvements in wallpaper production. The most important later refinements included devices for rapid drying of more quickly printed papers, embossing and gilding processes, and combining the steps of grounding, printing, and embossing in a series performed by machine. During the second half of the century, the previously abandoned engraved-cylinder process was readopted to print oil-based pigments on smoother surfaces. Papers made this way were washable and were called "sanitary" papers. They were promoted for use in bathrooms and kitchens in England as well as in America.[21] During the second half of the nineteenth century, photographic and lithographic printing processes were developed for wallpapers and patented as well.[22]

GROWTH OF THE AMERICAN INDUSTRY

During the late nineteenth century, the American wallpaper industry boomed, and was able to effect steady decreases in the cost of its products. The public responded by purchasing in increasing quantities. A writer for New York's magazine *The Manufacturer and Builder* asserted in 1869 that, "In every mansion, house, and hut in the land, the work of the paper-stainer now confronts the inhabitant."[23] The public also demanded new patterns, not only annually, but also seasonally. An article in *Scientific American* for July 24, 1880, revealed that "Old patterns are, nowadays, entirely unsalable, and the rule is that each year's patterns must be entirely new and distinct from those of the preceding season."[24]

14-7

Figure 14-7 A Philadelphia "Paper Hangings Warehouse" of the 1840s appears in intriguing detail in this lithograph signed by the artist W. H. Rease. In the windows are accurate depictions of Pignet's scenic wallpaper *Les Huguenots* (see *figure 9-22*). Floral stripes and panel papers typical of mid-century styles are unfurled around the central scene; they are displayed over a border that is not the simple diamond pattern it appears at first glance, but rather rolls of wallpaper on shelving. Such "diamond shelving" was a standard fixture in wallpaper sales rooms. Courtesy of the Library Company of Philadelphia

The growth of the wallpaper industry will not be statistically documented here, because the basic facts of phenomenal multiplication of numbers of factories, of the production capabilities of each, and of the consumption of paper hangings are available in many popular sources. Industrial histories and trade publications made a number of boastful statistical claims that could only be verified by laborious research, but that bear witness to a dramatic boom. Numbers of journalists reported on their visits to various individual factories and detailed production statistics, frequently in terms of the number of times the annual production of a given factory could be wrapped around the earth.

While J. Leander Bishop reported in his *History of American Manufactures* of 1868 that the annual output of wallpaper as of June 1850 was valued at $107,040,[25] by 1900 the U.S. Census reported wallpaper production for that year at $10,663,-209.[26] *The Decorator and Furnisher* in September of 1890 recorded that in 1840 two million rolls of wallpaper had been produced in America, a figure that had risen by 1890 to an annual production of a hundred million rolls.[27]

Some indication of the price reductions is given by one of many accounts that can be cited. In 1879, London's *The Furniture Gazette,* reporting on "The American Trade in Paper Hangings," stated that "The people of the United States spend $8,000,000 per annum for wallpaper, their requirements being about 57,142,860 rolls or 457,142,400 yards." The article also detailed:

> Wallpapers which were sold wholesale five years ago at from 12½ to 13 cents a roll, are selling at from 6 to 6½ cents a roll. Those which sold five years ago at 45 to 50 cents have now come down to 24 and 25 cents.[28]

By 1895, the New York firm of Alfred Peats was offering, at retail, papers ranging from 3 cents a roll to 30 cents a roll.[29]

In 1895 there were thirty-one manufacturers of paper hangings listed in the New York City directory, five in the Boston directory, ten in Philadelphia, and many others are known to have been manufacturing and/or jobbing wallpaper in Chicago, New England, New Jersey, and Pennsylvania.

15 1840 TO 1870: THE IMPRINT OF MACHINES, THE ELABORATION OF STYLES

Wallpaper printing machines stamped many papers of the mid-nineteenth century with a look—in the physical matter of paper and coloring materials—that can be readily distinguished from earlier block-printed papers. At the same time, aspiration to the elaboration achieved in the most expensive French designs (while these in fact still were being printed by hand) had an equally marked effect on the look of wallpapers used in America. Many patterns hung from the 1840s through the 1860s bear the imprint of these two, not always congenial, influences on their design.

Some of these wallpapers appear disastrously overelaborated to modern eyes. In fact, the majority of patterns commercially produced during this period provoked the censure of English critics as soon as their densely scrolled and flower-strewn designs began to dominate wallpaper offerings in shop windows. Although they began to voice their objections during the 1840s, not until the 1870s were these critics able to transfer to a wide reading audience of Americans their disgust with the whole vocabulary of ornament derived from, and elaborating upon, Renaissance, baroque, and rococo decorative styles. During the many intervening years, American consumers were enchanted by the gorgeous floral and pseudo-architectural *décors* created by the great French wallpaper factories of the Second Empire period (1852–70). In this era, Desfossé et Karth and Delicourt dazzled visitors to every international exposition. Impressed with the splendors of the French floral displays and other extravagant patterns, home owners who could not afford the imported versions of wallpaper flowers bought instead cheap American copies.

The critics saw all the scrolls and the realistic flowers and foliage as "falsifications" —attempts to deceive the eye in patterns founded on false principles of design. Under English tutelage, by the 1870s and 1880s, many Americans had learned to scorn the French and French-inspired patterns as "vain displays" and to prefer the simpler English styles, which often were modeled on medieval and Japanese designs.

But in the meantime, elaborate machine prints brought a store-bought look to the parlors of mid-century Americans—a look that found favor as soon as it became a possibility. Farmers' sons and daughters were moving to towns and cities. Establishing little shops, joining the ranks of factory and office workers, or finding a newly evolved world of feminine domesticity in the suddenly urbanizing society they confronted, they hastened to gloss over their country origins. The pasted-up ornaments and patterns from wallpaper sample books looked citified indeed, and those printed by machine made papering as cheap as painting. The simplicities of what we, looking back, would call folk art were not what the urban immigrants wanted to take to town.

A new middle class was seduced by the appeals of generous curves in intricate scrollwork patterns, bedazzled by the magnificence of cabbage-roses-by-the-yard, and delighted to have its choice among a wide variety of decorative styles. Wallpaper patterns made elaborate ornamentation available in quantities no painter or stenciler could have produced even with weeks of labor.

TECHNOLOGY'S INFLUENCE ON DESIGN OF THE MID-NINETEENTH CENTURY

Machine-printed wallpapers differ in appearance from many of their block-printed predecessors and contemporaries in the small scale of their design and in the thin-bodied, nearly transparent, colors with which they were printed. Requirements of the machines dictated these changes. For instance, repeat lengths had to be relatively short—under eighteen inches—to fit cylinders that were usually no larger than six inches in diameter.

Another limitation on the size of patterning was imposed by the technique for making printing cylinders. On wooden blocks for printing, raised flat printing surfaces could be carved as large forms to print broad areas of coloring. But on the cylinders, large areas of felt could not be relied upon to stay in place within the confines of small metal walls if the forms outlined by those metal walls became too large (see page 307). The felt carried the color for printing solid shapes, and when wet with the color, it would simply become too distorted and perhaps even detach itself from the wooden core in the middle of large shapes, where the metal outlines ceased to be effective as retaining walls. Therefore, any forms to be printed in solid colors on machines had to be designed in relatively small dimensions, encouraging a generally smaller scale of design than was permitted for block prints.

Coloring materials also had to be adapted to the machines; they had to be thinned down and chemically adapted to dry more quickly than the thick-bodied distempers of the block printer. This fast drying was particularly important on the later, more sophisticated machines for printing many colors, since the series of cylinders applied those colors in rapid succession as the paper was carried around the central cylinder (*figures 14-4, 14-6*) and fast drying helped to cut down on blurring and streaking during this process.

The hues of the machine colors also tended to differ from block-printed examples. They were generally brightened during this mid-nineteenth-century period by the standard use of new, cheaper synthetic pigments that had gradually become available during the early years of the century. These colors included chrome yellow, which, though discovered in 1797, became cheap and widely available only after quantities of chrome ore were discovered in the United States in 1820.

A piercing and distinctive bright blue was introduced to a great many wallpapers of the 1850s and 1860s, produced from artificial—"French"—ultramarine. The color was not commercially produced before 1828, but when it was adopted by the wallpaper trade, its brilliance was exploited, especially in contrast with shades of brown.

Two sources of bright piercing green, which were later found to be poisonous, brightened the surfaces of many mid-century papers: Sheele's green (copper arsenite) and Schweinfurt green (copper aceto arsenite). Both colors had been discoved earlier but were not widely used for wallpapers until this period.

By 1857 reports of poisoning by "arsenical" wallpapers were alarming the British public. Medical opinion in both England and America split over the question of whether arsenic in the pigments coloring wallpaper patterns posed a threat to health. Doctors and researchers continued to argue about its effects through the turn of the twentieth century. However, for consumers, sensational revelations of horrors suffered by people who slept, dined, or simply sat in rooms papered with arsenic-bearing patterns eliminated all doubt about its deadly effects. Two specimens of arsenical paper were bound into the third annual report of the Massachusetts State Board of Health for 1872. Dr. T. W. Draper, who submitted these specimens, concluded: "Cases of poisonings by this means constitute a mass of evidence which cannot well be refuted."

The mass of evidence in other publications recounted the sufferings of a chemist's children due to "an excessive quantity of arsenic" in the paper of the schoolroom of his house and the "partial poisoning" of a doctor in Northampton "through inhaling the particles brushed off by a bright green wallpaper." The doctor, George Johnson, testified about his symptoms before the National Health Society, describing himself as:

> . . . a physician, middle-aged, active, healthy, and perfectly well, until in an evil hour some months ago, he hung a garden-room with a bright, new, green paper, the apartment being one in which he was accustomed to sit after dinner. He began almost immediately to be strangely out of sorts—the symptoms of his indisposition being like those of "hay-fever." A violent influenza-like affliction oppressed him constantly . . . Asthma supervened . . .

He analyzed the wallpaper and found it to be highly arsenical. Its removal relieved his symptoms. In conclusion, the doctor warned that

> . . . it is not Scheele's green alone that should be suspected. Roseate tints, greys, browns, and yellows may be killing people quite as possibly as the lively greens, and this in spite of the fact that arsenic does not readily volatize.

Appleton's Journal reported an American "cry for legislation" against "poisonous wallpaper" in its "Miscellany" for June 29, 1872. In 1874, Robert Clark Kedzie of the State Board of Health of Michigan submitted eighty-six samples with his six pages of "facts and inferences prefacing a book of specimens of arsenical wall papers." He borrowed from a fiction thriller's lexicon to entitle this official report "Shadows from the Walls of Death." After such publicity, it is no wonder that wallpaper manufacturers of the nineteenth century often featured the words "non-arsenical" in their advertisements.[1]

Perhaps more remarkable is the fact that the fear of arsenic did not curtail the popularity of wallpaper. Rather, that popularity grew unabated. In the mid-century, whether identifiable as patterns derived from a particular architectural style, as rococo revival patterns, or as "naturalistic" patterns, many of the wallpapers frequently bear other more easily recognized marks of their era. In addition to indications that they are machine-printed, and to the presence of new pigments (which can be detected by chemical analysis), a few distinctive palettes mark many wallpapers as products of this period. Among the myriad combinations of bright colors found in these wallpapers, a few combinations recur with such consistency that they stand out from the rest as particularly characteristic of their time. These include light bright green with gray on a glossy white satin ground; deep rose reds in combination with brown, and the bright-blue-with-brown combination mentioned above.

FRENCH, AND FRENCH-INSPIRED, REALISM

In contrast to the machine prints, with their limited colors, many of the expensive hand-printed French imports incorporated dozens of thick-bodied, chalky distemper colors that were applied by hundreds of intricately carved wood blocks. Although some mechanical devices were introduced to speed the block-printing and to reduce costs, French masters of the art of hand-printing strove more for refinements in painterly effects, delicately shading their patterns and producing elaborate, large-scale, non-repeating designs that no machine could have made. Even their repeating patterns had many of these characteristics. Expensive French papers were designed to proclaim at first glance their distinctiveness from the cheaply made products of the big factories *(figure 15-5 and color plates 70, 73 through 77).*

Textile Imitations. The French preference for naturalistic wallpaper patterns and for papers imitating fabrics was long-lived. As late as 1874, in the face of English pronouncements branding as "dishonest" all but abstracted flat-patterned wallpapers, the French manufacturers had countered with the pronouncement given on page 179 that "falsification" was the "first destination" of wallpaper and that, above all, wallpaper falsifications, "the exact imitation of stuffs, and even of their floss," was one of the industry's most profitable branches.

That branch continued to bring profit to American retailers selling French wallpapers up through the 1860s. In 1850, a St. Louis wallpaper manufacturer and importer, George B. Michael, detailed fabric weaves and colors when advertising wallpapers in his stock. The textile vocabulary he applied to papers included: "Crimson velvet tapestry, foulard, satin crape, copper-plate, and watered."[2] "Velvet" had long meant "flocked" in American wallpaper advertisements. "Foulard" was a silk fabric printed with colored dots on a white ground. "Crape," a variant spelling of "crepe," denoted imitations of fabrics with textured surfaces. "Copper-plate" were imitations of textiles printed from copper plates with patterning formed of thin, engraved lines. "Watered" was an adjective used to describe the effect achieved by subjecting silks, cottons, and linens to heavy pressure, heat, and water or steam. This gave a lustrous uneven sheen, like the glint of light on water. *Figure 15-4* shows how this effect was translated to wallpaper patterning. The variety of textile imitations so described in St. Louis in 1850 probably included cheap machine prints as well as expensive block prints.

A. J. Downing, perhaps the most influential mid-nineteenth-century American taste maker, was more subject to contemporary fashion than he was guided by

Color plate 68 In this wallpaper, probably American-made and dating between 1840 and 1850, marbleized effects have been machine-printed. In an enlargement of a 2-inch portion of the paper, the imprint of the metal outlines used to form printing surfaces on rollers is visible as a line of thicker color around each printed shape. The thin-bodied pigments are relatively transparent, and give a grainy texture in which all the streaking runs in a vertical direction, the direction in which the printing rollers were turning. The sample is 20 inches wide. Cooper-Hewitt Museum, 1956-171-6; gift of Mrs. Adrienne Sheridan

68

69

Color plate 69 The wallpaper stripes shown here are formed of flowers block-printed in the chalky distemper colors that produced the realistic painterly effects so admired by the French. The paper and borders survive where they were hung during the mid-nineteenth century, in the drawing room of a Hudson River mansion built in 1803 near Rhinebeck, New York. Flocked or "velvet" borders with flowers matching those of the stripes were placed at cornice level and just above the baseboard.

Color plate 70 (opposite) This wallpaper sample was tipped into an album produced by the Parisian firm of Jules Desfossé for presentation to the officials of the Exposition Universelle held in Paris in 1855. Reacting to developments in machine-printing, French block-printers refined their craft. This example is a *tour de force* of block-printing, and a model aspired to in the miles of machine-printed floral papers produced up to the turn of the twentieth century. It epitomizes as well the illusionistic, painterly patterns scorned by avant-garde English critics and pattern designers. Labeled *Dessin à fleurs, 18 couleurs,* the sample measures 18¾ by 13½ inches. Cooper-Hewitt Museum library

71

72

Color plate 71 (left) Flowers spring from scrolls and stripes over a polished satin ground. Probably English, of about 1835–45—although it could have been made in France or in this country—this is an unused remnant of the paper hung in an early-nineteenth-century house in Lynchburg, Virginia: the home of John Early. The paper is machine-made and the patterning was printed both from blocks and by machine. It is 22 inches wide. Cooper-Hewitt Museum, 1969-144-1; gift of Mrs. John Early Jackson

Color plate 72 (right) A Gothic stripe has a strong admixture of scrollwork taken from baroque and rococo sources. Made in France or England, the paper dates from the 1840s. It is block-printed over a polished white satin ground on machine-made paper 19½ inches wide. This paper was found in the Whipple house in Wentworth, New Hampshire. Cooper-Hewitt Museum, 1968-53-1; gift of Edwin R. Humiston

73

Color plates 73, 74, 75
Elements from *"Décor Chasse et Pêche,"* also called *"Renaissance,"* one of the most spectacular mid-nineteenth-century French panel sets, or *décors,* are shown in these three plates. They include a frieze or cornice, a horizontal border for baseboard or chair-rail level (these two elements were printed on one piece, to be cut apart), a pilaster (10 feet high, with bits of cornice and bottom border printed behind its capital and base), and a central console showing hunting dogs attacking a deer, plus putti in the guise of huntsmen (7 feet high). These reminders of the hunt, deemed especially appropriate for dining rooms, were designed in 1846 by an artist whose name is recorded in French sources only as "Wagner." It was first issued by S. Lapeyre in 1847, then reissued by Jules Desfossé in 1859–60. The firm of Desfossé et Karth took over the blocks and continued to print this paper after 1865. The sharp blue shown here, especially in combination with brown, was a favorite during the late 1850s and 1860s in America. A set of this paper that was hung in the mid-nineteenth century survives in a house in Portland, Maine. Cooper-Hewitt Museum, 1955-12-10, -11, -12; purchased by Friends of the Museum Fund

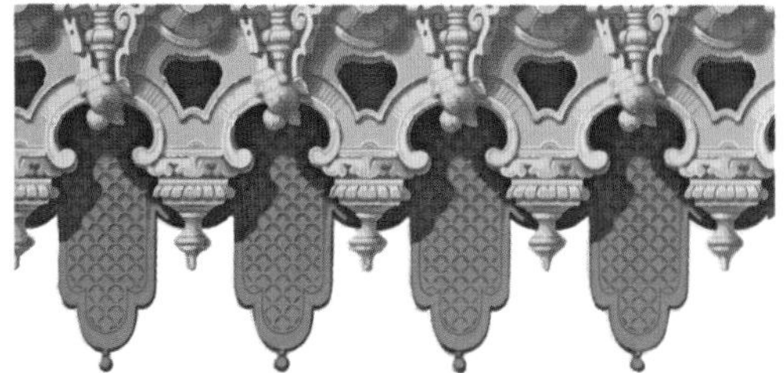

75

74

76

Color plate 76 In the mid-nineteenth century, the walls of the dining room in a Hudson River mansion of 1803 were divided into panels using stylish "fresco papers." Naturalistic flowers, very like those used in the drawing room in the same house *(color plate 69),* were block-printed in France in thick distemper colors, which have stood up amazingly well with some restoration over time. The flowers were combined in these papers with imitations of architectural moldings to frame vertical rectangles in the favorite scheme for wall division during this period.

77

Color plate 77A, B The top and the bottom of a wallpaper pilaster are shown as they were printed against an off-white background that is shaded as if the pilaster were casting a shadow. From the panel set "Regence," it was block-printed in Paris by Jules Desfossé or by Desfossé et Karth between 1851 and 1865. Including the elements designed to be joined to borders top and bottom, the piece of wallpaper shown here is 22 inches wide. Vertical elements very like this pilaster were featured in the decorative scheme shown in *color plate 76,* where realistic multi-colored landscapes appear at the top and at the bottom of the verticals. But here it is particularly bizarre to see the piercing of the architectural capital and base, which have been so carefully designed to simulate the solidity of carving and plasterwork, with realistically rendered vistas that plunge into a distant expanse of countryside. Cooper-Hewitt Museum, 1955-3-1; gift of A. Germain

78

79

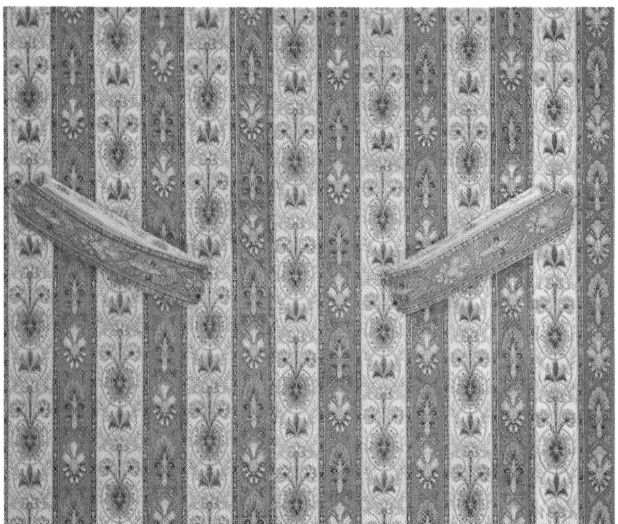

80

Color plate 78 (top left) In the trade literature of the mid-nineteenth century, this would have been called a "curtain paper." Printed by machine with rollers of the type used for printing wallpapers, this 35-inch-wide paper shade has the brown ground coating on both sides. Cooper-Hewitt Museum, 1961-175-1; gift of Elinor Merrell

Ruskinian design principles when it came to choosing wallpapers. He cited "flock papers, made to imitate woven stuffs, such as silk or worsted hangings" as "in the best taste."[3] During the middle of the century, "velvet" and "satin" continued to be the two most often-used textile words in American wallpaper advertisements. Whether they placed their advertisements in Galveston, Richmond, or Pittsburgh newspapers, or in papers of any other southern or western city, sellers of wallpaper used the same words found in the wallpaper advertisements of city newspapers from the northeast. These dealers and their customers apparently shared Downing's taste for the flocked imitations of velvet as well as for satined surfaces on their wallpapers.

By the 1840s, those "satin" papers so much advertised in America were very highly polished, and had a much harder sheen than the satin papers of earlier decades. The Zuber factory called their glossiest finish for wallpapers *"lissage."* The process for achieving this newly hard and brilliant sheen in satin papers was described by Messrs Desfossé et Karth in their article on wallpaper that appeared in London's *The Furniture Gazette.* They explained that what the English called satining was in French called *lissage,* something distinct from what the French themselves called *satiné. Lissage,* they explained, was

> . . . Used especially for marbles and imitations of Persian calicoes, [it] is executed by polishing the papers with a hard stone, simple flint or agate, fixed at the end of a counterweighted rod, and worked in a pear-wood groove. To give more brilliancy still, they use wax soap, either as a mixture in the paste, or to be rubbed gently on the paper in the same way as a polisher rubs his furniture.[4]

Color plate 79 (opposite, top right) Another "curtain paper," this commemorates New York's Crystal Palace Exhibition of 1853. It illustrates the glass building that once stood on the site of New York's present Public Library at Forty-second Street. The pattern is machine-printed over a "rainbow" ground. It is 34½ inches wide. Cooper-Hewitt Museum, 1944-66-1; purchased in memory of Eleanor and Sarah Hewitt

Color plate 80 (opposite, bottom) Draped curtains were made of this paper, which, though thicker and stronger, has the texture and pliancy of modern paper toweling. In the margin of this machine-printed pattern appears a printed mark from which the pattern can be identified as one registered with the British Patent Office by Jeffrey and Company of London on July 29, 1872. Matching tiebacks are shown in the photograph. This paper was found in Paris, but similar wallpaper novelties were doubtless used in this country as well. Cooper-Hewitt Museum, 1967-11-1, gift of Mrs. Lester S. Abelson

Satin effects described as having a "high polish and glancing appearance" were achieved at the Howell factory in Philadelphia as well. A report of 1873 reported that:

> . . . a simple brushing apparatus accomplishes the work. A number of cylinders armed with Tampico grass are made into large brushes. These are caused to revolve with immense velocity as the paper is carried over them and the result is the high polish and glancing appearance that such paper displays.[5]

Woodgrain Imitations. In addition to fabric imitations, French eye-fooling specialties included imitations of woodgrains. The English and American manufacturers made them as well. Since they turn up in every kind of American house of the nineteenth century, it is clear that they enjoyed widespread popularity. Even the sternest English critics, who warned against imitating any surface different from that of the material itself, were charmed by woodgrains. The English publication of the 1840s, *The Journal of Design,* was an early voice for those who advocated "honest" and simple flat patterns. Yet in the October issue for 1849, the editors included a little sample of an English wallpaper imitating Pollard Oak, manufactured by Robert Horne. Apparently recognizing their own inconsistency, the editors commented:

> Our readers know that imitations are hardly reconcilable with our principles, but we cannot withhold our approval of the admirable manner in which the artist has here imitated the wood. Hung upon a perfectly smooth surface and well varnished, it would deceive even a practiced eye. The "getting up" altogether is excellent.[6]

In this instance, what has been called "pride in ingeniousness"—fascination with technique—seems to have seduced even the most vehement of mid-nineteenth-century crusaders into abandoning their own principles of good design. This seems symptomatic of what the modern British art historian Sir Nicholas Pevsner (who published pioneering works about "Victorian" decorative arts) found at the core of aesthetic failures of the decorative arts in general during this period. He suggests that this pride took the place of aesthetic appreciation in the Victorian mind.[7]

Although English critics of the 1840s might have felt twinges of guilt when they admired well-executed wallpaper fakery, the French continued to enjoy it unabashedly. Both the English and the French purveyed their eye-fooling concoctions to middle-class and rich Americans, and American manufacturers followed suit. In the 1840s, Josiah Bumstead's Boston firm was stocking "Imitations of OAK, MAPLE

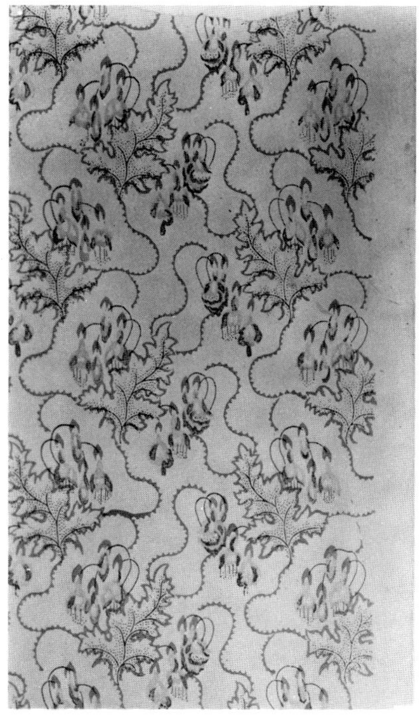

15-1

15-4

15-2

The Weekly Junior Register.

15-3

Figure 15-1 (top left) A simple machine print, patterned in three shades of blue and gray on white paper that has no coating of ground color, dates from the mid-nineteenth century. This is not a full width of wallpaper, but a fragment only 7 inches wide. The small scale of the patterning and of each element in its motifs is typical of cheap mid-nineteenth-century machine prints. Cooper-Hewitt Museum, 1938-62-79; gift of Grace Lincoln Temple

Figures 15-2, 15-3 (bottom: recto and verso, a single sheet) As happened in a number of Southern towns during the Civil War, the local newspaper in Franklin, Louisiana, ran out of paper, and someone scared up rolls of wallpaper to carry this issue, dated 30 October 1862. The patterning *(figure 15-2)* was printed in three colors over a white ground: the gray background first, reserving white leaf shapes; then tan veining for the leaves and green for the grapes. Again, the scale of the cheap machine-printed patterning is small; this paper is only 16¼ inches wide. Courtesy of the New-York Historical Society

Figure 15-4 (above, right) A realistic imitation of a gray moiré silk forms the background for this floral pattern, which bears the mark of Desfossé et Karth. Realistic roses are block-printed in shades of lavender in a bouquet with blue flowers and green foliage. Details are picked out in glossy, oil-based colors. The rest of the patterning is block-printed in chalky distemper colors. The sample is 19 inches wide. Cooper-Hewitt Museum, 1972-42-205; gift of Josephine Howell

15-5

15-6

15-7

and other woods, Varnished, Figures in Oak on rich Crimson and enamelled grounds, after the English style."[8] These papers were often varnished after they were hung. Perkins, Smith and Co. of New Bedford, Massachusetts, specified that their wallpaper "Imitations of Marble, Oak, and Mahogany" were "for varnishing."[9] A. J. Downing prescribed papers that "present the appearance of the graining of oak wainscot" for "the entry or living room of a cottage, or to whole interiors of cottages in the Gothic style."[10]

Papers printed in oil-based, non-water-soluble colors, and advertised as washable or "sanitary" wall coverings, frequently imitated woodgrains. A closely related novelty introduced into the trade during the 1860s involved the lamination of extremely thin veneers of wood to paper for use as wall decorations. In 1868, a writer in *The American Builder and Journal of Art,* a Chicago publication, reported:

> The recent attempt to supersede [paperhanging's] use by wood-hangings has scarcely succeeded. The wood shrinks after being placed upon the wall, and peels off, sometimes bringing the plaster with it . . .[11]

Yet the attempts continued. At Philadelphia's Centennial Exhibition, Charles W. Spurr of Boston exhibited his "wood hangings and marquetries." Many of his trade cards, printed on samples of the wood hangings, survive in library collections, souvenirs of his display at that great world's fair.[12]

Floral Papers. Of all their eye-fooling productions of mid-century, the French wallpaper manufacturers' realistic portrayals of flowers were their most magnificent and also their most popular *(color plates 69 through 71).* French wallpaper designers arranged flowers in stripes, in spotted patterns, or in meandering masses. They strung them out in leafy lines to form the borders for panels dividing a wall into a series

Figure 15-5 (opposite, left) About 1850, Desfossé et Karth printed this paper by Charles Muller. Still available in 1913, it was called by George Leland Hunter in *Home Furnishing* "the best 'roses' paper in the world." The flowers are pink, soft green, and tan. Cooper-Hewitt Museum, 1909-23-1; gift of Wolf and Carillo

Figure 15-6 (opposite, top right) This 1830 rococo revival pattern is probably English. Block-printed in blue, brown, and gray over a light gray-brown ground on machine-made paper, 21½ inches wide, it was the earliest paper in the hall of the Lyman house, Exeter, New Hampshire. Cooper-Hewitt Museum, 1948-61-4, gift of Mrs. J. Templeman Coolidge

Figure 15-7 (opposite, bottom right) Scrollwork typical of cheap rococo-revival papers, c.1850, this was machine-printed in green, red, and white on ungrounded paper—20 inches wide—from the "Mary Waters Homestead," Dodge, Massachusetts. Cooper-Hewitt Museum, 1940-74-60; gift of Mrs. Rollin Stickle

15-8

15-9

15-10

Figure 15-8 (opposite, left) A Sarony and Major lithograph of about 1848 shows a scrolled and striped wallpaper within the context of an idealized middle-class American setting. The generous curving forms of the pattern fit into the prototypical parlor, plentifully endowed with ornament, fabric, and plump, rounded progeny. Courtesy of the Smithsonian Institution

Figure 15-9 (opposite, right) This Gothic Revival pattern when hung would have formed stripes made up of "imitations of magnificent carved work" and "pinnacles," the very elements in wallpaper decoration Downing had warned his readers against. This paper was used in the hall of a house built on Martha's Vineyard in 1844, and was probably hung when the house was quite new. The 19-inch-wide pattern was block-printed in brown, black, and peach (now faded to white in most places) on machine-made paper with a white ground polished to a satin finish. It is probably French. The border used with this paper simulated orange, green, and brown lace gathered up with small bunches of blue and pink flowers over a black ground. A near duplicate of the pattern, but with a fountain under the arch instead of a cherub bearing flowers was illustrated by Nancy McClelland in *Historic Wall-Papers,* where she referred to it as an eighteenth-century pattern. However, both its style and the fact that it is printed on seamless paper betray its nineteenth-century origin. Cooper-Hewitt Museum, 1974-45-1; gift of Mrs. George deLys

Figure 15-10 (above) The casually asymmetrical foliage of this mid-nineteenth-century pattern parts occasionally to reveal Gothic scenes and bits of stonework. The paper was block-printed in white and darker browns on a tan ground during the 1840s. The sample is 19 inches wide, a width closer to French and American standards than to paper sizes on which English papers were usually printed. Cooper-Hewitt Museum, 1938-62-10; gift of Grace Lincoln Temple

15-11

Figure 15-11 (opposite) Medieval castles are viewed through a network of foliage typical of many inexpensive wallpaper patterns of the mid-nineteenth century. Machine-printed in green, gray, and red on ungrounded paper, this was probably made in the United States during the 1850s. This sample is 16 1/3 inches wide. A duplicate of the pattern in shades of blue on white was found in a mid-nineteenth-century house in Louvale, Georgia. Another duplicate, in shades of brown, is preserved on a bandbox in the Cooper-Hewitt collections. Cooper-Hewitt Museum, 1938-62-7; gift of Grace Lincoln Temple

Figure 15-12 Simple alternation of two views of buildings along the paper length, with bits of statuary and a few trees to mark a separation between the two scenes, was apparently deemed design enough for this wallpaper commemorating the Paris Exposition of 1855. This French paper, 19 inches wide, was block-printed in shades of brown. It was found in the Credle house in Washington, North Carolina. A duplicate pattern, printed in bright blue and brown on white, was found in the hall of the Benjamin Bowers house in Fall River, Massachusetts. Cooper-Hewitt Museum, 1962-139-1; gift of Sophia Credle

15-12

of vertical rectangles *(color plate 76).* They showed flowers growing artfully on imitation trellises and pillars. They entwined flowers with scrollwork in endless variations—a favorite theme for the American market *(color plate 71).*

Patterns like that of *figure 15-5,* introduced in the middle of the century, were produced and sold for years thereafter, through the turn of the twentieth century. Machine-printed American derivations from the French florals were less beguiling, but sold in larger numbers because they were cheaper.

The mimetic skills of designers wielded the same power to tempt English critics away from their principles when they imitated flowers as did their skills in faking woodgrains. Commenting on a sample of "trellis-work paper" pasted into *The Journal of Design* in 1849, the editors again confessed how much they liked a floral paper because it was skillfully done:

> Without binding ourselves to agree to the principle on which this pattern is made, we decidedly commend the excellent execution of it. . . . We feel bound, however, to declare, as a broad principle, that we cannot sympathize with any shams, and the imitation of light and shade, in this instance, we are afraid, brings it within the category of affectations.[13]

With the widespread use of mass-produced derivations from them in the mid-century, the French floral papers themselves began to suffer some loss of fashionable appeal. Overexposure of a style inevitably leads to changes in fashion. Because they were particular targets of the critics, these floral extravagancies also lost some of their appeal to the stylish buying public. Nevertheless, some Americans continued to buy realistic floral papers from an industry that continued to produce them right through to the twentieth century. Among those buyers must have been many who simply loved flowers and whose admiration of the seemingly real botanical specimens on these wallpapers surpassed any concern about keeping abreast of decorative styles that were currently chic.

ROCOCO REVIVAL PATTERNS

The decorative vocabulary of the mid-nineteenth century was richly embellished with scrollwork, with curving foliate and floral forms *(color plate 71, figures 15-6 through 15-8).* The "C" and "S" curves, the asymmetrical cartouches, and the fantastic acanthus leaves of late-seventeenth- and eighteenth-century baroque and rococo ornament were revived during the 1830s for wallpaper patterns as well as for every other kind of furnishing. By the 1850s, the style was at its most florid. The scrolls and flowers in rococo revival wallpaper patterns formed a harmonious background for similar ornaments on the best-known American manifestations of the style: the "Belter chairs"—and beds and tables—the most fashionable of any furniture made in New York between the mid-1850s and the mid-1860s. John Henry Belter (1804–63) was an immigrant German cabinet maker who made elaborately curved forms from layers of laminated wood, which was steamed and bent and then pierced with carving; in some examples the wood looks almost lacelike in its curves and realistic depictions of flowers, fruit, and leaves.

Belter's chairs and rococo revival wallpapers exemplify the "universal replacement of the straight line by the curve," which Pevsner recognizes as a "chief characteristic" of mid-Victorian design. This curve in wallpapers is the same that Pevsner describes: "As against [that in] other styles favoring curves, the Victorian curve is generous, full, or . . . bulgy."[14] Its appeal to a prospering, well-fed, self-confident class of Americans parallels the same appeal to a corresponding class in England.

Even on striped wallpaper, so strongly dominated by vertical linearity that curves would seem forever confined to complete subordination, the mid-nineteenth-century predilection for that generous curve tempted the ornament outside the confines of the vertical lines *(figure 15-7).* The ornament surges beyond the borders of other shapes as well, blurring the discrete design elements in the patterns, so that they seem enmeshed in continuous, relatively amorphous, masses of growth. In *color plate 72* the bulgy curves of rococo revival ornament impinge on elements of a style more chastely geometric in origin, if not in mid-Victorian execution: the Gothic.

Scrollwork, the favorite ornamental device in dense and luxuriant patterning on wallpapers of the 1840s through the 1860s, became a particular target of the design reformers. In 1841, Augustus Welby Northmore Pugin had sounded the call to battle:

> Glaring, showy, and meretricious ornament was never so much in vogue as at present; it disgraces every branch of our art and manufactures, and the correction of it should be an earnest consideration with every person who desires to see the real principles of art restored.[15]

More than a quarter century later, Pugin's earnest tones resounded through Charles Locke Eastlake's particular comments on scroll ornaments in wallpapers. In his *Hints on Household Taste* of 1868, Eastlake addressed the question of choosing wallpapers:

15-13

15-14

Figure 15-13 (left) "Landscape figures" that floated as asymmetrical vignettes on mid-nineteenth-century wallpapers could fall within any style, depict almost any scene. This example probably dates from the 1830s or 1840s, and was printed in France. The landing party in the multi-colored miniature landscape was block-printed in bright red. The paper was found in Marblehead, Massachusetts. The sample is 19 inches wide. Cooper-Hewitt Museum, 1938-62-83; gift of Grace Lincoln Temple

Figure 15-14 (right) In a cheap, machine-printed pattern of the mid-nineteenth century, a variety of romantic figures are dispersed among grotesquely large and blurred scrolls. Probably American-made, the multi-colored patterning is printed on ungrounded paper, 18 inches wide. Cooper-Hewitt Museum, 1938-62-31; gift of Grace Lincoln Temple

> I will not venture to lay down any definite rule for the choice of patterns, but I would earnestly deprecate all that species of decoration which may be included under the head of "scroll" ornament.[16]

Sentiments so charged with high moral tone bear witness to the diehard appeal of mid-century scrolls on wallpaper.

GOTHIC REVIVAL PATTERNS

Augustus Welby Northmore Pugin (1812–52), an English architect and designer of decorative arts, was a convert to Roman Catholicism. He had the zeal of a convert, and a vision of Gothic as the only true Christian style. In 1841 Pugin advocated a Gothic revival in his book *The True Principles of Pointed or Christian Architecture.* He championed honesty of materials and the simplicity of medieval design and scorned all showy ornaments. After paragraphs of vituperation hurled against "those inexhaustible mines of bad taste," Birmingham and Sheffield, where metalwork "abominations" that turned monumental crosses into lightshades were manufactured, Pugin paused to mention some other "absurdities":

> I will commence with what are termed Gothic-pattern papers, for hanging walls, where a wretched caricature of a pointed building is repeated from the skirting to the cornice in glorious confusion,—door over pinnacle, and pinnacle over door. This is a great favourite with hotel and tavern keepers.[17]

His objection was based on their breach of design principles. He commented: "Again, these papers which are shaded are defective in principle; for, as a paper is hung round a room, the ornament must frequently be shadowed on the light side." And he elaborated:

> The variety of these miserable patterns is quite surprising; and as the expense of cutting a block for a bad figure is equal if not greater than for a good one, there is not the shadow of an excuse for their continual reproduction. A moment's reflection must show the extreme absurdity of *repeating a perspective* over a large surface with some hundred different points of sight: a panel or wall may be enriched and decorated at pleasure, but it should always be treated in a consistent manner.[18]

The highly emotional, moralistic tone of such writing was to color much of the comment on ornamental design, including wallpaper, for most of the remaining years of the nineteenth century.

Andrew Jackson Downing, popularizer of the Gothic style in America, had read his Pugin. But his advocacy of Gothic architecture was delivered in a calmer voice and was unconcerned with many of Pugin's principles. It does not take the form of polemic in his *Architecture of Country Houses,* published in 1850. Only after Downing's book made the architectural style popular in America did the Gothic become really popular for wallpaper patterns. Although Gothic designs had occasionally been offered for sale beginning in the eighteenth century (see page 65), and although instances of their being advertised earlier in the early nineteenth century can be cited —as when a New Haven firm in 1833 offered "200 pieces paper-hangings, Gothic style with borders to match"[19]—they were not to become significant in the American wallpaper trade until the middle of the nineteenth century. Gothic patterns introduced American paper hangers to the pointed arches, crockets, trefoils, and rose windows derived from Gothic buildings in forms Pugin would doubtless have judged wretched caricatures *(color plate 72, figures 15-9, 15-10).* In fact, Pugin's more authentic medieval patterns never gained much popularity in America. (For an example of paper in a style derived from Puginesque models see *figure 16-4.*)

Nor would Downing have approved of most of the Gothic patterns that his book helped to popularize. The Gothic papers he admired were "plain, with only panels and cornices printed on them—giving the room in which they are placed a simple and elegant effect."[20] He also praised the "enhanced architectural effect which may be given to a plain room" by some Gothic style wallpapers, singling out for praise those that imitated the graining of oak wainscot.

In Pugin's shadow, he wrote that "good taste will lead us to reject all showy and striking patterns," and warned that:

> A good deal of taste is requisite in the choice of paper-hangings, in a house where there are rooms of importance. All flashy and gaudy patterns should be avoided, all imitations of church windows, magnificent carved work, pinnacles, etc.[21]

The very words Downing used in warning could describe many surviving examples of mid-nineteenth-century Gothic wallpapers used in American houses. That good taste he expected of his readers was apparently elusive.

Downing's "good taste" would not itself have met the standards of his mentors.

He had nothing to say of the absurdities of repeating perspectives or of the bad principles behind "shadowed" patterns that had so troubled Pugin. Nor did he appreciate the subtleties of thought about Gothic wallpaper patterns articulated by an even closer English mentor, John Claudius Loudon, for whom Downing wrote some articles, and whose ideas on other subjects are closely repeated in Downing's book. For Downing, good taste in the choice of Gothic patterns did not exclude realistic picture papers, like *figure 15-10,* as it did in Loudon's view. Loudon appreciated wallpaper ornaments that were conventionalized and abstracted from nature, not imitative: Loudon suggested that the motifs in patterns should be

> fanciful compositions of artificial forms and lines, or of plants and animals imagined in imitation of nature's general manner, but not copied from any of her specific objects.[22]

Downing's writings reflect none of this appreciation of the abstract and stylized in pattern design. They reveal no concern for the analytical thinking about wallpaper that was leading designers in England, including Pugin, to the flat patterns of the medieval period as sources of their motifs for wallpapers.

The papers that have survived in a great many American houses are far removed from the medieval sources, although a very few flat patterns derived from more authentic models have been found *(see again figure 16-4).* Although Gothic wallpapers were deemed most fitting for Gothic cottages, they also offered the owners of houses in classical architectural styles or owners of very plain structures with no architectural ornament, an inexpensive means of experimenting with a new, fairly radical revival style without undertaking architectural alterations. Downing recommended Gothic papers as well as those imitating ornaments from other architectural styles, particularly because they were a cheap means of embellishing the simplicities of cottage interiors. He remarked that "Within a couple of years, cheap patterns of paper have been introduced, exactly suited to the walls of cottages, in various styles of architecture—such as Gothic, Italian, Grecian, etc."[23] However, he did not illustrate them.

MID-CENTURY "LANDSCAPE FIGURES"

Small scenic vignettes, self-contained little views that were descended from the "landscape figures" of early-nineteenth-century wallpapers, were freed from the confines of geometric frameworks and ordered only by the mechanical necessities of machines that repeated the same elements at regular intervals. These small scenes seemingly floated in a kind of picturesque disorder, very different from the crisp geometry that had surrounded landscape views in earlier patterns *(figures 15-10 through 15-14).*

Usually the mid-century landscape vignettes were surrounded by meandering foliage, flowers, or scrollwork. The foliage and flowers were naturalistically rendered, with shading and delineation calculated to give the illusion that a thick growth of vines or endless arrays of flowers covered a wall, parting at intervals to give, most illogically, glimpses of distant buildings, figures in a group, or tiny landscapes in endless repeat. Pevsner's characterization of mid-nineteenth-century design in general notices that "all outlines are broken or blurred."[24] This comment seems especially applicable to this group of wallpapers.

Occasionally, a wallpaper designer would gather scrolls, flowers, and foliage into more regular picture frames of specifically baroque or rococo caste to surround landscape figures in more orderly ovals or other regular shapes *(figure 15-6).* These framed vignettes were often arrayed in regular drop repeats setting up a diamond grid when hung in large expanses on walls. Others, however, depended on asymmetrical ordering more closely akin to the wallpapers with landscapes floating among amorphous infills.

FRESCO PAPERS

Wallpapers of a distinctive, formulaic type were called "fresco papers" during the mid-nineteenth century. A. J. Downing explained, in *The Architecture of Country Houses,* that these "fresco papers . . . give the same effect as if the walls were formed into compartments or panels with suitable cornices and mouldings."[25] In effect, they gathered the earlier nineteenth-century wallpaper ornaments imitating dados, columns, and friezes into configurations that unified the separate architectural elements and compartmentalized the major part of the wall space above the chair rail, creating a series of vertical rectangles—illusionistic paneling.

Although they sometimes still framed landscapes, these elaborate fresco papers often framed a solid color or a relatively monochromatic repeating pattern *(figures 15-15, 15-19).* The most spectacular examples surrounded gorgeous displays of flowers, statues, fountains, and garden ornaments overflowing with magnificent growth. A number of American houses preserve examples of fresco papers in which the framing elements of the panels surround a relatively large expanse of plain color,

81

82

Color plate 81 (left) William Morris designed what he called the "Daisy" pattern for wallpaper in 1862. After unsuccessful attempts by his own firm to print it, he turned it over to Jeffrey and Company, who issued the pattern in 1864. The first of his wallpapers to reach the public, it was based on floral designs of the Middle Ages. The simple, even naive design was popular through the turn of the century, and was often emulated *(figure 16-10).* This sample, printed during the 1930s from Jeffrey's old wood blocks in green, red, orange, and brown on a cream field, is 22½ inches wide. Cooper-Hewitt Museum, 1935-23-7; gift of Cowtan and Tout, Inc.

Color plate 82 (right) This "Trellis" design of 1862, produced in collaboration with the architect Philip Webb (1831–1915), who drew the birds, was actually the first wallpaper design on which Morris worked. However, Jeffrey did not issue it until 1864, after the introduction of "Daisy." By incorporating a trellis as the framework for the patterning, Morris, whose first loyalties were to medieval designs, has used a motif often seen in wallpapers of the 1870s and 1880s that were based on Japanese sources *(color plate 93 and figure 16-23).* Printed in the 1920s from the old wood blocks, this sample is 22½ inches wide. Cooper-Hewitt Museum, 1941-74-19; gift of Robert W. Friedel

83

Color plate 83 (opposite) An intricately constructed swirling pattern of 1876—"Pimpernel"—is unusually dense, complex, and naturalistic compared to other Morris designs. Since many gradations of hue were featured in "Pimpernel," involving the use of many wood blocks, it was one of the more expensive Morris patterns. The sample, printed in about 1934, is 22½ inches wide. Cooper-Hewitt Museum, 1935-23-18; gift of Cowtan and Tout, Inc.

Color plate 84 This machine-printed pattern is probably American-made, dating from the 1890s. It is a remnant of a paper used in The Manse, Deerfield, Massachusetts. Both in coloration and in the use of swirling floral patterning it reveals the influence of Morris wallpapers on the cheaper products of the industry. It also incorporates mannerisms in the drawing of leaf forms taken from Art Nouveau sources *(color plate 100 and figures 17-8 through 17-10).* Cooper-Hewitt Museum, 1972-51-10; gift of Deerfield Academy, Deerfield, Massachusetts

84

85

Color plate 85 These remnants of paper used in the bedroom of George Peabody Wetmore at Château-sur-Mer in Newport, Rhode Island, were given to what was then the Cooper Union Museum in 1939 by his daughters. The swallow frieze, just over 11 inches wide, was designed in 1875 by Brightwen Binyon for Jeffrey and Company of London as part of the first set of wallpapers featuring matching frieze and continuous dado (not shown here) printed by Jeffrey. The fill pattern relates closely to a pattern of orange boughs designed by Charles Locke Eastlake for Jeffrey, and in overall effect shows the influence of Morris designs. Cooper-Hewitt Museum, 1939-45-4, -8; gift of Edith and Maude Wetmore

Color plate 86 (right) At "Château-sur-Mer," a grand house built on the sea at Newport, Rhode Island, in 1852, and remodeled by the architect Richard Morris Hunt in 1872, the bedroom of its owner, George Peabody Wetmore, survives much as he had it furnished in the 1870s. The trend-setting English wallpaper and frieze shown opposite are used above a dado of ebonized—black stained—wood paneling. The woodwork and furniture are also ebonized, and picked out in gold. On the ceiling, a copy of the wallpaper pattern has been painted and it is outlined with elaborate painted borders. The recessed bed alcove is papered in Morris's "Willow" pattern *(figure 16-6).* Courtesy of the Preservation Society of Newport, Rhode Island

87

Color plate 87 A remnant of another high-style wallpaper by a leading English architect and designer of the 1870s, William Burges, was given to what was then the Cooper Union Museum by Edith and Maude Wetmore. Fragments of this 11-inch-wide frieze, two widths of which are shown here as they were printed side by side, were found in the room shown opposite, and the pristine samples at the museum were used to reproduce the border. Its presence there provides further evidence of the influence of leading English designers on American taste during the period. This Burges pattern of 1875, like many of those designed by Pugin and Morris, was derived from medieval sources. Cooper-Hewitt Museum, 1939-45-5; gift of Edith and Maude Wetmore

Color plate 88 (opposite) Reproductions of the Burges border and of an appropriately medieval flat pattern of the 1870s were hung during 1980 in this second-floor bedroom at Château-sur-Mer. The painted ceiling decorations are faithful restorations of the ones that were uncovered when the room was being restored. Courtesy of the Preservation Society of Newport, Rhode Island

89

Color plate 89 The restoration of the bed-sitting room of Mrs. George P. Wetmore on the second floor of the Newport house included reproducing still other original Château-sur-Mer wallpapers, which Edith and Maude Wetmore gave to what was then the Cooper Union Museum in 1939. While the wallpapers at Château-sur-Mer illustrated in *color plates 87 and 88* were medieval in style, these stylized passion-flower motifs represent a taste for the Japanesque. The paper here used as a frieze was designed to be used as a dado as well. The ceiling is a painted restoration of the original. Courtesy of the Preservation Society of Newport, Rhode Island

or of unobtrusive faint patterning, centering on an elaborate ornament or statue *(figures 15-16 through 15-18).* Josiah Bumstead's printed billhead gives a quite detailed description of such paper, a description fully consistent with Downing's:

> FRESCO PAPER forms the wall into panels of plain style, or of richly decorated figures in colors, and gold; some are intended for small-sized rooms.

The vertical orientation of the panels created by the elements included in sets of fresco paper *(color plate 77, figures 15-16 through 15-21)* is distinct and consistent, though it is not emphasized in contemporary descriptions of the wallpaper fashion. However, this verticality was to contrast pointedly with the horizontality of papering styles that immediately succeeded the fresco papers. (Contrast horizontal schemes of wall division in the 1870s and 1880s, shown in *color plates 86, 88, 89* and in *figures 16-11, 16-22, 16-39, 16-42* with the vertical panels of mid-century shown in *color plate 76* and *figures 15-16 through 15-21.*) In retrospect, the contrast emphatically distinguishes the decorative schemes of the two eras.

In 1850, Downing recommended:

> If the fresco papers (which may now be had in New York, well designed, of chaste and suitable patterns for any style of architecture) are chosen, they will produce a tasteful, satisfactory and agreeable effect in almost any situation.[26]

The Boston billhead already cited furnishes documentation that fresco papers were also available in that city during the 1840s, and a St. Louis advertisement of 1850 confirms that the wallpaper styles available in the east were quickly taken west. George Michael offered "gold and silver, foulard, landscape, intaglio, medalion, crimson, and drab frescos."[27] And a Richmond, Virginia, advertisement of 1845 documents their availability in the south.[28]

As mentioned in Chapter 9, by mid-century the great panoramic scenic papers featuring distant vistas had given way to the *décors,* which emphasized the interior, the ornament within the room, subordinating the scenery to the architectural and decorative embellishments of the framing elements. These *décors* became the preoccupation of the famous French factories from the 1840s through the 1860s. Americans used the term "fresco papers" in much the same way the French used *décors.*

Americans also grandly characterized the most elaborate French specimens of the *décor*–fresco genre "tableau decorations." At New York's Crystal Palace Exhibition of 1853, Jules Desfossé won his bronze medal "for large and fine Tableau Decoration

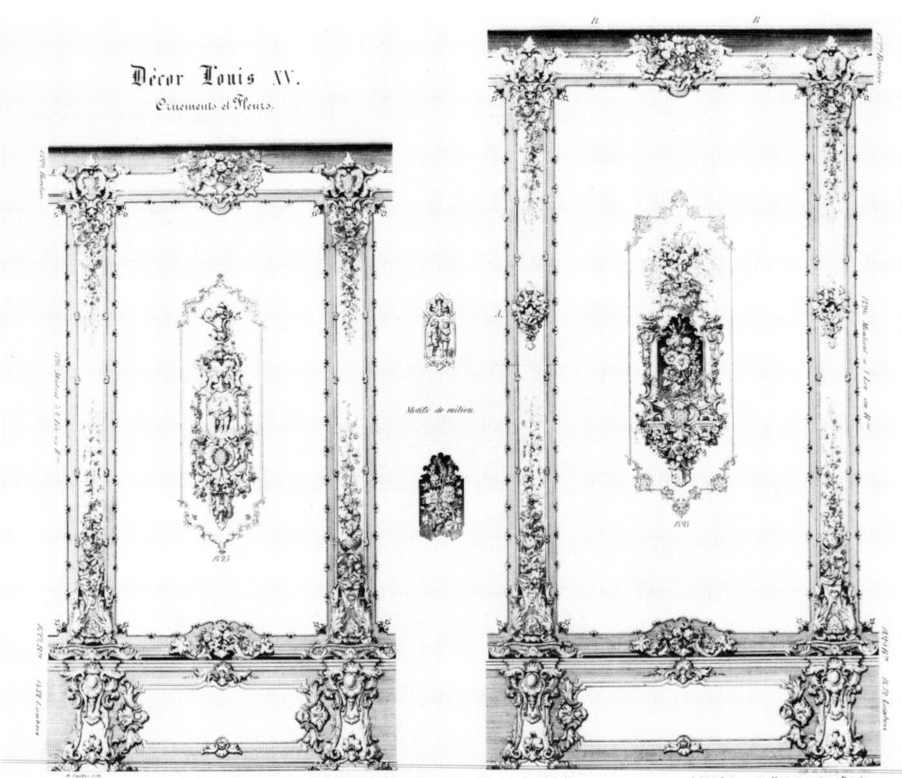

15-15

Figure 15-15 (top) This panel set—what Americans called "fresco papers"—is from an album of engravings illustrating papers produced by Étienne Delicourt (active in Paris, 1838–60). Delicourt's papers and, later, his designs as printed by his successors were available through New York dealers Emmerich and Vila from 1850 past the turn of the century. Fragments of the dado paper shown here were found in the Jesse Fay Hosmer house in Camden, Maine, where it was used with a repeating pattern of scroll-like foliage printed in gray and white with gilded accents. Cooper-Hewitt Museum library

15-16

15-17

15-18

Figure 15-16 (opposite, bottom left) At Landsdowne, in Natchez, Mississippi, the walls were papered during the mid-nineteenth century with French papers. The multi-colored floral border papers (by Zuber), dominated by maroon flocking and pink and red flowers, frame alternating vertical panels, one with a repeating pattern of gray and white foliage (by Delicourt, 1853), the other a sculptural figure on a plain white background. Courtesy of the Mississippi Department of Archives and History

Figure 15-17 (opposite, bottom right) These mid-nineteenth-century French panel papers with pilasters are preserved in a third-floor stair hall of a house built in 1847 by Dr. Lewis Windle in Cochranville, Pennsylvania. Rust brown and deep green shades predominate. Compare with *figure 15-18.* Courtesy of Ron Goldstein

Figure 15-18 (above) These papers hung until the 1960s in the library at Rose Hill, a frame house built in 1802 in Yanceyville, North Carolina. The French wallpaper cornice, pilasters, and fruit borders framing the panels duplicate the patterns shown in *figure 15-17* in a Pennsylvania house. Courtesy of Mrs. J. Williamson Brown

15-19

Figure 15-19 A plate from the album *"Collection d'Esquisses, Delicourt et Cie."* shows a "landscape view" as updated with mid-century wallpaper embellishments. The flower-covered trellises, balustrades, and cornices would have divided the walls into panels, using the neo-Grec architectural elements admired for their stylish effects during the 1850s and 1860s. The Delicourt firm was associated with that of Campnas et Garat between 1838 and 1859, so both names appear on the print. Cooper-Hewitt Museum library

15-20

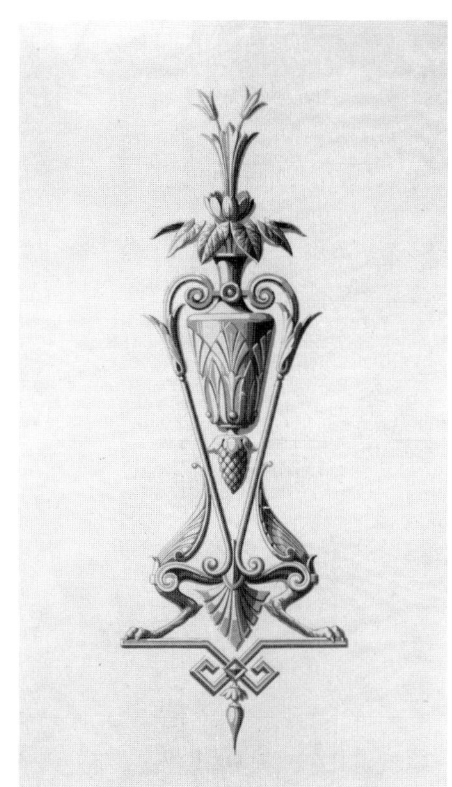

15-21

Figure 15-20 (left) At Elm Farm, built by Richard Little about 1850 in Salisbury, Connecticut, panels filled with an imitation of tufted green velvet with yellow buttons were used in the parlors. Around the tufting was the frame of brown flocking embellished with gold leaves and scrolling meanders that is shown here. Surrounding these elements, plain brown flocked paper was hung. The fragmentary portion illustrated here is 24 inches wide. Cooper-Hewitt Museum, 1955-87-2; gift of Mrs. Allen E. Smith

Figure 15-21 (right) This stylized motif, block-printed in shades of tan and yellow over a glossy ground of tan combed to imitate woodgraining, was probably intended as the center for a panel in a set of "fresco papers." Made in France during the mid-nineteenth century, the paper is 19½ inches wide. Cooper-Hewitt Museum, 1959-160-7; gift of Josephine Howell

15-22

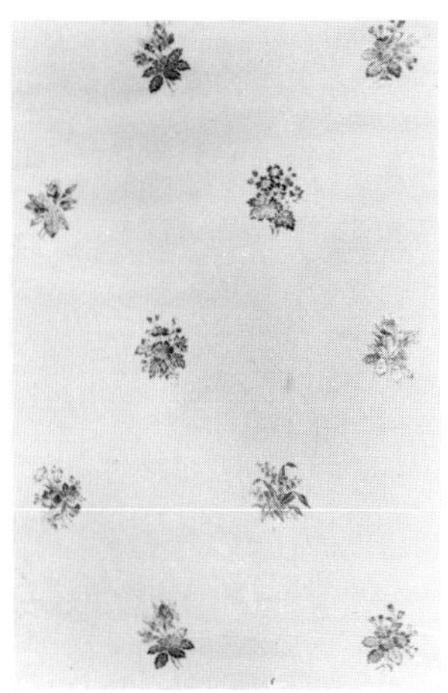

15-23

15-24

Figure 15-22 (top) The family group in this anonymous watercolor of the 1870s sits in a parlor papered with a pattern the artist has depicted in brown on tan. In reality, it probably was a pattern gilded and embossed, as were the two wallpaper samples illustrated below. The scale and spacing in the placement of the motifs are typical of a great many wallpapers of the 1850s and 1860s, as is the dark pattern of the border and its narrow width. Collections of Greenfield Village and the Henry Ford Museum, Dearborn, Michigan

paper."[29] Equally large and fine was Desfossé's *"Décor Chasse et Pêche,"* elements of which are shown in *color plates 73 through 75.* A set of this paper installed in the nineteenth century has survived in one New England house.[30] Similar papers by Delicourt, who also won a bronze medal at New York's Crystal Palace, must have been hung in numbers of American houses, purchased through Delicourt's New York distributor.

Not all fresco papers or panel sets were so grandiose as that shown in *figure 15-15.* In less exuberant interiors, more modest repeating patterns like that of *figure 15-23* filled panels framed with plain-colored flocked paper or with more sober imitations of moldings. "Velvet" or flocked paper was described in *The American Builder and Journal of Art* for March 1869 in such a panel arrangement as

> dark, rich, and heavy and seldom used entirely,—most frequently for borderings or strips which separate panels of lighter tinted paper. These strips are called styles, and are edged on each side with a narrow gilt bordering, which forms a very effective and elegant design. The panels in a room are sometimes of two colors, alternating with each other, or different shades of the same color.[31]

Fresco papers also imitated woodgrained paneling. An example of woodgraining arranged in the fresco style with red figured borders outlining vertical panels survives in deteriorated condition in a house near Natchez, Mississippi, called Roseland.[32]

GILDED AND EMBOSSED PATTERNS

During the 1850s and 1860s, in apparent reaction against the strong colors and dense ornamentation of wallpapers laden with scrolls, flowers, and foliage, many Americans began to hang sparsely ornamented patterns on which light touches of gold were

Figure 15-23 (opposite, bottom left) This gilded and embossed pattern, printed on gray, fits the specifications for papers in "good taste" among people of "moderate means," as described in 1869 (see page 360); 19½ inches wide, it is a remnant from paper still in the morning room at Acorn Hall in Morristown, New Jersey. Cooper-Hewitt Museum, 1974-27-5; gift of the Morris County Historical Society

Figure 15-24 (opposite, bottom right) Many of the gilded and embossed papers of the mid-1800s had motifs like this one. Its style might have been called "Renaissance." It was found at Elm Farm in Salisbury, Connecticut, which was built about 1850. This single motif is about 3 inches tall. Cooper-Hewitt Museum, 1955-87-6; gift of Mrs. Allen E. Smith

spotted across delicately tinted ground colors—grays, beiges, eggshell shades, and white. Most of these patterns featured small embossed gold motifs on satin grounds. The gold was real gold leaf that had been subjected to the pressure of intricately detailed metal dies on heavy machines that stamped their imprint into the paper while bonding the gold leaf to it. The stamping technique was a specialty of two German factories—that of Karl F. L. Herting at Einbeck and of Englehard at Mannheim. The Englehard papers were distributed in France by Desfossé, and probably came here through American distributors for that French firm.[33]

Americans also made such papers. The patent office preserves a patent of 1867 "for making stamp-gilt paper hangings," and accounts of visits to American factories describe a variety of methods for applying motifs in gold and embossing them.[34] One such account of 1873 details the making of gilt paper at the Janeway factory in New Brunswick, New Jersey:

> after grounded or tinted it is conveyed to a long room and passed over tables, while girls at the proper intervals lay on a stencil and brush shellac powder upon it; thence it passes to other girls, who lay the gold leaf over this, whence it is passed to a press in which plates are set with the design traced upon their surface, the whole being kept at a temperature of 130°. As the paper comes beneath the press, the heat melts the shellac and makes it glutinous and at the same time the dies are pressed upon the gold leaf so that the parts which the dies touch adhere and the superfluous leaf may be brushed away.[35]

The patterns that were the most popular vehicle for this technique were variations on the types illustrated in *figures 15-23 through 15-25.* On similar examples, delicate sprigs, tiny birds, leaves, rosettes, and decorative devices—like fleur-de-lis and stiff little cartouches—were distributed in the most basic and regular of drop repeats, widely spaced. The painting illustrated as *figure 15-22* shows a room wallpapered with one of these patterns and gives a sense both of the scale of the motifs and their spacing, typical of the period. An article of 1869 on "The Parlor Walls and Wall Ornaments" in Chicago's *The American Builder and Journal of Art* surveyed a variety of styles in wallpaper, to conclude:

> . . . that which obtains among people of moderate means and good taste, is a plain ground of stone color, in all its different shades, with a small gilt leaf, or cluster, or clusters, of delicate tracings; to which is sometimes added a little color, such as black and crimson.[36]

Americans hung them as the simplest of patterning covering the whole space of a wall or as centers of panel arrangements, as part of what they called "fresco decorations." The paper shown in *figure 15-23* was used in this way.[37]

BORDERS

In the mid-nineteenth century, borders were the most important elements in those fresco decorations dividing walls into panels. However, when borders were used more simply as the edgings for walls covered with a single wallpaper pattern, the borders chosen were rarely as wide or as visually important as they had been earlier in the century.

Nevertheless, borders continued to be *de rigueur.* Most often, they were hung at cornice level as in *figure 15-22.* "Velvet"—flocked—borders, generally darker in tone than the fill patterns, were preferred over all others, as they had been earlier in the century. Many took the form of scrolls like those used in the patterns covering walls, but strung out in a long line. Trailing bands of ivy, of grapes among grape leaves, and of blooming vines were favorite naturalistic motifs for borders, since the natural patterns of growth of those plants were easily adapted to a border format. Gilding, which had been prominently advertised as a feature of borders as far back as the eighteenth century, became even more important on these mid-century borders. Offerings of "French Gilt Velvet and Plain (or Common) Borders in Great Variety" headlined many wallpaper advertisements of these years. Many of the by-now standard architectural motifs and imitations of architectural moldings remained in the stock of American wallpaper dealers during this period.

PATTERNS PREFERRED FOR SPECIFIC ROOMS

Between 1840 and 1870, while French taste continued to dominate American wallpaper design, the newly industrialized factories making wallpapers in this country enjoyed their first real boom period. Advertisements boasted of the cheapness of paper hangings, and even a few surviving writings by consumers, who had no vested interest in promoting the product, comment on the economy of papering walls. James Watson Williams of Utica, New York, wrote a veritable testimonial for paper hanging in a letter of 1850 to his wife. Most of the letter was concerned with the house they were building, and Williams included:

> . . . One word as to painting walls. If they crack, the paint does not hide the defect. Papering does. Very fine and appropriate paper is made now, and more ornamental than any painting that is not quite Expensive. I rather admire some of the papers particularly for dining room and library. The cost of painting and papering is on the whole rather in favor of the latter.[38]

While Williams preferred the dining room and library papers, wallpaper dealers promoted the use of wallpaper everywhere. And by the middle of the century, wallpapering was not restricted to the confines of the home; it was being used in public and commercial buildings as well. The Newark, New Jersey, firm of Rutan and Birdsall advertised in the local city directory for 1859. Their offering of "French, English, and American Designs" was followed by an impressive list of places deemed appropriate settings for these papers: "Parlors, Halls, Dining Rooms, Offices, Banks, etc." Not to be outdone, the Newark firm of C. Abbott boasted that it had papers for "Parlors, Halls, Dining Rooms, Vestibules, Cottages, Banks, Lodges, Division Rooms, &c."[39] In addition to this variety of suggested settings, churches were specified as suitable places for wallpapering by J. F. Bumstead and Company of Boston. The firm mailed out a circular in 1853 that described the way a church interior might be decorated with its products:

> A church can be papered in a becoming style, at an inconsiderable expense. The walls of marble blocks; or panelled with columns intervening; and the ceiling in plain or watered figure, with moldings and center pieces;—all of a very chaste and church-like character.[40]

A wallpaper seller of Salem, Massachusetts, offered papers for kitchens in 1846. It is not clear today what form these might have taken. Perhaps they were printed in oil-based pigments, and were of the type advertised in England as "Sanitary Papers." Or perhaps they were like the varnished wallpapers described by Thomas Webster and Mrs. Paine in *The American Family Encyclopaedia of Useful Knowledge* (1845 and 1859) as "useful in places where it is liable to be much soiled," although they had "an unpleasant appearance . . . not fit for elegant apartments."[41]

As in earlier periods, mid-century advertisers of wallpaper were more specific when suggesting which papers should be used for halls than they were when advertising papers for other rooms. Architectural papers, imitations of stone work, and of oak and marble were those most often recommended. Papers simulating stone surfaces, like those illustrated in *figures 12-32 through 12-34,* were described in Bumstead's circular of 1853. It featured marble and granite, adding:

> These, cut into large or small blocks, are much used in halls, entries, & c. in producing a substantial looking effect and withall very durable in wear.

In 1852, one Philadelphian, advertising similar wallpapers for halls, specified that his imitated "SEANA AND OTHER MARBLES."[42]

The use of architectural papers and of fake wood and stone was not restricted to halls. In 1853, S. H. Gregory and Company of Boston advertised "Architectural Decorations" for a variety of rooms and buildings including "Parlors, Library and Dining Rooms, Halls, Saloons, and Churches."[43] Oak and marble papers, favored for use in halls, were also suggested for dining room use in a Galveston, Texas, advertisement of 1857.[44]

For dining rooms, panel papers centering on trophies of the hunt and trophies of the vineyard were deemed appropriate and fitting for rooms in which game and wine were enjoyed. One Englishman commented on dining room papers in an issue of *The Art Union* for 1844:

> Flock papers have held their ground in popular estimation longer than any other, and we cannot ourselves get rid of the old association which connects them with the dining room . . . But flock paper is fit only for large and lofty apartments, in small rooms its effect is weighty and oppressive.[45]

Perhaps Americans followed the English example and used flocked papers for dining rooms, though specific references to this practice have eluded present research. For dining rooms, the Bumstead company, in that same circular of 1853, recommended "deep and dark" papers as appropriate. While they did not specify that these papers be flocked, they went on to detail that "Rich crimson, maroon, green, and blue satin grounds, with figures in oak and brown colors" were "suitable for dining rooms, where the effect of warmth and comfort is desired."

For parlors, light-colored papers with gold accents were generally preferred. Many mid-century American advertisers described their best offerings as "finest parlor papers," and this phrase may well have designated the gold and white papers. In 1846, a Philadelphia company, Howell and Brothers, published an advertisement in which its recommendations were more specific in describing papers for that most archly Victorian of room types: "Splendid Decoration Frescoes, with or without gold, for Parlors" were offered. Bumstead's listed on its printed billhead of the 1840s:

15-25

Figure 15-25 A typically narrow border of the mid-nineteenth century is shown here with a wallpaper on which a repeated motif is embossed and has a metallic finish. The border, only 3 inches wide, is predominantly red and blue. Each leafy wreath, of which only one example is included in the illustration, is 5 inches high, embossed in silver against a light gray ground. The paper and border were hung during the mid-nineteenth century at Elm Farm, in Salisbury, Connecticut. Cooper-Hewitt Museum, 1955-87-5; gift of Mrs. Allen E. Smith

"Gold, Silver, and Cloth Papers, on embossed and enamelled grounds of exquisite finish and beauty for Drawing rooms and Parlors."[46] By 1853, this same Boston firm advised its customers even more emphatically that "Gold and white embossed" were "not the most costly" but "are deemed by many to be in the best taste for drawing rooms, especially where a light prevailing hue is desired." "Richest Gilt-Velvet and Pure White Parlor Papers" were among the wallpapers offered for sale in New Bedford, Massachusetts, in 1853 by Perkins, Smith and Company.[47]

In their book of 1869, *The American Woman's Home,* Harriet Beecher Stowe and Catharine E. Beecher advised the budget-conscious housewife, anxious to make her parlor look "furnished," to put her money in cheap wallpaper rather than in expensive Brussels carpeting. They suggested "good satin papers" in "a lovely shade of buff, which will make the room look sunshiny in the day-time, and light up brilliantly in the evening." They added that "a maroon bordering, made in imitation of the choicest French style which can not at a distance be told from it, can be bought for six cents a yard."[48]

Bedrooms seem to have been left to individual taste, and that often meant flowered papers were hung. Downing noted that in bedrooms, "Paper hangings are largely used for the walls in most of our country houses, and should always agree in general tone with that of the furniture in the apartment."[49] The Bumstead company in its 1853 flier headlined its most appropriate bedroom papers as "Light and Bright" and noted that "white satin grounds, with figures in brilliant colors, taken from the flower garden, herbarium, &c." gave "to chambers and bedrooms a sprightly and cheerful look."

The growing popularity of wallpaper is evident in writings of the middle third of the nineteenth century. In 1850 Downing, whose opinions about interior decoration in *The Architecture of Country Houses* were so widely read, commented:

> Paper-hangings offer so easy, economical and agreeable a means of decorating or finishing the walls of an apartment, that we strongly recommend them for use in the majority of country houses of moderate cost.[50]

By 1861, the writer of *Eighty Years Progress of the United States* included wallpaper in his survey of American paper production, noting:

> The use of paper-hangings, which has become so common in the past ten years, superseding hard finish and painted walls for city dwellings, absorbs a large amount of paper.[51]

In *The American Builder and Journal of Art* for March 1869, an author identified only as HSB observed in an article on "The Domestic Economy of Architecture: The Parlor Walls and Wall Ornaments":

> . . . upon the whole, for convenience ordinarily, and as a matter of taste extraordinarily, we find that papering obtains among all classes.

The article concluded:

> Taken as a whole, there is nothing for wall hangings that can, for general use, supersede paper. . . . There is something about a papered wall that gives one a sense of protection and comfort, whatever may be said of the taste of substituting paper for a plain uniform wall, pure white or tinted.[52]

WINDOW SHADES AND PAPER CURTAINS

By the middle of the century, another wallpaper product, already in use, had achieved wide popularity. The wallpaper manufacturers used their wood blocks and printing machines to make what we would now call window shades. These were known in the trade as "curtain papers." Examples shown at New York's Crystal Palace in 1853 caught the attention of the British commissioners, who described them in their general report of the New York Industrial Exhibition:

> A peculiar article in paper-hangings is largely manufactured for the Western States. This is about thirty-five inches wide, and is known as "curtain paper." An ornament, within a panel, is printed, extending to the length of about 1½ yards and those are cut off and used as substitutes for roller blinds, by a large class of people in the West.[53]

While numbers of advertisements for curtain papers do call them to the attention of "country merchants," their use was not limited to the west. They were advertised in every major city. A thirty-five- or thirty-six-inch width was fairly standard. Beaty and Curry of Philadelphia advertised in 1846: "Country merchants supplied with curtain paper 36 inches wide."[54]

Curtain paper was not new to the 1840s and 1850s. Plain green and blue papers were advertised for use as curtain papers by New Bedford paper stainers in 1810[55] and continued to be mentioned by those firms in advertisements of 1834.[56] In 1812, the advertisements of John Perkins of New Bedford featured "Patterns for Window Curtains and Chimney Boards."[57] In 1828, Perkins offered "ready made Paper Curtains (plain and figured and warranted not to fade)."[58] Isaac Pugh of Philadelphia had "bright colored paper for curtains" in 1846.[59] There are also examples of window shades surviving from the nineteenth century indicating that sometimes wallpaper patterns were cut out and pasted on plain paper..

Philadelphia-made curtain papers seem to have enjoyed a particular cachet. Large numbers of them appear to have been shipped all over the country. In addition to advertisements in Philadelphia publications, there are other references to Philadelphia curtain papers found farther afield. In an Albany, New York, city directory for 1849–50, William Richardson mentioned "the splendid and elegant patterns" of his "SHADED PHILADELPHIA CURTAIN PAPERS, just the kind to suit customers." "Shaded" refers to the rainbow, or *irisé,* technique described on pages 273–275 *(color plate 79).*

While most of the products that wallpaper sellers called "curtain papers" were similar to the straight, flat, window shades of today, it is possible that sometimes these words may have been used to describe papers intended to be draped from rods like fabric draperies *(color plate 80).*

16 THE 1870S AND 1880S: A MAJOR CHANGE IN TASTE

During the 1870s, Americans continued to cover their walls with ever-increasing quantities of paper. However, their taste in wallpapers was changing. They were less often attracted to the French designs because they were beginning to heed English design reform theory and to buy English-made wallpapers in which application of that theory was visually evident. Ironically, at the very time of this shift in American taste in decorative arts from France to England, a shift in taste in the fine arts went the other way: from English painting to the French work of the Barbizon School.

The elaborate and realistic ornamentation of French and French-inspired wallpaper lost its hold on the American wallpaper market only after this country had been inundated by English design theory and English designs, which had been introduced in several ways. The theories had been widely circulated in John Ruskin's book, *The Two Paths* (1859), important to subsequent American thinking about design and decoration, and in Charles Locke Eastlake's *Hints on Household Taste* (1868, with a first American edition in 1872), which mixed and popularized ideas derived both from Ruskin and from Owen Jones, author of *The Grammar of Ornament* (1856). Reforming theory had also been circulated by publication of speeches of William Morris. Even more important, the look inspired by his theory was introduced in Morris's wallpaper patterns, which were sold in America. And only after a great many more of the new reformed English designs were showcased at the Centennial Exhibition of 1876 in Philadelphia was the change in taste fully effected.

THE ENGLISH BACKGROUND

As noted earlier, English critics and designers had been trying since the 1840s to change popular taste in wallpapers. This was part of a much larger and more significant movement to improve the design of all manufactured furnishings and decorative objects, as well as the whole system that divided the labor required to design and make these objects for everyday use. Seminal thinkers in the movement like Thomas Carlyle (1795–1881), Augustus W. N. Pugin (1812–52), and John Ruskin (1819–1900) attributed many of the woes of the nineteenth-century English workingman to industry's destruction of his role as a true artist-craftsman. The man who designed an object no longer made it. This "unnatural" separation was blamed for the fact that design was "bad"—not appropriate to the materials and production techniques used to make objects. It was also blamed for the reduction of the formerly creative craftsman to the status of wage-earning slave of the machine.

Some reformers decried factory production altogether, urging a return to hand

16-1

Figure 16-1 Illustrations of the "True" (left) and the "False" (right) principles in wallpaper design as included in Wilmot Pilsbury's article "Principles of Decorative Design" in London's *The Furniture Gazette* for August 30, 1873 (see page 394). Courtesy of Avery Library, Columbia University

16-2

16-3

Figure 16-2 (left) This pattern, found in the early-nineteenth-century house of Captain Daniel Norton in Suffield, Connecticut, would certainly have passed Mr. Pilsbury's standards and would have ranked as a good example of the true principles of wallpaper design during the 1870s. On a brown ground, the curving framework is printed in brown and the stylized bouquet in orange-red and purple with green. The sample is 9¾ inches wide. Courtesy of the Stowe-Day Foundation, Hartford, Connecticut

Figure 16-3 (right) This sample, produced by Jules Desfossé and tipped into the album his firm presented to the judges of the 1855 Exposition Universelle in Paris, could have served as the model for Mr. Pilsbury's illustration of the "false" in wallpaper design. The pink roses are realistically block-printed in subtle shadings in a bouquet of green foliage. The polished white satin ground in combination with pale gray overprinting creates an effective imitation of moiré silk. Cooper-Hewitt Museum library

16-4

Figure 16-4 This sample, about 1860–70, found in the Norton house, Suffield, Connecticut, provides a rare example of American use of a wallpaper based on the most chastely two-dimensional medieval flat patterns. Each motif was machine-printed in red, green, and metallic gold over a tan ground on textured (ribbed) paper. The sample is 9 inches wide. Courtesy of the Stowe-Day Foundation, Hartford, Connecticut

Figure 16-5 (opposite) Morris registered this "Marigold" pattern as a wallpaper and again as a chintz in 1875. The combination of densely patterned foliage and flattened, stylized flowers inspired American imitations *(color plate 91).* A relatively inexpensive Morris paper, only one printing block was used for the background—blue in this 22-inch-wide sample made in 1934 reserving the pattern shapes in the off-white ground color. Like all Morris patterns, it was available in many colors. Cooper-Hewitt Museum, 1935-23-4; gift of Cowtan and Tout, Inc.

16-5

16-7

16-6

Figure 16-6 (left) Morris's "Willow" pattern, registered in the British Patent Office in 1874, was apparently very popular in America. It survived in many late-nineteenth-century houses *(color plate 86)* and was imitated by American manufacturers. This sample, 22½ inches wide, was block-printed about 1934 in green with black veining on a cream ground. Cooper-Hewitt Museum, 1935-23-1; gift of Cowtan and Tout, Inc.

Figure 16-7 (right) In a brochure issued around the turn of the century, Morris and Co. noted that this "Sunflower" pattern (designed in 1879) was "Probably the most popular of all William Morris's designs for monochrome wallpapers. It is printed in a great variety of shades, and can also be had in gold and bronze colors." The sample shown here, only part of a repeat, is preserved in an early Morris and Co. sample book now at the Cooper-Hewitt Museum. Cooper-Hewitt Museum, 1969 purchase

16-8

Figure 16-8 In an early photograph of Chicago's Glessner house, built in 1885 and designed by H. H. Richardson, Morris's "Sunflower" pattern is seen on the walls of a bedroom. The curtains, hung in the most approved manner of the day—from straightforward brass loops on a rod—are made of Morris's "Rose and Thistle" chintz, designed in 1882. On the chair, yet another Morris chintz was used, the "African Marigold" pattern of 1876. Courtesy of the Glessner house, Chicago

craftsmanship so that the workman, like his medieval predecessor, could enjoy the experience of designing as well as executing his work. In this group were Pugin and Ruskin, who thought that with training in art, the craftsman would then create designs surpassing those of an artist or designer who did not understand the true nature of his materials and the processes of forming them into useful and beautiful objects. Conversely, the artist or designer who perfected craft techniques could also produce superior work.

The products of the true artist-craftsman would exhibit in their very construction his understanding of the character of the medium in which he worked. "Honest" construction—forthright expression of the character of materials and of how they were formed and put together—became articles of faith with many nineteenth-century design reformers. It was popularly codified: "As a rule, all imitations, in whatever material, of a totally different surface from that which characterizes the material itself, are false."[1] Wood was to look like wood, to be formed into furniture made up of straight, solid pieces, cut to follow the wood grain, and joined with mortise and tenon plainly visible, not "deceptively" glued together and disguised with veneers. Such furniture represented a rejection of what were ridiculed as the structurally unsound forms of French rococo chairs and tables, or Belter's carved and laminated elaborations on those forms, with curvaceous legs and backs exhibiting a total "misunderstanding" of the nature of wood. The reformers also rejected some of the nineteenth century's pet examples of ingenuity such as iron cast to look like logs and twigs in rustic furniture and papier-mâché formed to imitate wood. By the same logic, they rejected paper hangings printed to simulate wood—or marble, or a vast landscape, or any one of the variety of *trompe l'oeil* subjects surveyed in Chapters 9, 10, and 12.

This principle, carried one step further, would have rejected the idea of using paper hangings at all, since they were a "totally different surface" imposed on the plaster or on whatever material was used to form the wall itself. However, a nineteenth-century love of ornament kept the reformers from going so far. That last step was to be taken only in the twentieth century.

Models for structurally honest furniture construction were found in medieval forms, as were models for two-dimensional, abstracted flat patterns used in textiles and wallpapers. Pugin, as mentioned on page 341, was an architect and designer of decorative arts who zealously advocated the Gothic style. His writings were the first to popularize its revival in English architecture. Pugin's trend-setting wallpapers for

London's new Houses of Parliament were produced during the late 1840s. Drawn from medieval sources, they incorporated conventionalized two-dimensional motifs. Although these papers influenced English taste in the middle of the century, such papers were very rarely seen in America. The few that were imported *(figure 16-4)* had little effect on American wallpaper design.

In contrast to the Englishmen who aspired to become artist-craftsmen—calling for a revival of craft techniques, rejecting the industrial advances, and emulating designs of the Middle Ages—was a quite different group of English designers. This second group also advocated reform in the design of everyday objects. But they aspired to improve and change the process of designing for the new machines, which they accepted with enthusiasm. They too recognized the importance of understanding the nature of materials. With a slight twist of the craft-oriented logic of Pugin and Ruskin, this group urged that a new generation of industrial designers be trained as true artists who would also acquire a full understanding of machine production and create designs appropriate to the capabilities of the machines.[2]

In contrast to the Ruskinians, they insisted that the design of decorative objects presented an entirely different set of problems from those of the fine arts. Like the Ruskinians, they did feel that good decorative design must be derived from nature, but only by abstracting and conventionalizing its forms. Ruskin, on the other hand, permitted the fine artist to decorate objects with forms as realistically rendered as the portraits in easel paintings.[3]

The multiplication of "Art Manufactures"—one of the period's terms for furnishings, decorations, and other objects that were ornamented—as well as the multiplication of markets for art manufactures, attracted the attention of a new public in England. This public ranged from Prince Albert himself to manufacturers anxious to rival the French products and to newly taste-conscious consumers. The ranks of these consumers extended farther and farther down the social and economic scales while machinery continued to reduce the cost of all furnishings, including wallpaper. In England, periodicals like *The Art Union* (1844–8) and *The Art Journal* (1849) popularized these two sets of ideas—those of the Ruskinians and those of the industrial designers—about the education of craftsmen and designers and the nature of good design. And they focused attention on the quality of the design of English decorative goods made in the factories. *The Journal of Design,* published between 1849 and 1852, called for design reform. It included tipped-in wallpaper samples judged worthy of use and emulation.

One of the industrial designers whose ideas were publicized in these magazines was Owen Jones (1809–74). Jones was one of the leading English designers of the mid-century who tried to improve the design quality of factory-produced goods. Believing that creative contemporary design should build on the accomplishments of the past, he studied ornaments of all cultures and all ages, and in 1856 published selections from each in *The Grammar of Ornament.*

In the preface to his *Grammar,* Jones stated that the first of the "main facts" he had "endeavoured to establish" was "that whenever any style of ornament commands universal admiration, it will always be found to be in accordance with the laws which regulate the distribution of form in nature."[4] His search for universally applicable "natural" laws of design led him to analyze historic ornamentation of every culture. He judged the relative success of each and in the best he found common characteristics, the bases for formulating thirty-seven "propositions," or laws, with which he prefaced his book. He titled this list "General principles in the arrangement of form and colour, in architecture and the decorative arts, which are advocated throughout this work." Several propositions are especially important for a study of wallpaper. The ones that were most often applied in patterns with distinctive characteristics marking them as designs by followers of Jones are:

> Proposition 8: All ornament should be based upon a geometrical construction.
>
> Proposition 13: Flowers or other natural objects should not be used as ornaments, but conventional representations founded upon them sufficiently suggestive to convey the intended image to the mind, without destroying the unity of the object they are employed to decorate.[5]

The influence of these propositions is apparent in the strong geometric framework and abstract motifs of many English patterns of the mid-nineteenth century and in a few wallpapers found in American houses *(figure 16-2).*

Another of Jones's propositions is particularly important in the context of a study of wallpapers, because it influenced the coloring of many patterns. It is Proposition 23, which stated: "No composition can ever be perfect in which any one of the three primary colours is wanting, either in its natural state or in combination."[6] "Natural state" meant pure hue, and "in combination" meant secondary and tertiary hues. The combination of red, blue, and yellow (or its substitute, gold) in many English and American wallpapers of the mid-nineteenth century can be traced to this source.

Owen Jones produced wallpaper designs for several British manufacturers. They were often derived from cultures distant in time—that is, historical styles—or from cultures distant in space, especially those of the Middle East. An admiring English critic in 1876 described Jones's wallpapers as "abstract ornamental forms based upon principles of natural growths but totally devoid of imitative character" that "kept strictly within the bounds proper to mechanically produced ornaments" and had a strongly "intellectual and abstract character."[7] Constance Cary Harrison, an American who published a book called *Woman's Handiwork in Modern Homes* (1881), was less analytical in describing some of his wallpapers:

> Owen Jones' designs are chiefly small berries, fruit, hips, and haws, etc. with flowers and foliage, simple and unpretending, yet most attractive to the eye.[8]

Jones's abstracted and conventionalized style also included patterns made up of simple shapes, like basic triangles, repeated in purely geometric patterns. His style was uncompromising in its adherence to his own principles, and, to some eyes, harsh in its strength of geometry and sometimes of color. Although his wallpapers did not immediately enjoy wide success in America, his *Grammar of Ornament* and its propositions were to have great influence on the wallpaper trade in America during the 1870s and 1880s.

Owen Jones's book of 1856 excited the attentive opposition of the important English critic John Ruskin, whose published work had previously been preoccupied with painting and architecture. However, Ruskin's *The Two Paths* was devoted exclusively to the arts of design and decoration, and much of it responded directly to propositions of Owen Jones. *The Two Paths* appeared in nineteen American editions between 1859 (the same year it was first published in England) and 1891, more editions than for any other book on decorative arts published in America during this period.[9] In five lectures on aspects of decorative design, Ruskin subordinated questions of form and visual principles to moral issues. As Ruskin had "helped convince the majority of Americans . . . that architectural questions had an important bearing on their moral life,"[10] so, in *The Two Paths,* he helped convince them that their standards for designing and purchasing decorative objects had a similarly important relationship to their souls. This book was the basic source for much of the subsequent thinking about applying moral standards to the applied arts.

Ruskin's impact on attitudes toward all the arts was perhaps even greater in America than it was in England. It was Ruskin who first popularized art interest in

16-9

Figure 16-9 Charles Locke Eastlake included this design for a wallpaper among several similar patterns in the first edition (1868) of his *Hints on Household Taste.* Eastlake's conviction that only abstract and two-dimensional flat patterns were acceptable for wallpapers, as well as his strong impulse to simplify (in reaction to the fussy realistic papers of the mid-nineteenth century), are rather clumsily demonstrated in this adaptation of a fifteenth-century pattern. Cooper-Hewitt Museum library

Figure 16-10 (opposite page) Repeatedly published in books and journals celebrating Philadelphia's Centennial Exhibition, this illustration of Walter Crane's elaborate wallpaper composition did much to popularize the English styles in America. Crane's design is here shown as Walter Smith illustrated it in his book *Household Taste: The Industrial Art of the International Exhibition.* Judges at the exhibition found it deserving of a medal for "its great excellence and chastity of design, connected with exceedingly harmonious colouring." The fill pattern "La Margarete" owes an obvious debt to Morris's earlier "Daisy" pattern *(color plate 81).* Smith, like many writers who illustrated this design, explained at length that it was inspired by Chaucer's lines praising the lowly daisy as "she that is of alle floures flour/ fulfilled of all virtue and honour/ and ever alike fair and fresh of hue." He went on to detail Crane's visual allusions to the wifely virtues symbolized by the daisy and by Alcestis, wife of the God of Love, who is depicted in the frieze. The lilies and doves of the dado were recognized as additional allusions to wifely purity. Author's collection

TO WHOM DO YE OWE YOUR SERVICE & WHICH WILL YE HONOUR TELL ME I PRAY THIS YERE THE LEAF OR THE FLOWER
SI DOUCE EST LA MARGARETE

16-10

16-11

Figure 16-11 Another trend-setting illustration of an exhibit at Philadelphia's Centennial Exhibition, which was also frequently reprinted, shows a room designed by the fashionable New York decorators and furniture makers Kimbel and Cabus. Their Americanization of the English Modern Gothic style features a patterned dado, an overly large fill pattern, and a very wide frieze. This illustration is taken from Harriet Prescott Spofford's *Art Decoration Applied to Furniture,* published in New York in 1877. Author's collection

16-12

Figure 16-12 In 1877 Christopher Dresser registered this design in the United States Patent Office. Trained as a botanist, Dresser abstracted from nature, producing patterns as visually flat as his teacher Owen Jones would have had them, but rendered in a personalized style quite markedly by the same hand that illustrated his books (see Chapter 16, footnote 37). Rendering by Roy P. Frangiamore of the Patent Design Photograph at the United States Patent Office

this country, especially during the 1860s, by "incorporating the unfamiliar, art, within the framework of the familiar, religion and nature."[11] In contrast to Jones and his followers, who regarded decorative, ornamental art as something quite distinct from fine art, Ruskin expounded on the principle that furnishings and the decorative arts were legitimate branches of art.

In the first lecture in *The Two Paths* Ruskin proclaimed:

> I have but one steady aim in all that I have ever tried to teach, namely—to declare that whatever was great in human art was the expression of man's delight in God's work.[12]

The full title of this first lecture conveys the import of the thesis Ruskin went on to explore: "The Deteriorative Power of Conventional Art Over Nations." Ruskin equated *sin* with *geometry, formula,* and *legalism* in ornamental design. His prose apparently cajoled good church-going Americans, attuned to the fact that "The Wages of Sin is Death," into accepting his thesis. Ruskin's allusions to Owen Jones could not have been more thinly veiled. Ruskin illustrated his notion that these qualities had a morally deteriorative effect on those who produced and used objects cast in the formulaic molds by pointing to the corrupt and evil "Hindoos," whose sophisticated designs were based on geometry and law, and were devoid of natural feeling. He opposed these people, whom he depicted as a nearly criminal race, to the Scots, whom he held to be models of Christian virtue, and whose crude arts were in direct harmony with nature.

The "two paths" Ruskin offered to aspiring students of design presented a "choice, decisive and conclusive, between two modes of study which involve ultimately the development, or deadening, of every power he possesses . . . the way divides itself, one way [that of studying, following, and celebrating nature] to the Olive mountains, —one [that of adhering to arbitrary rules of geometry and imitation of past models of decorative design] to the vale of the salt sea."[13] Clearly, Ruskin was warning his hearers and readers not to follow Owen Jones into the salt sea. To follow the true path, each student of art must himself experience close, direct confrontation with, and observation of, nature. In the training of a decorative designer, Ruskin gave priority to the kind of training he felt was also the only training appropriate for a fine artist: learning to draw directly from observed nature.

Ruskin saw this approach to teaching design students as an approach diametrically opposed to that of the newly established state-supported schools of design in England. He perceived that their system, strongly influenced by Jones, taught design not as a branch of art, but as a branch of industry. He condemned their teaching methods as false, as based on regularized and hollow theories of design, grounded in geometry and mathematical formulas, and too often derived from historical models rather than from nature.

The wallpaper designer who most successfully took the path to the Olive mountain, while not too drastically offending the laws of Jones's followers, bound for the salt sea though they were, was William Morris (1834–96). The stylized naturalism of Morris's floral patterns *(color plates 81 through 83, figures 16-5 through 16-8),* while relatively abstracted and two-dimensional when compared to full-blown mid-nineteenth-century French renderings of cabbage roses, was much softer and closer to nature than the rigidly conventionalized abstract patterns of Jones and his followers.

William Morris was the most powerful and appealing spokesman for the English Arts and Crafts Movement, which traced its roots to Ruskin. Morris urged artists and architects to apply their minds to the design of objects for everyday use. Not only did he speak to them through lectures and publications; but he also inspired them by the example of his life. A skilled weaver, dyer, illuminator, and printer as well as designer, he drew his designs as well as produced many of them. He encouraged other architects and artists to learn crafts and to design ornamental objects. Many heeded his plea. His poetry, his socialism, and his personality, as well as his patterns and his crusade for the recognition of decorative arts as equal, not "lesser," arts, with those of the painter and sculptor won audiences in America that continued to grow until the turn of the century and beyond.

Morris went directly to nature, as Ruskin would have had him do, to find inspiration for most of his designs. He drew the species of flowers and foliage in forms that could be recognized immediately. He studied the curves that spring from natural growth rather than curves that the mathematician creates with the compass, and incorporated these as gracefully undulating stems and branches that carried flowers and foliage up the lengths of his patterns. The leaves and blossoms were rendered as forms of flat, solid color, usually within strong outlines. While he made limited use of shading to create some sense of depth in his patterns, it was shading produced with stylized lines. The flowers and foliage seem suspended in a very shallow space —a matter of inches on the surface of the wall.[14] In a lecture on pattern design, Morris himself analyzed one basis for the popular appeal of his patterns. He called it his ability "to mask the construction of our patterns enough to prevent people from counting repeats, while we lull their curiosity to trace it out."

In the tradition of Pugin and Ruskin, Morris found in medieval designs the sources for some of his patterns, including his very first wallpaper, "Daisy," designed in 1862 *(color plate 81)*.[15] For inspiration occasionally he also turned to textile patterns in the collections of the Victoria and Albert Museum, but in his best-loved and most characteristic wallpapers he created a personal style derived from natural forms, translated to pattern through his own abstracting process rather than the application of formulaic devices of traditional, historic ornamental styles *(color plate 83, figures 16-5 through 16-8)*. An English critic writing on "Modern Wallpapers" in *The Furniture Gazette* for May 18, 1876, judged Morris's wallpapers lacking in "the intellectual or abstract character of Jones' designs," but granted that they were "sufficiently conventionalized in principle."[16]

During the early 1870s, Morris patterns were known to only a few in America. *Scribner's Monthly* in June of 1873 reported that the Boston firm of Bumstead and Co. had become the American agent for Morris's firm, Morris, Marshall and Co., and offered their wallpapers and other decorative goods for sale. *Scribner's* noted, however, that "these articles are very expensive, and, going exclusively into rich houses, do not influence the popular taste to any perceptible degree."[17] But during the 1870s, the popularity of Morris patterns grew dramatically. In his book of 1878, *Modern Dwellings in Town and Country,* the American architect H. Hudson Holly reported a wallpaper dealer's perception of the beginnings of change:

> I remember going to a paper hanging establishment, a short time since, the proprietor of which, while showing me designs from the famous Morris Company of London, mentioned that the taste of the public was at so low an ebb that it offered but little inducement for their importation. He remarked that the Americans were improving in this direction however; for only a few years ago the worst designs of the European market passed current.[18]

Others had recognized an even more drastic shift in the public taste in wallpapers. As early as 1877, Mary Elizabeth Wilson Sherwood, "daughter of a congressman, wife of a lawyer, sometime novelist, and indefatigable socialite," credited with "giving the stamp of social approval" in America to the new English styles of the 1870s,[19] noticed the popularity of Morris wallpapers as part of a "tasteful Renaissance which seems to have started with the pre-Raphaelite painters in England, and which, like all great reforms, had a kind of simultaneous sunrise in all lands." She judged that Morris's wallpapers had come as part of a "wave" that "has widely bathed the shores" of "that New World which is the Old," a wave which "threw up . . . much valuable flotsam and jetsam" and "taught us properly to appreciate the sincerity and purpose of true Art."[20]

Even that broad public which was deaf to the critics' pleas for change in the *principles* of design responded to Morris *patterns,* if only by buying imitations of his designs. By 1884 the New York magazine *Carpentry and Building* could report:

> It is quite remarkable how quickly the supply for cheap and truly artistic papers has responded to the demand. The paper manufacturers have employed the best artists and have given prizes for good designs. They have taken hints from Morris and his followers. . . . There was a time when if one wanted a good paper for his wall he must pay the enormous prices asked by William Morris and Co. of London. Now he can find quite as good designs as Morris ever made by looking over the stock of any first-class American papers at not more than one-third the price of the no better papers from England.[21]

Morris wallpapers changed the look of wallpapers produced in Europe as well as America. In 1876 *Scribner's Monthly* stated: "The interest excited by the beautiful designs and novel colors of the so-called 'Morris Papers,' for interior decoration, has inspired European manufacturers to the highest achievements in the field of household art."[22]

Morris patterns provided respectable bourgeois Americans with a happy solution

Figures 16-13, 16-14 Christian Herter registered these designs for a wallpaper and a frieze at the United States Patent Office in 1878. No color descriptions were included. The use of light-colored figures against a dark ground is closely related to the color schemes in Herter furniture, where light wood inlays were set off by the ebonized black wood of their backgrounds. The studied and stylish flat-patterned abstraction of this design clearly reflects knowledge of English design theory, but Herter has introduced an ease and sophistication in mixing classical and contemporary sources that is quite distinctive. No examples of wallpaper printed from the patent designs have yet been located. Rendering by Roy P. Frangiamore of the Patent Design Photographs accompanying Patent 11,001, Jan. 28, 1879, at the United States Patent Office

16-13

16-14

16-15

Figure 16-15 E. W. Godwin, one of Britain's leading architects of the 1870s, designed the Japanesque wallpaper pattern shown in the dining room of the Hartford, Connecticut, house built in 1871 for Major James Goodwin by the architect Francis Goodwin. The house was demolished about 1952. This photograph, included in a private publication of 1933, *Rooftrees, or the Architectural History of an American Family,* by Philip Lippincott Goodwin, clearly illustrates the wallpaper pattern printed by the English wallpaper manufacturers Jeffrey and Co. in 1872. The pattern enjoyed fashionable status well into the 1880s, as evidenced by its illustration in the January 1884 issue of New York's *The Decorator and Furnisher.* The original design for the paper in red and black on a light golden-yellow ground is preserved at the Victoria and Albert Museum. Courtesy of the Stowe-Day Foundation, Hartford, Connecticut

16-16

Figure 16-16 Japanese influence is apparent in a wallpaper pattern designed by Louis C. Tiffany. The spiderweb motif seen here also appears in Tiffany's better-known metalwork. This illustration was published in 1880 in the promotional booklet *"What Shall We Do With Our Walls?"* (see frontispiece) that Clarence C. Cook wrote for the manufacturers of Tiffany's wallpaper designs, Warren Fuller and Co. (see Chapter 16, footnote 83). Author's collection

ARTISTIC WALL-PAPERS.

MESSRS. J. S. WARREN & CO. have opened a RETAIL DEPARTMENT in their Manufactory, 129 East 42d Street, near Grand Central Depot, New York, for the sale, at the lowest prices, of their productions of artistic WALL-PAPERS of high excellence.

The designs and colorings are carefully considered, and are not excelled by those of any Foreign Makers; they include the modern styles, with Dadoes, Borders, and Friezes to match, after MORRIS, DRESSER, EASTLAKE, and other authorities on Decorative Art.

They offer their productions at less than one-half the prices that are asked for the same quality of imported goods.

Purchasers are invited to witness the process of Manufacturing.

N. Y. Elevated R. R. Station in front of the Manufactory.

16-17

to the conflict between their love of luxuriant naturalistic detail and their new consciousness of the moral superiority of the more austere abstract wallpaper patterns. Marvelous testimony to middle-class American reaction to these papers appeared in a sentimental novel, *Maude Mohan,* written in 1872 by Annie Hall Thomas. She described

> . . . those wonderful greyish-blue backgrounds on which limes, lemons, and pomegranates with their respective foliages intertwine luxuriantly. How eye comforting and perfect they are in their wonderful admixture of grey and blue "undertone" on which blooming fruits repose, that look as if they were executed by Nature.

And her heroine went on to instruct:

> You have got to learn how to make your walls artistic without the aid of pictures. I am wrong there, for these "Morris Papers" as they are familiarly called, after their inventor, the author of the "Earthly Paradise," are pictures in themselves.[23]

When Annie Hall Thomas used words like "artistic," she was reflecting the Ruskinian doctrine popularized by Morris himself that the decorative arts were equal to the fine arts. When she called the patterns "pictures in themselves," she was revealing not a response to the abstract, conventionalized aspects of the patterns that marked their propriety in respecting the rules for decorative art (as articulated by Jones and his followers), but a response to their more traditional representational qualities. Most of her enthusiasm was devoted to admiration for the colors Morris used—his grayed tones (secondary and tertiary hues), often in palettes that were closely related in value, hue, and tone. These produced that eye comfort she so appreciated, especially in contrast to the bolder palettes of some of the patterns designed by Jones, and the bright colors of the naturalistic French floral patterns.

Figure 16-17 This wallpaper dado designed by Louis C. Tiffany was repeatedly featured, as here, in advertisements published during 1880 and 1881 by New York wallpaper manufacturers. In several other advertisements in which the illustration was used, it was specifically attributed to Tiffany. The Japanesque taste displayed here was part of what a writer in the March 1886 issue of *Dixie* (Atlanta, Georgia) called "the Mikado Style of decoration," adding that it was "the latest craze among wealthy people of aesthetic taste, in the north." Courtesy of Avery Architectural Library, Columbia University

16-18

Figure 16-18 Designed by an Englishman, Luther Hooper, and manufactured by the English firm Carlisle and Clegg, this frieze and dado for a staircase were published in London's *Cabinet Maker and Art Furnisher* in June of 1881 (the source of this illustration) and later in a number of American magazines. Courtesy of Avery Architectural Library, Columbia University

An appreciation of Morris papers that demonstrated closer sympathies with the standards of Owen Jones was published in 1877 by another American writer, Harriet Prescott Spofford. In her taste-making book *Art Decoration Applied to Furniture,* she particularly admired "one Morris paper of idealized jasmine flowers and leaves" for its colors and for the "conventional treatment" that made of the natural model "this new creation, this leading up of beauty into higher reaches by the new combination of the old elements."[24]

Morris wallpapers, reflecting ideas of both Ruskin and Owen Jones, appealed to the students and partisans of each of these writers. If the Morris designs provide visual evidence of forms in which the combination of these ideas met with popular success, Charles Locke Eastlake's book *Hints on Household Taste* furnishes written evidence of the way the gospels of the several English design reformers were summarized and reconciled in a published format that enjoyed phenomenal popular success. Published in England in 1868, America in 1872, the book had a persuasive moralistic tone, derived from Pugin and Ruskin, that struck resonant chords in the souls of an American middle class whose Puritan inheritance was strong. But Eastlake was certainly following in the line of Owen Jones when, in introducing the book, he announced his ambition "to suggest some fixed principles of taste for the popular guidance of those who are not accustomed to hear such principles defined."[25] In doing this, Eastlake professed to speak to the common sense of his contemporaries. He delved into the whole issue of "truth" and "falsity" and devoted a full chapter of *Hints* to "The Floor and the Wall," and illustrated some of his own wallpaper designs *(figure 16-9).* He emphasized that "flat should be decorated with flat ornament," explaining:

> Indeed, common sense points to the fact, that as a wall represents the flat surface of a solid material . . . it should be decorated after a manner which will belie neither its flatness nor solidity. For this reason, all shaded ornament and patterns, which by their arrangement of color give an appearance of relief, should be strictly avoided. Where natural forms are introduced, they should be treated in a conventional manner, i.e. drawn in pure outline, and filled in with flat color, never rounded.[26]

Eastlake's book ranks as the most important practical guide to house furnishing in America during the 1870s and 1880s. It went through seven American editions between 1872 and 1883. Among books dealing with house furnishing and the decora-

16-19

16-20

16-21

tive and industrial arts, none but Ruskin's *Two Paths* was so often published, republished, and brought out in pirated editions in America during this period. In her book of 1877, Harriet P. Spofford described the impact of Eastlake's *Hints:*

> The book met a great want. Not a young marrying couple who read English were to be found without "Hints on Household Taste" in their hands, and all its dicta were accepted as gospel truths. . . . The book occasioned a great awakening, questioning, and study in the matter of household furnishing.[27]

Where Ruskin had provided theory, Eastlake filled in with practical examples and illustrations. That their respective roles as theorist and practitioner were recognized is apparent in such statements as this by an anonymous writer in *Appleton's Journal* for 1873:

> It is the principle of Ruskin, and the practice of Eastlake, that the essential and necessary structure of an object should never be lost sight of, nor concealed by secondary forms or ornament.[28]

The other theorist whose ideas had an equally important part in shaping the thought in Eastlake's book, Owen Jones, was not mentioned by the writer in *Appleton's.* By taking from Pugin as well as from Ruskin both bombastic tone and moral principles, and by combining those ingredients with rules of design derived from Owen Jones, Eastlake was at last able to persuade masses of American wallpaper consumers of the error of their ways. As the popularizer of ideas that had been gathering momentum in England for twenty years, he played a crucial role in weaning American decorators from the French wallpapers. No longer could Americans in good con-

Figures 16-19, 16-20, 16-21 Wallpapers in the parlor of the Bush house, built in 1877–8 in Salem, Oregon *(figure 16-19),* include a frieze, 18 inches wide, machine-printed in pastel blue and pink and in metallic gold and silver over a creamy white ground on paper textured like paper toweling *(figure 16-20).* The matching side-wall pattern, like the frieze, is marked by Leissner and Louis (active New York 1872–82). Courtesy of the Salem Art Association, Salem, Oregon. *Figure 16-21,* Cooper-Hewitt Museum, 1972-21-3

science fail to regret their mistake in falling prey to the seductions of the French craft that was in the eyes of its own masters "before all else an art of falsification" which "should never give lie to its first destination."

During the 1870s, Eastlake was not the only popularizer of the terms and standards for evaluating design that had been used by a community of English critics since the 1840s. In 1873, Wilmot Pilsbury, a less important English critic, published an article titled "Principles of Decorative Design" in *The Furniture Gazette.* In that article, he included two pictures of wallpaper designs *(figure 16-1)* that serve as unusually clear demonstrations of the standards of "truth" and "falsity" as they were regularly applied to wallpaper in popular publications of the 1870s and 1880s.

He analyzed in detail the one on the left as a demonstration of the "true" principles of "suitability" (or modification of motifs to ornamental purposes) and "conventionalization" (in a conventional style, abstraction, or stylization of motifs). He praised the designer of the paper on the left for his use of symmetry, and for "adapting the treatment to the purpose of decorating a flat vertical surface . . . there is no indication of projection or relief . . . there is a design and intention throughout, the treatment is purely ornamental, and suited to the purpose to which it is to be applied." The pattern on the right, Pilsbury's figure 4, was lambasted as embodying all the false principles:

> A few flowers, which appear accidentally grouped together into a bouquet, have been copied as literally as the nature of the process and cheapness would allow. The purpose of a wallpaper seems never to have been considered, hence every essential condition is disregarded. The forms are rounded with light and shade, so as to appear to project from the surface of the paper . . . The flowers look as well sideways as upright . . . It is evident that in fig. 4, the leading aim has been literal imitation. It contains no design, and strictly speaking, it is not ornament at all.[29]

By the 1870s, thinking like this had become part of a common body of critical principle, and it was being brought to America not only in publications, but also by visitors and immigrants. A transplanted English design school principal, Walter Smith, became professor of art education in the City of Boston Normal School of Art, and state director of art education. In 1872, he published a book called *Art Education, Scholastic and Industrial.* It was filled with teachings current in the English design schools, where the principles of Owen Jones were basic, but not untouched by Ruskin. A sample of such teaching appeared in Smith's advice:

> An imitative natural treatment of design will be found to gratify the tastes of the young and the ignorant, as it does also that of the savage and the *roué* worn-out taste of a frivolous or luxurious age. Nature, when simply copied as ornament, suggests the incapacity of the designer, as well as his ignorance of historical methods; nature conventionalized is evidence both of knowledge and originality, and in all industrial art will be found adapted to its requirements and satisfying the most refined perceptions.[30]

As such analytical thinking about "art manufactures" gained currency in a suddenly design-conscious popular press, its application to wallpapers manufactured in America became widespread. The abstractions and conventionalizations of wallpaper motifs might follow the style of the Middle Ages, the Renaissance, or of the ancient Greeks, the Japanese, the Moors, or still other cultures and times. But whatever the style, the public was taught to demand "conventionalized" abstracted patterns for wallpapers. Such patterns, it was told, gave evidence of the imposition of the human mind on the forms found in nature. For, as Harriet Spofford put it in phrases derived from Ruskin: ". . . the natural representation, taken and transformed . . . conventionalized—has gone through an organic process; it has gone through this organic process in the human brain, and in becoming idealized is as much finer, loftier, and nobler as the living soul is better than dead matter."[31]

Conventionalized wallpaper patterns designed by artists who had been indoctrinated with such principles scored a long-lasting triumph at one of the major celebrations of American life in the 1870s: the Centennial Exhibition at Philadelphia in 1876. British exhibitors revealed their new consciousness of the aesthetic importance of "art manufactures" or "objects of everyday use" in a range of products including embroideries designed by William Morris and Walter Crane, executed by London's Royal School of Art Needlework, and wallpapers exhibited by Jeffrey and Company *(figure 16-10).*

Jeffrey mounted an exhibit that stood out among the international showing of wallpapers, and it received awards and high praise in official reports. The papers were described as "works of art in paper decoration, which display the highest and purest taste. Such productions deserve special recognition, and tend to elevate paper as a decorative article."[32] Among the designers represented in Jeffrey's display were William Morris and Walter Crane. An American who wrote a volume of *Reports and*

Awards at the Centennial commented that this notable exhibition from Great Britain "might well serve for a suggestive model to our designers of decorative paper."[33]

This British wallpaper success assured the popularization of tastes and design ideas that had gained only limited recognition in the American wallpaper trade prior to their exposure at the exhibition. The English wallpapers did indeed serve as "suggestive" models that transformed the look of many wallpapers made in America.

THE ANGLO-JAPANESE STYLE

One important English designer who visited the Centennial Exhibition was Dr. Christopher Dresser (1834–1904). Trained as a professional botanist, Dresser recognized Owen Jones as his mentor in the arts of design. Dresser's enthusiasm for scientific progress and for the machine stood in sharp contrast to the abhorrence of the machine among participants in the Arts and Crafts Movement.

When he visited Philadelphia in 1876, Christopher Dresser was en route to Japan to pursue his interest in Japanese design. He had begun to collect Japanese prints after seeing examples at London's international exposition in 1862. While Owen Jones had been more taken with designs from the Middle East—the "Hindoo" work so derided by Ruskin—Dresser was one of a growing number of artists and designers in London who were intrigued by the abstract qualities in Japanese art and decorative arts. Among the important figures who were also collecting and studying the formal qualities of things Japanese were the painter James McNeill Whistler (1834–1903) and the architects Edward William Godwin (1833–86) *(figure 16-15),* and William Burges (1827–81) *(color plate 87),* as well as the designer and writer Bruce J. Talbert (1838–81). During the 1870s, Godwin and Burges began to produce handsome Japanesque wallpapers that were the inspiration for a style commercialized and known in the American wallpaper trade as "Anglo-Japanese" or, more flippantly, "Anglo-Jap."

While visiting the Centennial Exhibition in Philadelphia, Dresser perhaps worked out a business arrangement with the local wallpaper manufacturers Wilson and Fenimore. Whenever the arrangements were made, in the year following the centennial, the name of this Philadelphia firm appears as "assignor" to thirteen wallpaper design patents (numbers 9,975–9,987) registered in the United States Patent Office by Christopher Dresser of London. Each design demonstrates Dresser's interest in the basic geometry of plants, which he showed in configurations organized within simple and bold geometric structures *(figure 16-12).* Visually, in their flatness, sharp linearity, and conventionalization, the descent from Owen Jones is evident. Dresser's reliance on Japanese motifs is apparent in the wallpaper shown in *color plate 92.* Dresser's name became widely known in this country during the 1870s and 1880s, not only in the East Coast centers, but also in centers far to the west. In St. Louis during 1878, one wallpaper dealer thought the name of Dr. Christopher Dresser was well known enough and commanded enough admiration to warrant its use in a promotional description of his store, featuring "Dr. Dresser's Art Designs."[34] The American authors H. Hudson Holly and Constance Cary Harrison praised his wallpaper designs—Harrison coupled the name of this industrial designer with that of the great artist-craftsman when she mentioned that "wallpapers . . . of Dresser and of Morris are familiar in our houses."[35] In 1888 the New York trade magazine *Decorator and Furnisher* carried an article entitled "Ceilings, Walls and Hangings: Some Modified Views by Dr. Dresser."[36] His books on the principles of design were sold and republished in this country.[37]

The Japanese designs Dresser found so fascinating and suggestive also influenced American designers. One of America's leading decorative artists, Christian Herter (1840–83), turned his attention to the design of wallpaper during the 1870s. German-born, Herter had studied in Paris with Pierre Victor Galland (1822–92) before assuming responsibility for Herter Brothers of New York, makers of furniture and decorators of houses (active 1865–after 1900). After Christian Herter became head of the firm in 1870, it reigned supreme as the leading manufacturer and purveyor of fine decorative goods in New York. Herter even produced architectural designs as well as decorative schemes for a pair of houses built for William H. Vanderbilt on Fifth Avenue in New York. Other super-rich clients of the 1870s included J. P. Morgan, Jay Gould, the Carters of Philadelphia, and Mark Hopkins of San Francisco.

There is a good record of the firm's wallpaper patterns because in 1878 Christian Herter registered twelve designs for wallpaper at the U.S. Patent Office (Patents 10,990, 10,992–10,998, 11,001–11,003), and illustrations of those patterns are still on file. All the designs were abstract flat patterns, showing the influence of the new English designs. Some were Japanese in style, one a design of peacock feathers. Two of these designs at the patent office show fill paper and frieze with light-colored flat patterning of floral and animal motifs on a dark ground. The patterns relate closely to the inlaid wood patterning that is distinctively characteristic of the furniture made by the firm *(figure 16-13, 16-14).*

Another American whose decorative designs in other media brought him fame, the artist and glass maker Louis Comfort Tiffany (1848–1933), also produced wallpaper patterns in the Japanesque style *(figures 16-16, 16-17).* Some were quite successful, but others, like the dado pattern shown in *figure 16-17,* look like standard exercises in what was fast becoming a hackneyed commercial formula by the late 1880s, "Japanesque" being almost any arrangement featuring asymmetry, particularly as it was imposed on an otherwise horizontal and vertical patchwork of shapes that was suddenly interrupted by diagonal slashes. The inclusion of representations of Japanese fans, vases, and kimono-clad figures also marked a paper as "Japanesque" within this commercial vocabulary. The worst examples of its use completely disregarded the features of Japanese art that London collectors, painters, and designers had studied with such appreciation. In addition to the fact that the originators of the Japanesque style had gained a new appreciation of asymmetry from this non-Western source, they found that in Japanese design all the principles Jones and Eastlake had spelled out as essential for decorative design had been understood and demonstrated through ages of Japanese fine and decorative arts. These included the reliance on natural forms for design inspiration, the valuation of abstraction over imitation of nature, the flattening of space to create strongly patterned flat surfaces, the application of color in flat, unshaded solids, the importance of strongly outlining shapes to prevent any *trompe l'oeil* illusions imitating or attempting to reproduce the forms or spaces seen in nature. Eastlake's dicta about wallpaper pattern design seemed to describe precisely countless textile patterns that were standardly produced in Japan, and from which the West began to borrow freely. Eastlake had written in 1868:

> all shaded ornament and patterns, which by their arrangement of color give an appearance of relief, should be strictly avoided. Where natural forms are introduced, they should be treated in a conventional manner, i.e., drawn in pure outline, and filled in with flat color, never rounded.[38]

The English instigators of the interest in Japanese art advocated study and application of Japanese design *principles,* and warned against slavishly copying the actual designs. However, these English designers did not practice what they preached, and they themselves not only bought Japanese textiles, but also directly copied them in wallpaper designs, reproducing Oriental patterns, or elements from them, line for line. American wallpaper manufacturers also made some direct copies or close adaptations of the Japanese textile prints *(color plate 90, figures 16-18 through 16-21),* but more often motifs were borrowed; asymmetry was introduced into the layout for a pattern; and admixtures were then derived from almost any source *(color plates 95 through 97),* whether or not their forms were congenial with those taken from the Japanese. The Americans learned to appreciate Japanese designs not by importing them across the Pacific, but by seeing their imitations in avant-garde British design of the 1870s, which interpreted and adapted the Japanese forms in ways far removed from their Japanese sources. Therefore, American writers of the 1870s and 1880s were quite accurate when they called the style "Anglo-Japanese."

THE WALLPAPER CRAZE IN PRINT

A great many writers were describing these wallpapers in the astonishingly large number of books, articles, and pamphlets on interior decoration that were published during the late 1870s and 1880s. All featured advice on wallpapering. In daily newspapers and popular journals, in books on architecture, in a proliferation of architecture and building trade journals, as well as in the new journals devoted to interior decoration—like *The Decorator and Furnisher,* whose first issue appeared in New York in October of 1882—wallpaper was a subject given an almost unbelievable amount of space.

The first publication in America of Eastlake's *Hints,* an event of 1872, has already been singled out as a milestone. Another important precedent was the publication of Charles Wyllys Elliot's series of articles on "Household Art" in *Appleton's Art Journal* during 1875 *(figure 16-22).* That series was followed by similar ones by other authors in journals of more general interest, like *Appleton's Journal* and *Harper's.* Elliot's articles were among the first on such subjects later to be expanded to book form. In 1876 much that was included in those articles appeared in *The Book of American Interiors* by Elliot. Almost every major American decorating manual of the 1870s and 1880s had first been published in one of the leading journals. Elliot introduced another feature that subsequently became important in American publications on house furnishing: he illustrated American rooms furnished and papered in the new English manner.

After he had first produced it as a series of articles, run in *Scribner's Monthly* between November 1875 and April 1876, Clarence Chatham Cook (1828–1900), a pioneer in the professional criticism of art in America, published his influential book,

16-22

Figure 16-22 In New York's *Art Journal* of 1875, Charles Wyllys Elliot published this simple dining room. Although Elliot judged flat tints of color more desirable than patterned wallpaper, he used this illustration to demonstrate that "Most pleasing effects may be had at little cost by papering the space above the picture-moulding with a gay pattern of birds or flowers." Courtesy of the Avery Architectural Library, Columbia University

Figure 16-23 (opposite) A frieze and a trellis pattern dominate a room illustrated in Clarence Cook's *The House Beautiful* (New York, 1878). Strongly Japanesque in style, they reveal the influence of the Japanese love of abstracting from nature while still representing her beauties in two-dimensional, conventionalized forms. In the 1880s the taste for such papers became amazingly widespread. Author's collection

"A bed is the most delightful retreat known to man."
No. 86.

16-23

16-24

Figure 16-24 (opposite) In 1883, Warren, Fuller and Lange (successors to Warren, Fuller and Co.) published a new edition of "*What Shall We Do With Our Walls?*" and inserted a new frontispiece, the illustration shown here. The $1,000 prize design by Mrs. Candace Wheeler shows a frieze, fill, and dado in each of which bees are incorporated. A honeycomb provides the background for the fill, and hives among clover dominate the dado, while a few stray bees buzz among the trailing foliage of the frieze (see page 426). Courtesy of the Art and Architecture Division, New York Public Library, Astor, Lenox, and Tilden Foundations

Figure 16-25 The layout for this advertisement demonstrates the favored scheme for wall division during the 1870s and 1880s: Frieze at cornice level, fill paper below that, and dado below chair rail. Each major pattern area was finished with patterned borders. The advertisement appeared in the New York magazine *The Decorator and Furnisher* of May 1883. The contemporary importance of English and Japanese designs and of "relief decorations" is indicated in the text. Courtesy of the Avery Architectural Library, Columbia University

16-25

The House Beautiful (1878). The frontispiece Cook used was an illustration by the British artist Walter Crane (1845–1915) of a stylish wallpapered interior in the new manner of the Aesthetic Movement. This was the movement to integrate art and life, to bring art into the house, to interest artists in the crafts, craftsmen in the fine arts, and the general public in both. It began in England during the 1860s. The English designers and critics mentioned above were all part of this Aesthetic Movement, and Clarence Cook knew their work. The spirit of the movement was reaching these shores when Cook published illustrations like that shown in *figure 16-23* and when he observed:

> among the smaller facts that must be taken note of in drawing the portrait of these times is the interest a great many people feel in everything that is written on the subject of house-building and house-furnishing. There never was a time when so many books, written for the purpose of bringing the subject of architecture . . . down to the level of popular understanding were produced as in this time of ours. And, from the house itself, we are now set to thinking and theorizing about the dress and decoration of our rooms . . . and books are written, and magazine and newspaper articles, to the end that on a matter which concerns everybody, everybody may know what is the latest word.[39]

In the same year Cook's *The House Beautiful* appeared, H. Hudson Holly's *Modern Dwellings in Town and Country* (1878) was full of praise not only for the wallpapers of Christopher Dresser cited above, but also for those of Pugin, Eastlake, and J. W. Talbert. Hudson noted an improvement in American taste in wallpapers—which apparently meant an acceptance of the English styles.

By 1879, trade journals like New York's *Carpentry and Building* were reporting:

> Within a very few years . . . the status of wallpaper has changed as completely as it is possible for any art to change. From being the last resort for covering a defaced wall, it has come to be recognized as a decorative material of the highest value . . . in no direction is the growing artistic taste of the people at large more manifest than in the stimulus which the demand for tasteful paper-hangings has given to the manufacture of wall-papers.[40]

During the 1880s the popularity of wallpaper continued to grow along with the popularity of decorating literature that discussed wallpaper.

16-26

16-27

Figure 16-26 (left) *Carpentry and Building* for April 1879 carried this illustration, furnished by the Robert Graves Company of New York, describing it as "a characteristic representation of a wallpaper treatment as applied to a hall and stairway. In style . . . known as Anglo-Japanese, possessing some of the features of Japanese art, but in composition and treatment conforming to the ideas of English and American artists." The fill pattern is a close variant of Brightwen Binyon's design that Eastlake illustrated in the fourth (1878) edition of *Hints on Household Taste.* Courtesy of the Avery Architectural Library, Columbia University

Figure 16-27 (right) The M. H. Birge Company of Buffalo, New York, used this stair hall to advertise their papers in *The Decorator and Furnisher* of December 1884. The taste of the 1870s and 1880s for horizontal wall division with different patterns in each division is here seen carried to an extreme. Further use of patterns and borders on floors and ceilings might seem the figment of an advertiser's fancy, but they were present in many middle-class homes *(figures 16-35 through 16-37).* Photograph courtesy of the Art and Architecture Division, New York Public Library, Astor, Lenox, and Tilden Foundations

Many factors contributed to this outpouring of words, accompanied by illustrations, on a subject that had never before aroused much attention in the popular press. High-ranking among these factors were general prosperity and the rise of a new generation of house builders and homemakers who were moving in ever-increasing numbers into the cities. A writer in New York's *The Art Journal* observed in 1880: "In the ordinary dwelling, each new occupant of a house feels himself at liberty to choose the style of his wall adornments."[41] In that same year, 1880, Clarence Cook provided additional insight into the popularity of wallpapers:

> With our American migratory habits and love of getting out of one house and into another, it is a great recommendation that our new dwelling can be made to change its coat.[42]

Reduced publication costs, the development of new advertising techniques, and the great technological advances that had lowered the costs of all house furnishings coincided to encourage the multiplication of printed words about decorating. The ordinary householder was urged to add to his old, relatively simple concern for shelter, comfort, and personal pleasure in things he found pretty a new concern for fashion and taste, once matters of interest only to the rich. In addition, as discussed earlier, moral issues had been introduced to the discussion of house furnishing, providing additional incentive for reading this literature.

The 1880s also witnessed the publication of expensive books on decorating, produced for the not-so-ordinary housekeeper. *Artistic Houses,* one of the most ambitious of such books, was issued in 1883 in two expensive, oversized volumes, made up of exquisitely detailed photographs showing the most important and lavish interiors of America's financial, political, and social leaders, as well as of a few individuals

Figure 16-28 (opposite, top) *Carpentry and Building* included this illustration in the issue for December 1880. The text in the article "Wall Papers," in which it appeared, described it: ". . . two examples of how a ceiling may be treated, or rather it shows two different kinds of paper applied to the surface. One is a Japanese design consisting of feathery sprays of ferns, and the other of dots of gold, which form a small and regular pattern on the paper when seen near the eye. Both are very pretty, and both produce a very pleasant effect. In this cut the solid black portions represent the gold. It will be noticed that the ground work of both the centers and the borders is gold. The effect in the paper is quite different from that in the engraving, being exceedingly rich and brilliant." Courtesy of the New-York Historical Society, New York City

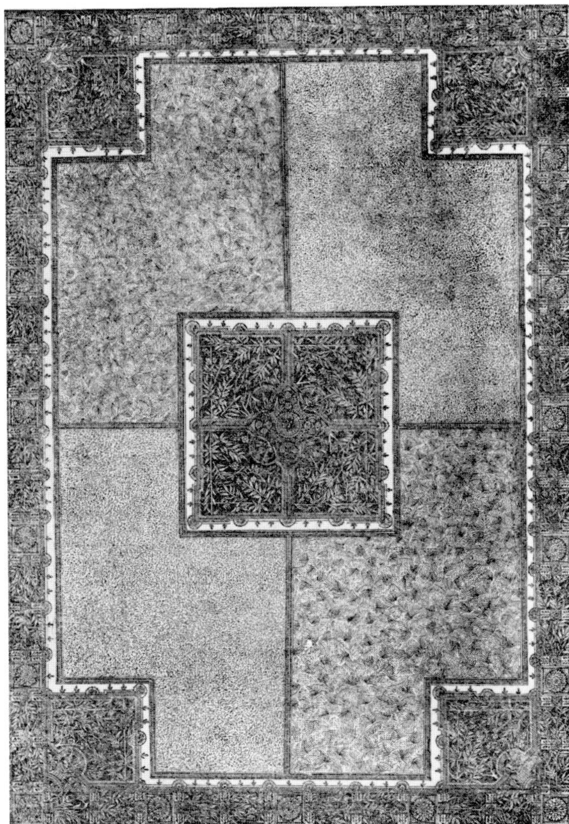

16-28

16-29

16-30

Figures 16-29, 16-30 The ceiling in the parlor of the Captain Edward Penniman house in Eastham, Massachusetts, on Cape Cod, is here illustrated in two photographs: *Figure 16-29* (top) shows about half of the wallpapered surface. *Figure 16-30* (bottom) shows a more detailed view of the left-hand corner and also includes portions of borders not visible in *figure 16-29.* Courtesy of the National Park Service, North Atlantic Region, Boston, Massachusetts

16-31

16-32

Figure 16-31 (above) Egyptian motifs, vividly colored in blues, violet, black, and gold, make this 1870s paper particularly appropriate for the setting in which it was found—Olana, the palatial Hudson River home built in 1870 by the artist Frederick Edwin Church (1826–1900). The house features a mixture of Middle Eastern architectural styles. Courtesy of the New York State Office of Parks and Recreation

Figure 16-32 (right) This 1880s pattern, machine-printed in metallic gold, maroon, turquoise, green, and blue shading to pink, typifies the generalized Middle Eastern style as it was commercialized in America. It is a far cry from the geometric flat patterns of the "Moresque," "Alhambran," Turkish, Indian, and "Hindoo" styles in Owen Jones's *Grammar of Ornament.* Courtesy of Sunworthy, Wallpaper Manufacturers of Canada

Figure 16-33 (below) Classical styles were not totally ignored during the era of the Aesthetes, but they were sometimes exaggerated and transformed, as here. This very wide frieze (20½ inches) has a late-nineteenth-century color scheme—a tan ground with deep red flocking, the pattern in shades of brown, pastel pinks, lavender, pale yellow, and metallic gold. This is an unused remnant of paper in the Bush house, Salem, Oregon. Made in New York by Beck and Co., it was hung in the 1880s. Cooper-Hewitt Museum, 1972-21-2; gift of the Salem Art Association, Salem, Oregon

Figure 16-34 (right) Floral patterns with scrollwork like this enjoyed some popularity through all the extremes and shifts in taste during the last quarter of the nineteenth century. This remnant of a pattern hung about 1880 in a house in Annisquam, Massachusetts, is 17 inches wide, machine-printed in white and cream on a mustard-yellow ground. It bears the mark of Howell and Brothers of Philadelphia, a company that began in Albany, New York, in 1813 and by the 1880s was the country's largest wallpaper manufacturer. Cooper-Hewitt Museum, 1940-89-1; an anonymous gift

16-34

16-33

from the world of art. The rooms included ranged from those of the artist Louis Comfort Tiffany to the splendid halls of the palatial residences of several Vanderbilts, of Marshall Field, General Grant, and Mrs. A. T. Stewart. Pattern had a prominent place on the walls of these trend-setting, densely ornamented establishments—patterns in leather, in silk, in carved wood, and in embroidery, and patterns on less expensive wallpapers as well. George W. Sheldon's accompanying text noted patterned wallpapers imported from Japan as well as papers designed by leading figures in the English art world, including Crane, Dresser, and Morris.[43]

Introductory paragraphs in the new wave of decorating publications echo each other in a chorus of self-congratulatory recognition of the birth of awareness of the "artistic" in decoration. They hail a "great reform," an English-inspired "Renaissance" that was sweeping away the bad old designs of the past. *Frank Leslie's Illustrated Newspaper* for May 27, 1882, proclaimed:

> The demand for high art in wallpaper has arrived at the financial dignity of a boom, while the utter cessation of the sale of what used, here to fore to be in frantic demand, proves in the most direct manner the progress of the art wave that is now happily passing over this vast continent.[44]

Wallpaper advertisements bristled with the word "Art." Adjectives like "quaint," borrowed from writings about Queen Anne architecture in England—the style most popular there during the 1870s—punctuated the sales pitches of wallpaper sellers. Phrases like "Wallpapers of Quaint Design and Colors by English Decorative Artists" characterized advertising text along with the names of specific English designers. English patterns and English insistence on abstract, flat-patterned wallpapers had finally swept aside the fashion for French wallpaper imitations of textiles and eye-fooling panoramas.

This Anglicization of the American wallpaper trade received a well-publicized boost when Oscar Wilde (1854–1900) made an eighteen-month lecture tour of America during 1882 and 1883. Wilde had absorbed many of the ideas of London's avant-garde decorative designers of the 1870s. A student of Ruskin and an admirer of Morris, Wilde adopted their ideas about the importance of art in the life of "everyman." He admired Japanese design, and took as his favorite decorative motifs the sunflower, the lily, and the peacock feather so frequently seen in Japanese work.

Wilde was the darling of the London salons. His affectation of a peculiar style of dress—knee britches, velvets, warm rich shades of green, brown, and gold—and his often repeated ecstasies over one or another of his favorite flowers, which he was wont to carry in hand as he went about London, made him an easy target for cartoonists. His association with celebrated theatrical stars won for him a prominent place in the English press. He had published a bit of poetry deemed stylish by some of the rich and famous, but his fame as a writer came later; *The Picture of Dorian Gray* was not to appear until 1891. When he came to America in 1882 he had gained a reputation as a critic and lecturer and as a source of highly quotable phrases. Sweeping through New York customs, he told officials, "I have nothing to declare but my genius." The papers repeated it the next day, and for several months spiced their pages with accounts of his further adventures and with drawings showing him in the wilds of America arrayed in full aesthetic garb. To such publicity Wilde would reply, "Caricature is the tribute which mediocrity pays to genius."[45]

The self-styled genius was not an artist or designer, but he played an important role in popularizing the English Aesthetic Movement, with its ideas about design and decoration summarized in the preceding pages. In America, he gave voice to the ideas of the more original thinkers and designers with whom he had associated in London. His fame drew crowds—many of whom doubtless viewed him as a freak but who nonetheless came to hear this superstar, 1880s style, in New York, San Francisco, Denver, Atlanta, and countless stopping places in between. He lectured to Americans on "House Decorations," "The English Renaissance," and "Art and the Handicraftsman." As he urged the public to take art into their homes, he praised the new English wallpapers.

The popular press in America recognized his impact as the evangelist of the English art movement. *Frank Leslie's Illustrated Newspaper* for May 27, 1882, featured "The Aesthetic in Wallpaper Manufacture" in a lead article. The London-born aesthetic ideas had indeed arrived at the most popular level in America when such a publication would print:

> Thanks to aestheticians and Oscar Wilde, the hideous wallpapers that used to make us feel horribly bilious, and cause one's strained and tortured eyes to ache again, are happily doomed, and, instead of abominable stripes out of all harmony, or aggressive squares, hard and uncompromising as tombstones, our homes will, in the near future, be beautified with tender half tones, with delicate, dreamy tints, while patterns born of the glories of the sun flower will gradually merge into the chaste and delicious nuances of the peach, the apple, or the hawthorne blossom. A New Era has dawned.[46]

During that "New Era" the saturation of the American wallpaper industry with English ideas about design was nearly total. Ranging from the most expensive English imports through the cheapest products of the American factories, the forms and styles of the 1880s wallpapers came from, or by way of, London. Sometimes the patterns Americans put on their walls were grotesque derivations from the London originals. But in no other era during the nineteenth century was there broader dissemination of distinctive styles to an enormous popular audience, which had been indoctrinated with a given set of design rules. It is fascinating to see the seminal designs diluted and distorted, yet still distinctive, as they seeped down through the trade and traveled across the great American continent as prints on cheap wallpapers.

Following the English precedent, major American figures in the art world were persuaded to design for American wallpaper manufacturers. Warren, Fuller and Company of New York capitalized most effectively on the English idea, purchasing the designs of Louis C. Tiffany *(figures 16-16, 16-17),* and of Samuel Colman (1832–1920), both of whom were well established as painters as well as decorators. Tiffany's stained glass was but one aspect of the decorating services he provided in association with Colman and others.

Warren, Fuller and Company featured the designs of Tiffany and Colman in the little book *"What Shall We Do With Our Walls?",* whose cover is shown in the front of this book *(figure 0-1).* Its author, Clarence Cook, praised their wallpapers by claiming that "in another field, the Messrs. Warren, Fuller and Co. are doing the same service for the decorative arts here in America that the Wedgwoods did in England."[47]

Cook's booklet for Warren, Fuller and Co. went through several editions after its initial appearance in 1880. In the 1884 edition, the text was illustrated with patterns that had received awards in a contest the company had sponsored during 1881. The company had offered $2,200 in cash awards to stimulate artistic efforts in the production of original American designs. Since she was already working as a partner with Tiffany and Colman in their decorating firm, Associated Artists, it is perhaps not coincidental that Mrs. Candace Wheeler won the contest *(figure 16-24).* The third prize in Warren, Fuller and Company's wallpaper design contest went to Mrs. Wheeler's daughter, Dora. Among the many published notices of this "contest" and its results was a full three-column article run in *The American Architect and Building News,* which reviewed all the entries, and noticed rather pointedly that of the top prizewinners "all . . . have been accustomed to design for the firm in question."[48]

Important American architects also turned their attention to the design of wallpaper, as English architects of the period were also doing. In Chicago, the *Inland Architect* in 1883 reported that the architect John Welborne Root had designed a set of wallpapers for the Chicago manufacturers of paper hangings, John J. McGrath Co., "comprising a dado and border, a wallpaper and frieze" deemed by the writer for the *Inland Architect* "characteristic, bold, and artistic."[49] No examples of Root's designs have been located.

American architects who were not designing wallpapers were recommending English patterns in the homes of their clients. Richard Morris Hunt used friezes by William Burges and Brightwen Binyon (dates unknown) at Château-sur-Mer in Newport, Rhode Island. Hunt enlarged and remodeled the house for George Peabody Wetmore beginning in 1872 and over the next several years. During this period Wetmore was, for the most part, traveling in Europe on an extensive honeymoon. So the choice of wallpapers almost undoubtedly reflects the architect's, rather than the client's, taste *(color plates 85, 86, 87, 88, 89).*[50] Peabody and Stearns used William Morris fabrics and wallpapers at Vinland, a brown stone villa built in Newport in 1883 for Miss Catharine Wolfe. At H. H. Richardson's Glessner house of 1885 in Chicago, in addition to the Morris patterns shown in the bedroom illustrated as *figure 16-8,* Morris patterns were used in almost every other room, either on walls as paper, at the windows as curtains, or on furniture as upholstery. The California architects Newsom and Newsom, who operated at a much more popular level, publishing plans for houses to be copied by large numbers of middle-class Americans, advised readers of *Picturesque Californian Homes* (1885) to make their homes beautiful with wallpaper, noting that a house was considered incomplete without the adornment of paper.[51]

WHERE AND HOW PATTERNS WERE HUNG

With wallpaper attracting so much attention and comment, authors published widely divergent opinions about which kinds of papers should be hung in which kinds of rooms. They disagreed in print about which historical styles were preferable (classical or medieval), about the superiority of Japanese or Moorish patterns, about whether flocking, imitation leather, embossed composition, or printed paper was better suited for use in living room, parlor, or library. Where one would-be arbiter

16-35

Figure 16-35 The music room of the Paddock mansion in Watertown, New York, was fashionably papered and draped when this late-nineteenth-century photograph was made. The figured wallpapers on the ceilings and walls provide only a background for the patterns on heavy portieres, curtains, carpets, table covers, and upholstery. Peacock feathers have been fashionably tucked behind a frame on the left-hand side of the arched doorway, and Spanish moss hanging over the arch provides the finishing touch. Courtesy of the Jefferson County Historical Society, Watertown, New York

16-36

16-37

Figure 16-36 (opposite, below) This is the parlor of the Ramsey House in St. Paul, Minnesota, as it was photographed in 1884. The ceiling is devoid of pattern, but pattern dominates the furniture, hangings, floor, and walls. While the room lacks some of the standard trappings of "aestheticism" such as a wallpaper dado, and while some of the mid-nineteenth-century rococo revival furniture would have been considered old-fashioned in 1884, the wide wallpaper frieze, horizontally banded curtains suspended by rings from rods, and a Japanese screen are sure signs of "aesthetic consciousness." Courtesy of the Minnesota Historical Society

Figure 16-37 (above) "In an age when a distinguished architect writes of the 'conscientious in decorative art' we may well speak of a conscientious room," Mary Elizabeth Wilson Sherwood wrote in 1879 in New York's *Art Journal.* This is just the kind of room she had in mind. It is the music or drawing room of the John Bond Trevor mansion, built in 1877 in Yonkers, New York, and photographed about 1894 in its full-blown splendor reflecting high-style decorative ideas of the 1870s and 1880s. A leading New York firm, Leissner and Louis (active 1872–82), is known to have executed the Japanesque painting on the ceiling, and perhaps made the papers as well. Courtesy of the Hudson River Museum, Yonkers, New York

IDIOSYNCRATIC FURNISHINGS.

THE FRIEZE.

THE FRESCO.

16-38

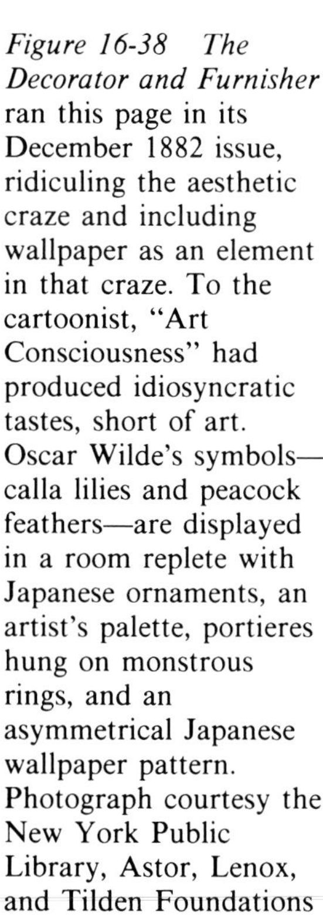

Figure 16-38 The Decorator and Furnisher ran this page in its December 1882 issue, ridiculing the aesthetic craze and including wallpaper as an element in that craze. To the cartoonist, "Art Consciousness" had produced idiosyncratic tastes, short of art. Oscar Wilde's symbols—calla lilies and peacock feathers—are displayed in a room replete with Japanese ornaments, an artist's palette, portieres hung on monstrous rings, and an asymmetrical Japanese wallpaper pattern. Photograph courtesy the New York Public Library, Astor, Lenox, and Tilden Foundations

of taste would insist that, as everyone knew, flocked papers were the only ones suitable for dining rooms, another would rail against the use of flocking in rooms where food was to be served, since flocking attracted odors and dirt more readily than other wallpaper surfaces.[52]

While disagreement on detail was widespread, certain general rules about wallpapers appear to have been nearly universal in America during the period. The idea that only flat, abstracted patterns were permissible was so pervasive that it could be the butt of a joke. Similarly, the Ruskinian notion that it was both possible and necessary to judge patterns in moral terms received such widespread acceptance that it became a platitude, susceptible to ironic comments by writers. This time had come in 1884, when a writer in *Carpentry and Building* reviewed the "tawdry" papers of mid-century, then observed with self-mocking humor: "We know better now. We are a people getting to despise imitations. We will be genuine, if we have bare and gray walls, or such paper only as our grocer wraps about bundles."[53] Clarence Cook, a writer who was basically a Ruskinian although he generally applauded the principle of using only abstract flat patterns, sometimes felt it necessary to assure his readers that it was acceptable, on occasion, to break the rule: "We are at liberty to choose where we will employ them, and where we will not."[54]

If during the 1870s and 1880s most taste makers, as well as manufacturers and consumers, agreed about the superiority of honest flat patterns to picture papers, there was even more general agreement about the advantages of dividing walls into horizontal sections, composed of a wide frieze at cornice level, a dado on the lower portion of the wall—below the chair rail and usually about three feet high, and a "filling" or "screen" of wallpaper between the two. Any and all of these elements were often furnished with additional border patterns *(figures 16-24, 16-25)*.

This horizontal scheme of wall division stands out in strong contrast to the mid-nineteenth-century schemes for dividing walls into vertical panels. The horizontal bandings of wallpaper imitated basic architectural divisions of walls, executed more expensively during this period in wood paneling, carving, painted decoration, and textile hangings.

The source of this scheme for wall division was again England, where Aesthetic trend setters were, by the early 1870s, using straw matting to cover the lower portions of walls—the dado areas, below chair-rail level. Eastlake, who felt that "the most dreary method of decorating the wall of a sitting-room is to cover it all over with an unrelieved pattern of monotonous design," had decreed in 1868:

> Paperhangings should in no case be allowed to cover the whole space of a wall from skirting to ceiling. A "dado," or plinth space of plain colour, either in paper or distemper, should be left to a height of two or three feet from the floor. This may be separated from the diapered paper above by a light wood molding stained or gilded. A second space, of frieze, left just below the ceiling, and filled with arabesque ornament, painted on a distemper ground, is always effective . . .[55]

For walls where there were no moldings, paper borders were quickly provided by the wallpaper manufacturers, and to save their customers the expense of having wide friezes painted, friezes were inexpensively printed on wallpaper.

Like Eastlake, the writer of a brochure for Morris and Co., describing the company's wallpapers displayed at Boston's Foreign Fair of 1883, suggested that if a dado of wood paneling was lacking, a plain dado be used with a patterned paper above. But he warned:

> Never stoop to the ignominy of a paper dado . . . If the wall obviously wants a dado, the effect may be at once got by fixing a rail of suitable section at the required height, and painting from rail to floor with one color.[56]

Morris and Company, feeling that wallpaper was simply to be a filling for the "natural architectural features of a room," added that "Our wallpapers therefore are simple fillings, they imitate no architectural features, neither dado, frieze, nor paneling."

However, the manufacturers who catered to a much larger trade than did Morris and Company continued through the 1870s and 1880s to manufacture miles upon miles of just such architectural imitations—dados and friezes—long since staples in the wallpaper trade, now dressed in new designs and given great prominence in interior decoration.

Christopher Dresser in his *Principles of Design* (1873) substituted for Eastlake's plain dado a more ornamental scheme:

> The dado of a room need not be plain; indeed, it may be enriched to any extent. . . . it may have a simple flower regularly dispersed over it; or it may be covered with a geometrical repeating pattern, in either of which cases it would have a border.[57]

16-39

Figure 16-39 This "Dining Room Decorated with Rottman, Strome and Co.'s Japanese Leather Papers" was illustrated in color in the *Journal of Decorative Art* published in April 1884 in London. The luxuriant pattern of fruit and trailing foliage in high relief was derived from Western rather than Japanese sources. It was enhanced by a generous application of gold. The gold of the scrolling leaf forms stands out against a neutral green-blue ground. The carpeting and drapery are red, the sideboard ebonized. Duplicates of the fill pattern have survived in the Gibson House in Boston, and in the Ballantine House in Newark, New Jersey. Photographs show that the same pattern was used in the Glessner house in Chicago. Cooper-Hewitt Museum library

16-40

Figure 16-40 Samuel Clemens celebrated the success he had achieved as "Mark Twain" by building a stylish mansion in Hartford, Connecticut, in 1874. Seven years later he commissioned Louis C. Tiffany and his firm, Associated Artists, to enhance its magnificence by redecorating. The firm installed Japanese leather paper in this, the dining room. That wall covering had disappeared but was replaced during the 1970s by a meticulously careful reproduction, shown in this photograph of the room. The reproduction was based on samples saved from the room and preserved at the Cooper-Hewitt Museum. Lilies, embossed and gilded in relief, stand out against a deep rich red ground. The stenciled woodwork design is attributed to a member of Associated Artists, Samuel Colman. Courtesy of the Mark Twain Memorial, Hartford, Connecticut

Figure 16-41 This is an unused remnant of the heavy paper wall covering, with patterning in relief, that survives as it was hung in the hall of the Bush house in Salem, Oregon. A Massachusetts banker and newspaperman built the house in 1877, and hung this paper during the 1880s. The patterning, colored tan and light orange, stands out in relief from the metallic gold ground. This sample is 10⅝ inches wide. It was hung with the frieze illustrated in *figure 16-33.* Cooper-Hewitt Museum, 1972-21-1; gift of the Salem Art Association, Salem, Oregon

16-41

16-42

Figure 16-42 (above) The *Journal of Decorative Art,* in London, included this advertisement, printed in full color, in the issue for March 1884. It shows the period's most popular composition wall covering with decoration in relief—Lincrusta Walton. The linoleumlike material was here used in the 1880s favorite combination of frieze, fill, and dado. The dado in brown imitated wainscoting; the leaf-patterned fill was shown in light brown; and the frieze was brown with gold highlights. Well into the 1920s, Lincrusta continued to be an important item in the American wallpaper trade. Cooper-Hewitt Museum library

Figure 16-43 This example of Lincrusta Walton was photographed in place where it survives in the hallway of the Edward C. Peters house, built in 1883 in Atlanta, Georgia. The house is used in 1980 as a restaurant. The Lincrusta is gilded and so closely resembles leather that it is difficult to dissuade its admirers from calling it leather. The design was patented by Paul Groeber of New York in 1884. A duplicate pattern survives in the dining room of the Benjamin F. Ferguson house built in Chicago in 1883. Photograph courtesy of the Atlanta Historical Society

16-43

16-44

Figure 16-44 Another of Paul Groeber's Lincrusta patterns, patented in 1884, survives in the Dexter cabin, built in 1879 in Leadville, Colorado. The log exterior belies the splendors of the interior: a multi-patterned room of the 1880s with its Japanese-inspired wall patterning in gold on green, ornate Lincrusta dado in black with gold ornament, curved Lincrusta cornice, and patterned ceiling. The cabin served as a hunting lodge and exclusive poker club for James V. Dexter, who had made a fortune in Leadville's mines. Courtesy of the State Historical Society of Colorado

"Enrichments" for dados were printed in endless variety by wallpaper manufacturers, led by Jeffrey's of London, who in 1875 introduced the first set of "combination" papers, featuring elaborate dado, related frieze, and fill paper. (The frieze from this first "combination" set issued by Jeffrey's is shown in *color plate 85.*) By 1880, they were even printing dados along the diagonal to go up staircase walls *(figures 16-26, 16-27).*

Motifs were frequently planned as parts of elaborate sets in which dado, filling, and frieze each carried patterning illustrative of three aspects of a theme. Walter Crane's design, which won prizes at the 1876 Exhibition, is the epitome of a literary approach to the selection of themes for such papers *(figure 16-10).* Taking a text from Chaucer, Crane depicted the God of Love and his queen Alcestis, the archetypical good wife, crowned with the daisy and clothed in its colors, with her attendants, the domestic virtues Diligence, Order, Providence, and Hospitality. The frieze is bordered with words from the text of Chaucer's poem "Flower and Leaf":

> To whom do ye owe your service,
> Which will ye honour
> Tell me I pray the yere
> The leaf or the flower

The motif of the daisy, introduced in Alcestis's crown, is taken as the theme of the fill paper "La Margarete" with a line from Chaucer's poem worked into the pattern: "*Si Douce est la Margarete.*" The wifely virtue of purity is symbolized by the lilies and doves of the dado. While this set of papers met with great success, it was not without detractors. One can well imagine a good nineteenth-century feminist deriding its message and intent, though such a criticism has not been located. However, a writer in *The Furniture Gazette* for May 18, 1876, criticized the papers on other grounds, citing them as "an instance of attempting too much with a paper" and of "putting too much meaning into a wall paper, and thus obtruding it beyond its proper place as a decorative background."[58]

Figure 16-45 This ornate pattern, replete with scrollwork and metallic gold glitter, was printed in brown over a beige ingrain ground. It was purchased in Philadelphia in 1885 for a house in Jersey Shore, Pennsylvania, where it survives in place with elaborate painted ceiling decorations of the same period. This unused remnant of the paper is 22½ inches wide. Cooper-Hewitt Museum, 1972-66-1; gift of Mr. and Mrs. William F. Williams

16-45

A more usual approach to the design of the frieze–fill–dado sets was to choose a theme in which visual relationships could be demonstrated in three segments, a theme not so highly charged with literary meaning and moral instruction. Candace Wheeler's "$1,000 prize design," shown in *figure 16-24,* presents a theme not so fraught with messages as well as an example of an American working within the English formula of three-part sets of wallpaper designed around a related theme. Although the idea is not spelled out in letters on the papers themselves, one suspects that the bees might have been intended to inspire industry on the part of a housewife.

The favorite part of the three-part scheme was the dado. Through the 1880s almost every mention or illustration of wallpaper in America included dados. By 1884, R. H. Pratt was moved to write an article in *The Decorator and Furnisher* entitled "Dados." In the article, he lamented:

> With our usual proneness to excess, we carried the dado idea too far; its use became fashionable, and those who considered or wished themselves in style, or to be considered so by others, had a dado in the hall, parlor and dining room, and sometimes in the bed chamber also, until the excess of dado became so tiresome that today, some people do not want it used in any part of their houses.[59]

However, Pratt went on in his article to recommend the use of wallpaper dados as protection for the part of the wall most likely to be soiled, and as especially appropriate where there was not much furniture—as in halls, dining rooms, and stairways.

Pratt, like most arbiters of taste who took their cue from Christopher Dresser, felt that the dado should be darker than the "side wall" or "fill" pattern, "giving a feeling of stability and strength to the decorative effect." Dresser had set the precedent for this opinion when he observed:

> The occupants of a room always look better when viewed in conjunction with a dark background, and ladies' dresses certainly do. The dark dado gives the desired background without rendering it necessary that the entire wall be dark.[60]

CEILINGS

Not content with the variety of patterns rising from dado level through fill to frieze height on walls, many wallpaper manufacturers printed patterns especially for ceilings during the 1870s and 1880s. These patterns sometimes added a fourth element to the thematic sets. They were not universal, however, largely because they were hard to hang. The authors of one decorating book of 1878 observed that for the space overhead "calcimining will be found invaluable as it is a difficult task to apply paper hangings to a ceiling."[61] Nevertheless, during the 1880s the wallpaper trade did energetically promote the sale of their products for ceilings *(figures 16-28 through 16-30). Carpentry and Building* noted in 1880: "Centers and borders for ceilings are especially designed to match each kind of paper, from the commonest to the most expensive grades."[62] In 1881, the same publication listed the prevailing—they described them as "really very few"—styles for filling the ceiling spaces between borders and ceiling centers: "A certain irregular grouping of forms is one of the most characteristic features . . . rayed stars are sometimes employed . . . and pale blue grounds find favor." The writer observed that these last two features of current ceiling papers with their analogies to the heavens were not the best, for "It is hardly good taste . . . to regard the ceiling as part of the immensity of space. It is better to look at the matter only decoratively . . ." Most of the commercially produced ceiling papers conformed to the suggestion of the leading design arbiters that patterns be flat and, for ceilings especially, that motifs not have dominant directional quality, creating a recognizable up and down. The abstracted and flattened motifs often took particularly bizarre forms on wallpapers for ceilings. Jagged lines, short streaks of lightning, meandering lines resembling nothing more than worms, little circles, snowflakes, and unidentifiable amoebalike shapes were strewn in disjointed combination over ceiling papers.

The coloring of ceiling papers was fairly standard. Almost all the cheap American ceiling papers during the 1880s seem to have incorporated gold, either for the ground or the figures, or other shiny metallic colorings, in copper, brass, or silver shades. Most of the popular decorating literature advised that from the dado to the ceiling, color and strength of patterning should grow increasingly lighter. *Carpentry and Building* noted in 1881 that the "prevailing tints in most of the ceiling papers are cream, light olive, blue, gray, and pale blue."[63] However, some of the leading English designers took exception to the prevailing popular wisdom in the question of dark- versus light-colored ceilings. Morris and Company, in their booklet of 1883, published for Boston's Foreign Fair, suggested that ceilings might be either lighter or darker than the walls.[64] Christopher Dresser was quoted in an 1888 issue of *The Decorator and Furnisher* as having found that dark colors on ceilings gave a room a "cozy" effect. He instructed that "more color may be placed on the ceiling than on the walls or floor, for it does not serve as a background."[65]

COLORS

Whether on ceilings or walls, the colors of wallpapers of the 1870s and 1880s were richer, warmer, and in overall appearance generally darker and more subdued than had been those of mid-century *(color plates 83 through 96).* Critics suggested that a close relationship between background and figure presented a more harmonious and desirable effect than did the strong contrast between a boldly colored figure and a white background. The subtle blending of colors, incorporating drab tones as well as stronger shades, had been masterfully achieved by William Morris. His color effects were praised particularly for the way in which they receded into the background and thereby played the role proper for most wallpapers. They created an "atmosphere," to use Harriet Spofford's word, rather than becoming the dominant element in the decoration of a room, calling attention to the paper itself at the expense of other furnishings. Morris color schemes were widely imitated, by both British and American manufacturers.

As might be expected, however, there were varying opinions about colors in the many publications dealing with this subject in great detail. Eastlake observed "the paper can hardly be too subdued in tone. Very light drab or green (not emerald) and silver gray will be found suitable . . ." *(color plate 94)* and "two shades of the *same color* are all sufficient for one paper" *(color plate 91).*[66] Eastlake perpetuated the mid-century preference for white and gold in drawing rooms with his suggestion that "embossed white or cream color, with a very small diaper or spot of gold, will not be amiss where water color drawings are hung." Eastlake also advised that "intricate forms should be accompanied by quiet color, and variety of hue should be chastened by the plainest possible outline."[67]

For all the words that English critics spun in praise of simplicity and modesty in furnishings, Victorian appreciation of luxuriance seldom failed to affect the coloring of wallpapers. Metallic gold was frequently incorporated in color schemes of friezes, fill patterns, and dados ranging from the most expensive to the cheapest papers *(color plates 97, 98).* Again and again, writers referred to the light, handsome effects achieved when flickering illumination was reflected in the gold that spotted wallpaper patterns or formed their background.

Morris and Company, who marketed the designs of the most widely imitated colorist of wallpapers during this period, William Morris, went into great detail about color in their booklet for the Boston Foreign Fair in 1883. They suggested that the householder take the color of the woodwork as a guide in choosing a wallpaper.

They considered acceptable woodwork that was colored either lighter or darker than the walls. The writer of the booklet, perhaps Morris himslef, observed about wallpapers:

> The revolt against crude, inharmonious coloring has pushed things to the opposite extreme, and instead of over-bright colors, we have now dirty no-colors. The aim was to get sobriety and tenderness, but the inherent difficulty was not less great than before.

After providing quantities of advice on taking into account the other furnishings, the booklet from Morris and Company concludes with *the* principle to follow when choosing wallpaper colors: "The best thing to say is that, when all is done, the result must be *color,* not colors."

Wallpaper color schemes recommended by American popularizers of the latest English ideas on decoration were more specific, giving very detailed recommendations of colors for given rooms, taking into account the exposure of the room in relation to the points of the compass, the time of day it would most often be used, and for what purposes.[68] Commonly these writers recommended darker, richer colors for dining rooms and libraries, lighter shades for drawing rooms.

Mrs. C. S. Jones and Henry T. Williams summed up some of the most often repeated ideas about appropriate wallpaper colors in their book of 1878, *Beautiful Homes, or Hints on House Furnishings:*

> . . . a paper should be of *secondary, tertiary,* and *grey* colors, and where the *primary* colors are introduced, let them appear only in small spaces or figures, for, by such a display of them sparingly applied, the effect is enhanced.
> All conspicuous contrasts, both of color and form, must be carefully avoided, and here we can understand the popularity of those lovely grays, pearl, sage, stone, or that exquisite "Ashes of Roses tint" that has been copied from the imported paper hangings of the celebrated Morris Company of London.[69]

Advice like this, derived from a variety of English sources, apparently influenced American wallpaper manufacturers to gray and deaden their palettes. Eastlake's recommendation of "very light drab or green (not emerald)" may have been one source for the predominance of olive shades in commercially produced wallpapers of the 1880s, with which metallic gold, shades of maroon, and touches of creamy yellow beiges were frequently combined, then accented with black. Examples of the productions of American manufacturers using this very distinctive combination of colors with gold survive in countless rooms papered during the 1880s and into the 1890s *(color plates 92, 94 through 97).*

ECLECTICISM

Not only are the surviving examples of cheap late-nineteenth-century American wallpapers marked by a certain range of color preferences, but they are also marked by rampant eclecticism in the choice of historical styles their designers imitated in creating them. The design-conscious of the 1870s and 1880s viewed themselves as the privileged heirs of all the cultures of all the ages, and they felt free to pick and choose their design motifs from any source, no matter what its origin in time or space. New York's *The Art Journal* characterized the attitude in 1877:

> The man who has conquered fortune steps into his new house, to which every country, every artist, every age, the eighteen Christian centuries, and the unwritten centuries of the great Confucius, and Buddah himself, have all made tributary.[70]

In 1875, another author had elaborated on the theme in *Scribner's Monthly:*

> The modern householder has one great advantage over his predecessors. He has the houses of the past as models. He may gather under one roof the styles and fashions of a dozen nationalities and centuries. The library may be Gothic, the dining room in the "Eastlake" style, the parlor French, one chamber Modern English, one chamber suggestive of India . . .[71]

This exuberant eclectic spirit was visually evident in wallpaper patterns printed during the 1870s and 1880s. Writing in 1882, a Philadelphia correspondent for the *Decorator and Furnisher* described wallpapers currently fashionable:

> In paper-hangings, the variety is simply exhaustless, and the most ultra-minded householder may give reign to his fancies without danger of violating any of the canons of existing fashion. So that he avoids plain panels, and sticks to this dado and frieze, he may do pretty much as he likes and still remain within the pale of permissible decorative art.[72]

Owen Jones's popular *Grammar of Ornament* gave manufacturers a whole range of decorative styles and suggestions for wallpaper patterns. By the 1870s, even the most remotely fashion-conscious were accepting Jones's insistence that forms be conventionalized, at least to some degree. Any decorative style that gave evidence of the impression of the human mind on nature, through abstraction, was permissible. The ornamental patterns of the Egyptians *(figure 16-31),* the Greeks, revived Adamesque (late-eighteenth-century English neoclassical) decorations, and traditional damask patterns, baroque in origin, which contrasted shiny and matte finishes in ground and figure, were all incorporated in patterns. These, however, played a less important part than the styles Ruskin, Jones, and Dresser advocated so spiritedly—the medieval, the Middle Eastern, and the Japanese.

"Properly" flat-patterned wallpaper interpretations of all these styles came to this country from England, and even from France, where manufacturers had also experienced some English-inspired reform.[73] American wallpaper factories also printed some of the purer reform-inspired designs. More often, however, they churned out garish popularizations of all the styles, particularly the Moorish *(figure 16-32)* and Japanese *(color plates 96, 97).* These commercialized American versions of the Eastern styles were designed in total misunderstanding of the reasons for English admiration for the originals. The abstraction, two-dimensionality, and refinement in treating natural and geometric forms were replaced by the commercializers with realistically shaded pictures of Japanese and Moorish objects—buildings, vases, fans—that appeared to reside in three-dimensional spaces. They drew realistic versions of the favorite Japanese flowers—chrysanthemums and cherry blossoms—and then forced them all into asymmetrical configurations. In these cheap American papers, Oriental motifs often appeared cheek by jowl with totally Western motifs, often in the form of little vignettes, descendants from the "landscape figures" of the early nineteenth century *(figures 12-7 through 12-9, 12-11, 12-12),* showing classical buildings, American scenery, or any other incongruous subject. In contrast to Japanese prints, in which blank spaces are so sensitively balanced with visually active areas, the unrelieved crowding of elements that make up cheap Japanesque patterns is ironic, especially since the prints inspired the earliest English interest in the style.

However, that crowding, that quality of dense confused patterning in the commercial Japanesque wallpapers, is closely related to a similar overburdening of other wallpapers of the period with scrollwork—one of the sins of mid-century the taste makers had not quite managed to completely drive off the American market. The "non-Aesthetic" housewife could still find a few cheap patterns tainted by scrolls *(figures 16-33, 16-34)* and naturalistic flowers. Nevertheless, the scorn of the stylish did manage to banish them, along with most stripes and spotty diaper patterns, from the shops and the major rooms of any who aspired to status among the fashionable.

Most of the taste makers advised the public to choose simple patterns that would assume a properly retiring position within the decorative scheme of a room. However, photographs of interiors, like those of *figures 16-35 through 16-37,* and surviving examples of wallpaper, like that shown in *figure 16-45,* give evidence of the popularity of complex, luxuriant ornament on the walls as well as in all house furnishings. Writing on "Modern Furnishing in America" in an article that appeared in *The Furniture Gazette* of 1877, Mr. Shirley Dare lamented:

> The artist had an idea of making an ample, dignified apartment that should have the state and the comfort of the latest English styles, but he grew bewildered, and the room is more like a fashionable furniture shop. The trouble is, that people want to put all that they find in decorative materials in one room, and one house, at once, by way of getting their money's worth probably. The result is what might be called the Babylonian style, an eclecticism of which there is a painful amount nowadays.[74]

WALL COVERINGS WITH PATTERNING IN RELIEF

In no area of the wallpaper trade of the 1870s and 1880s is there more ample evidence of rampant eclecticism and love of luxuriant, complex ornament than in imitations of baroque leathers and of elaborate architectural detailing executed in relief. The variety of composition materials used in them was bewildering. Late-nineteenth-century handcraftsmen as well as factories imitated panels with carved ornament in

Color plate 90 The abstraction of the flowers, combined with obvious references to Japanese design in the stylized knots punctuating the pattern, marks this American-made paper as one inspired by English interpretations of Japanese designs—"Anglo-Japanesque." It is a 22-inch-wide remnant of a paper hung in The Manse at Deerfield, Massachusetts, during the late nineteenth century. The background color was fondly called in the 1880s "ashes of roses." Cooper-Hewitt Museum, 1972-51-3; gift of Deerfield Academy

91

Color plate 91 (opposite) In this American pattern of the 1880s the manner of Morris and the Japanese taste have been blended. It is an unused remnant of paper hung late in the nineteenth century in a house in Carolina, Rhode Island. The darker blue background has been block-printed over a lighter ground, reserving—not covering—the lighter colored pattern shapes. This sample is 22 inches wide. Cooper-Hewitt Museum, 1972-39-13; gift of Elizabeth Albro

Color plate 92 This fragment of an "Anglo-Japanesque" wallpaper dado, 29 inches high, is marked in the margin "DR. DRESSER INV." Christopher Dresser, an Englishman, who is known to have designed wallpapers for nine British manufacturers, also designed for the Philadelphia wallpaper firm of Wilson and Fenimore, who patented thirteen of his patterns in the United States Patent Office in May 1877. This pattern, perhaps American-made, though it is not among the patented designs, is machine-printed. Cooper-Hewitt Museum, 1941-17-1; gift of Wilmer Moore

92

93

Color plate 93 An unused remnant of a paper hung in the Bush house in Salem, Oregon, during the 1880s also reveals a taste for Japanese design. This block-printed trellis pattern was hung in the sitting room. The sample is 18 inches wide. Cooper-Hewitt Museum, 1972-21-59; gift of the Salem Art Association, Salem, Oregon

Color plate 94
Anglo-Japanesque stylization is again apparent in this machine-printed paper of the 1880s, a remnant of patterning hung in the hall and sitting room of the Butler–McCook house (built in 1782) in Hartford, Connecticut. A key, or fret, motif forms an asymmetrical framework for the stylized flowers. Cooper-Hewitt Museum, 1971-82-5; gift of the Antiquarian and Landmarks Society, Inc., of Connecticut

94

96

95

Color plates 95, 96 A pristine sample (right) of a commercialized Anglo-Japanesque pattern of the 1880s or early 1890s is one of sixty-five wallpaper samples now at the Cooper-Hewitt Museum, all of which had been mounted on long strips of cloth for use on showroom display rollers. The pattern, 17 inches wide, is machine-printed, and its coloring represents the original appearance of a duplicate taken from the walls of a house in Franconia, New Hampshire, shown on the left. The color combination, featuring maroon, creamy yellow, black, and olive shades, is typical of a great many machine-printed wallpapers of the 1880s that have survived in American houses. Cooper-Hewitt Museum, 1970-26-4 Cdd; gift of Jones and Erwin, Inc. Cooper-Hewitt Museum, 1955-79-1; gift of Mrs. Philip Robertson

Color plate 97 The vogue for Japanesque wallpapers produced bizarre patterns like this concoction of the 1880s. Here, pictorial vignettes showing the Brooklyn Bridge, Niagara Falls, and other American views are forced within a framework considered Japanese because of its asymmetry. They appear cheek by jowl with Oriental vases and sprays of foliage. Another typical color scheme—maroon, brown, and yellow again, but with the addition of metallic gold—is representative of many commercial, machine-printed American papers of the 1880s. The sample, 19 inches wide, was found in the Bixby house in Salem, Massachusetts. Cooper-Hewitt Museum, 1938-62-18; gift of Grace Lincoln Temple

97

98

Color plate 98 Lincrusta Walton (see page 442) was patterned to suit every taste. The abstraction and stylization of floral forms in this example, with their references to Japanese designs, would have suited an aspirant to fashionable "aestheticism." It is a remnant of the Lincrusta Walton that was installed during the 1880s in the dining room of the John D. Rockefeller house at 4 West 54th Street in New York. It duplicates a Lincrusta pattern published in London's *Journal of Decorative Art* in March of 1884. The patterning is in relief on a sample that is 18½ inches wide. Cooper-Hewitt Museum, 1937-57-3; gift of John D. Rockefeller, Jr.

addition to moldings, ceiling centers, pilasters, and capitals. The repeating patterns stood out from the surface and the ornaments were fully as three-dimensional as the carved and plaster models for them, and were often accented by generous applications of gold. They partook of every decorative and architectural style that nineteenth-century designers could find in the many published sources available.

Japanese Leather Paper. "Leather Paper" in the late-nineteenth-century wallpaper trade was indeed paper, and not really made from animal hides. In the late 1860s, English and American dealers added Japanese Leather Papers to their offerings, so they could provide less affluent customers with the look of highly embossed leather remembered from Renaissance sources and currently enjoying a revived vogue in the houses of the super-rich, like those soon to be illustrated in *Artistic Houses* (see page 406).

The cheaper Japanese leather papers were not, in fact, products of technologically advanced large-scale factories, but of Oriental craftsmen who were highly skilled and worked on the papers for relatively low wages. To achieve a leatherlike effect, the Japanese used sheets of heavy paper, made near Tokyo, according to accounts published during the 1880s.[75] They dampened these, then subjected them to pressure and to the action of rolling wooden cylinders to reduce the length and width while increasing the thickness of the original sheets of paper. The rollers and weights impressed patterning, and gave a look of leather. They enhanced this simulation with coloring, oil, and varnish. Workmen also used hammers to produce patterning and textured effects, and to join one sheet of paper to the next.

Americans decorated bamboo and imitation bamboo furniture with these papers as well as walls. Many of the Japanese fake leathers came to America through the agency of an English firm, Rottman, Strome and Company. Rottman Strome entered the business of manufacturing leather papers in Japan in cooperation with the Imperial government, which had established a factory for the production of "leather paper" at Yokohama. Rottman Strome and the Japanese Imperial Factory increased their capacities with facilities at Tokyo. They exported the papers in increasing quantities as the taste for things Japanese grew stronger in England and America. Rottman Strome developed a process that eliminated the use of oil in producing the leatherlike effect, a technical improvement especially valued because it produced surfaces less sticky and subject to the accumulation of dirt. Though formed of individual sheets of paper, the Japanese papers were hammered together with great skill, making it difficult to detect the seams.[76]

Examples of leather papers in the patterns pictured in advertisements of the 1880s for Rottman, Strome and Company have survived in houses scattered across America. For instance, the pattern this firm illustrated in an advertisement of 1884, shown as *figure 16-39,* was used in large fine houses in Newark, New Jersey, Chicago, and Boston—not only have the papers survived but so have nineteenth-century photographs showing the patterns in place.

Lincrusta Walton. It is ironic that England, home of the most vehement opposition to elaborate imitations of luxurious finishes for walls, produced the one composition wall covering that most effectively faked the look of leather, plaster work, or carved architectural ornament. It was called "Lincrusta Walton." Its name was derived from that of its inventor, Frederick Walton, who patented it in 1877, and from the fact that, like linoleum, it was based on solidified linseed oil. It was also advertised as "The Sunbury Wall Covering" because it was manufactured at Sunbury-on-Thames. It was even more widely distributed in America than Japanese leather paper. Although it was not technically a wall *paper* it was, in the words of the writer of a late-nineteenth-century book called *Industrial Chicago,* "so much a part of the wallpaper trade that it may be considered a wallpaper specialty" *(figures 16-42 through 16-44 and color plate 98).*[77]

This thick linoleumlike compositon material was embossed while still in a semi-liquid state, then backed with heavy canvas. After 1887, it was given a light water-proof paper backing.[78] Paper hangers often put the embossed patterning in a plain, uncolored state on the walls and ceilings. They could even wrap it around corners and curves, because, as the manufacturer advertised, "when slightly warmed it becomes plastic and will cover a pillar as easily as a wall." Once affixed to the walls, it could then be colored. A favorite technique was to brush on color, then wipe it off, achieving effective highlights on the highly embossed areas. Gold was generously used in coloring Lincrusta patterns. Lincrusta could also be purchased with coloring already applied.

Introduced to America in 1879, it did not really become a major item in the wallpaper trade until 1882, when Frederick Beck and Company of New York organized a company to manufacture Lincrusta Walton under Frederick Walton's patents at Stamford, Connecticut, where it was made until the factory moved to Staten Island in 1916.

In 1884, Beck registered design patents at the U.S. Patent Office for several of the repeating patterns he made in Stamford. Additional testimony to the status of

Lincrusta as an accepted part of the wallpaper trade is furnished by the fact that the patent office registered and filed Beck's Lincrusta patterns under "wallpaper designs." The designers, whose names were given as Paul Groeber of New York, Augustin Le Prince, a French citizen living in New York, and Albert Leisel, a German citizen living in New York, exhibited eclectic tastes in their designs. The patterns ranged from what were described in the individual patent specifications as "modern renaissance style," "partly in the Egyptian and partly in the Eastlake style," "Japanese style," and "Venetian style of the last century," to "the Persian style."[79] One of the company's advertisements of 1884 for "the most beautiful of all decorations, indestructible, waterproof, sanitary," boasted of 150 new designs: "Renaissance, Egyptian, Moorish, Celtic, Florentine, Japanese, Greek, Byzantine, Eastlake, Mediaeval, Modern, etc."[80]

Numerous examples of Lincrusta hung during the 1880s have lived up to the advertising claim that they were "the indestructible wallcovering." They have survived in settings as diverse as old railroad cars, a log cabin in Leadville, Colorado *(figure 16-43),* a Gothic cottage in Woodstock, Connecticut, and the Manhattan mansion of John D. Rockefeller *(color plate 98).*

Lincrusta had many rivals and imitators. "Tynecastle," "Cameoid," "Anaglypta," "Subercorium," "Lignomur," "Corticine," "Cordelova," "Salamander," and "Calcorian" are among the brand names for paper, cork, rubber, wood fiber, and canvas wall hangings with decorations in relief that were manufactured in England and America during the late nineteenth century.[81]

INGRAIN PAPER

A novelty introduced to the trade in 1877 was to grow in popularity well past the turn of the century. On June 19, 1877, James S. Munroe of Lexington, Massachusetts, filed patent specifications for an "improved Wall Paper . . . of mixed cotton and woolen rags, dyed before pulping."[82] Distinctive characteristics included more thickness and strength than ordinary wallpaper and "ingrain instead of surface color, whereby it may be washed without marring or destroying its color."

"Munroe's Ingrain Paper" had a soft, textured, slightly mottled finish, a woolly feel. It was used both as a base on which patterns were printed and as a satisfactory wall hanging in plain, unpatterned colors. Similar papers were known as "Oatmeal Papers." These sometimes had even larger white flecks imbedded within the relatively thick paper pulp from which they were formed.[83]

17 1890 TO 1915: REVIVALS AND THE PURSUIT OF NOVELTY

After all the ambitious, self-conscious, and often successful strivings of the American wallpaper industry to improve the quality of the designs it printed, its performance around the turn of the century stands as a sad epilogue. Many American wallpapers of the 1870s and 1880s may have fallen far short of critical standards of their era, exhibiting an incongruous mixture of characteristics derived from English reform theory and features responding to the demands of American commerce and fashion *(color plate 97).* Nevertheless, a number of the designs were distinctive and sometimes beautiful; they met with great critical favor when they were introduced and remain visually interesting today.

Early in the 1890s, however, the design-conscious began to notice a change for the worse. On the pages of *Architectural Record* for 1893 John Beverly Robinson eulogized the "admirable designs of wallpapers" that had so recently been available, with their "half-conventionalized flower forms and tints and tones of color." He lamented: "It is rather curious, how rapidly the wallpaper designers have left the standard set for them some years ago, at the time of the aesthetic craze"—something Robinson went on to grant had been "much more than a 'craze' at its best." While the patterns of the past decade glowed in his memory, the current offerings of wallpaper dealers disgusted Robinson. "Now . . . wallpaper designers lead in the chase toward the rococo and the purely pictorial," he complained, adding:

> The character of wallpaper design has lapsed into the old magenta-roses-and-brown-leaves style, mere thoughtless collocation of naturalistically-drawn objects, offensive to good taste as would be a landscape painted on a floor.

Robinson found the patterned papers that were available in 1893 so bad that he could only suggest to his fellow architects that they resort to "plain cartridge papers [thick, stiff, solid-colored papers] and some flat ceiling designs which still persist." He concluded that the current crop of wallpapers reflected "undue pressures to produce vicious novelties."[1] His phrases are resonant with the same English reforming zeal that had animated similar critical commentary in the mid-nineteenth century, and that was to continue to be heard from architects and architectural critics involved in the Modern Movement of the twentieth century, long after a broader public had tired of it.

During the 1890s innovative architects were evolving new standards for interiors that precluded the use of wallpaper. Honest, structurally revealing construction in anything designed by man was becoming an article of faith in the movement toward architectural modernism. Avant-garde architects were arriving at the logical conclusion of mid-nineteenth-century statements like that of Eastlake suggesting that the wall, as the flat surface of a solid material, "should be treated in a way that would belie neither its flatness nor solidity." Now these architects perceived as a truth that the material from which the wall was constructed should provide its own decoration: there should be no applied skin of paper. And at the same time that theoretical concerns were turning them against wallpaper, these architects were also reacting against the decorating practices of their parents' generation. They ridiculed the way pattern had been omnipresent, in all the surroundings of their daily lives *(figures 16-35 through 16-37).* Frank Lloyd Wright (1869–1959), the most important American architect of his day, scorned the use of wallpaper in his new "organic" houses, along with any "house decoration" that, he was later to declare, "as such, is an architectural makeshift."[2] In this category he seems to have included any ornamentation that was not abstract, any introduction of extraneous forms or patterns that were not suggested by the structure and overall spatial organization of a building. In 1905, Adolph Loos, the Viennese architect, declared "Ornament is Crime" in a statement that was to become part of the canon of the Modern Movement.[3] In the light of such pronouncements, not only did American "Modern" architects cease to design wallpaper but they also stopped specifying its use in the houses they built.

These architects were not the only ones to disapprove the use of wallpaper during this period. The authors of books on interior decoration were also discouraging its use. The best known of these books were written for the ultra-rich by a variety of authors, who, as professional decorators, architects, or owners of their own mansions, had decorated grand houses in revival styles. For such people, paper wall coverings were now all too available in the meanest shops and were all too evident in the lowliest houses. For these snobbish writers and their clients, wallpapers had lost all cachet and had been relegated in their eyes to their original status as cheap, and poor, substitutes for more expensive materials.

In their book of 1897, *The Decoration of Houses,* the novelist Edith Wharton (1862–1937) and the fashionable architect Ogden Codman (1863–1951) dismissed the subject with scorn:

> It was well for the future of house-decoration when medical science declared itself against the use of wall-papers. These hangings have, in fact, little to recommend them. Besides being objectionable on sanitary grounds, they are inferior as a wall-decoration to any form of treatment, however simple, that

A MODERN DINING-ROOM

The May tree frieze is from a famous English wall-paper factory.

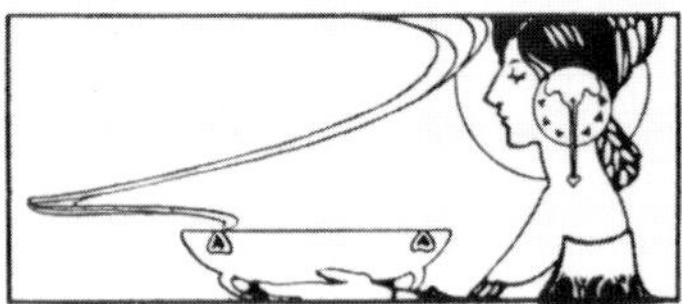

17-1

Figure 17-1 An American trade magazine, *The Wall-Paper News and Interior Decorator,* showed Walter Crane's "May Tree Frieze" in a modern dining room pictured in its January issue of 1906. The use of a wide frieze with a plain-colored wall, either papered or painted, was popular in this country around the turn of the century. Reliance on English imports for quality in contemporary wallpaper design was typical of the period. Cooper-Hewitt Museum library

17-2

Figure 17-2 "The Riverside" is one of many commercial versions of the abstracted, boldly outlined landscape frieze, with the look of storybook never-never lands, which were popular between 1905 and 1915. This 18-inch-wide sample with a color-shaded ground, blending from creamy yellow at the top to green at the bottom, was machine-printed in shades of brown, mustard yellow, green, and pink-orange. It is marked BSP. Cooper-Hewitt Museum, 1971-58-47; gift of the Philadelphia Museum of Art

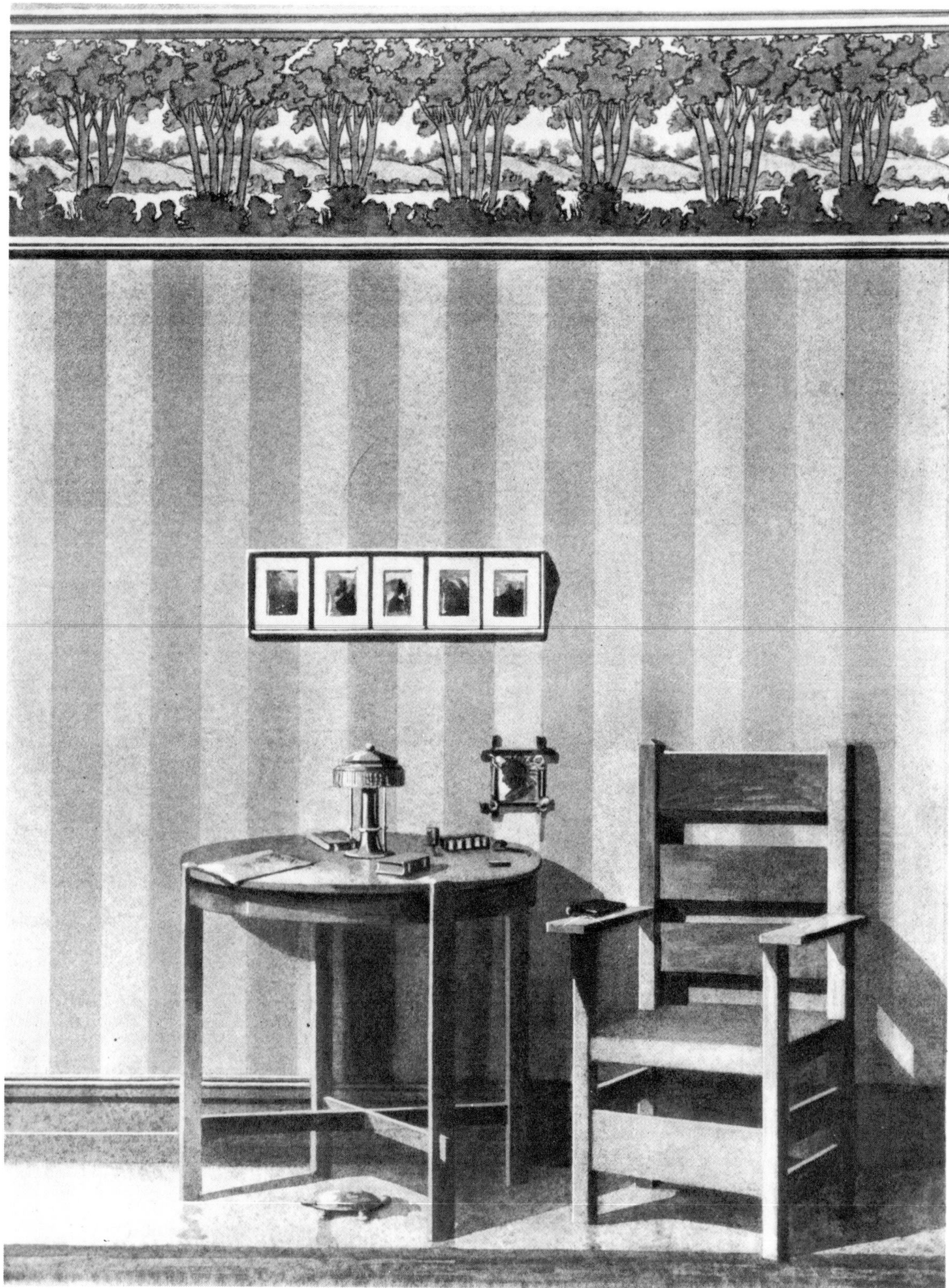

17-3

Figure 17-3 (opposite) The Robert Graves Company of New York illustrated its version of the theme so effectively used by Walter Crane—the wide, tree-filled landscape frieze—in the October 1908 issue of *The Wall-Paper News and Interior Decorator.* It was shown above a wide striped paper in a room with straight-lined furniture in the Craftsman style. Cooper-Hewitt library

Figure 17-4 The idea of the wide, tree-filled frieze has here been overwhelmed by a heavy dose of the scrollwork dear to the hearts of commercial wallpaper manufacturers at the turn of the century. This was published in *The Wall-Paper News and Interior Decorator,* January 1907, credited to the Gledhill Wall-Paper Co. Cooper-Hewitt library

17-4

17-5

17-7

Figure 17-5 (page 450) Walter Crane's nursery paper "The House that Jack Built" was featured by Jeffrey and Co. in the firm's exhibition at Chicago's World's Columbian Exposition of 1893. Designed in 1886, the 22½-inch-wide example shown here was machine-printed in greens and reds on a pale yellow ground. When Crane was in America in 1891 he told reporters that an American wallpaper manufacturer's pirating of drawings from one of his nursery books had led him to design his first wallpaper in 1875. Crane's nursery papers were popular in this country from the time Mark Twain used one in his Hartford, Connecticut, house during the 1880s to well past the turn of the century. Crane also wrote a series of books on decorative design that became standard texts in this country. Cooper-Hewitt Museum, 1938-62-57; gift of Grace Lincoln Temple

Figure 17-6 (page 451) The influence of Walter Crane's nursery papers is apparent in this design by Grace Lincoln Temple, who in fact gave the Walter Crane sample shown in *figure 17-5* to what was then the Cooper Union Museum in 1938. Her "Little Boy Blue" paper was printed about 1890–2 by the Thomas Strahan Company of Boston. The wallpaper, 20 inches wide, is machine-printed in yellow, green, blue, and brown on ungrounded off-white paper. Cooper-Hewitt Museum, 1938-62-58; gift of Grace Lincoln Temple

Figure 17-7 Jeffrey and Company of London embossed and gilded this paper about 1890. Over the gold, the background areas are printed with green. It is typical of the luxurious and traditional papers favored for houses built in classical styles around the turn of the century. The portion shown here is 23 inches wide. Cooper-Hewitt Museum, 1955-144-9; gift of Roger Warner

ART WALL HANGINGS AND FRIEZES

THE LATEST PRODUCTIONS OF SANDERSON & SONS, LONDON INCLUDING MANY MASTERPIECES BY NOTED ARTISTS

IMAGINE the charm and the character you could put into a child's room or a nursery with wall friezes like the above.

Such a room would be a continual delight not only to the little people but to the grown-ups as well, and you would be surprised to know at how small a cost such decorations may be had.

The subjects are varied so that any taste for any sort of room or hall is provided for. The work is thoroughly artistic and is reproduced from designs and originals of some of the best-known masters.

We invite correspondence from readers of the THE CRAFTSMAN, who are building or remodeling and who would like our help in the way of suggestions for interior decoration.

The productions of SANDERSON & SONS of London are unrivalled and their goods can be had only through

W. H. S. LLOYD COMPANY

Sole American Representatives of SANDERSON & SONS, London, and Other Foreign Manufacturers

26 EAST 22d STREET NEW YORK

Kindly mention The Craftsman

Figure 17-7A The Craftsman Magazine for October 1907 carried this advertisement. It suggests the popularity of friezes, of English design in general, of nursery papers, and—with its boldly outlined figures filled with flat color—of a graphic style that was used in other papers especially in friezes, not so specifically designed for children. Courtesy of Roy P. Frangiamore

17-7a

17-8

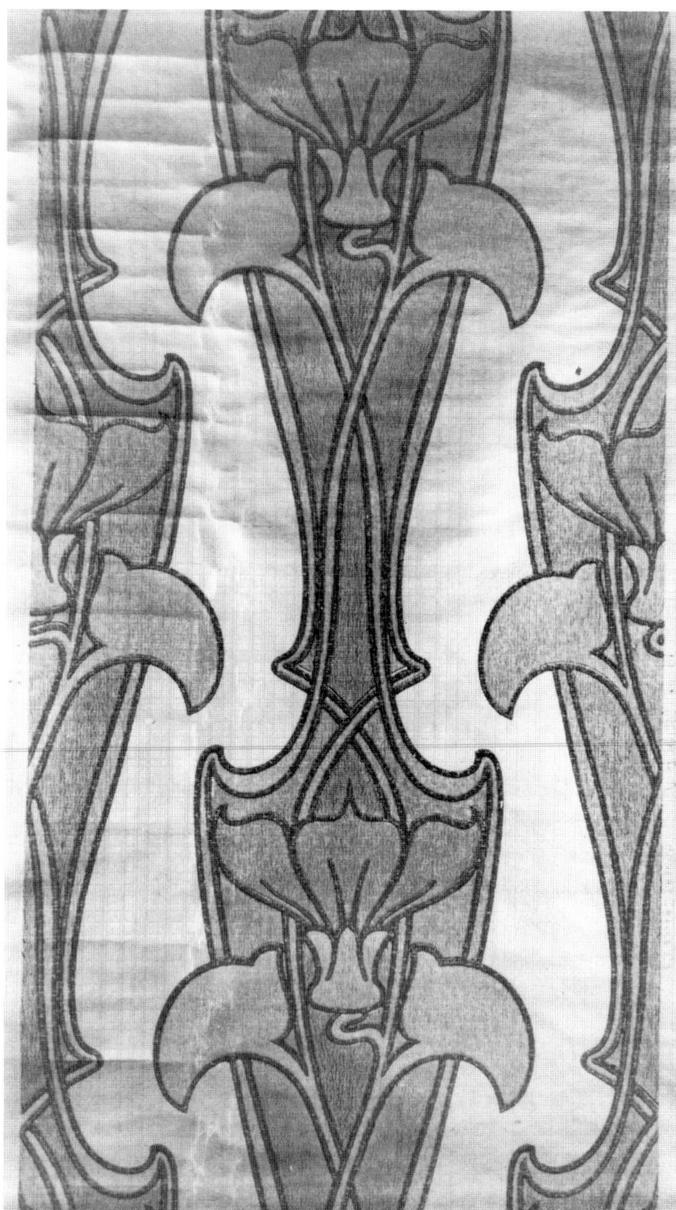

17-9

Figure 17-8 (above, left) This example of American Art Nouveau wallpaper is machine-printed in mustard and two shades of green on a yellowish ingrain paper stock, 19 inches wide. It bears the printed mark of one "Albert Ainsworth, Desgr" about whom no more is known to the author than an inclusion of his name, giving Hackensack, New Jersey, as his place of residence, on a list of the Society of Designers that appeared in *The Year's Art* for 1905. The Hackensack designer's awareness of English and European designers like C. F. A. Voysey is apparent in his design. Cooper-Hewitt Museum, 1956-42-101; gift of Mary M. Kenway from the estate of Sarah B. Russell

17-10

Figure 17-9 (opposite, right) An unused remnant of an Art Nouveau paper hung in 1904 in the library of the Hewell house, built c.1870 in Baltimore, this pattern is machine-printed: two shades of thin-bodied brown and one of green on a metallic gold ground, applied to paper embossed to simulate a plain-woven textile. The pattern covers 18 inches of the paper's width. It bears the mark of the Robert Graves Company, active in New York from 1860 into the 1920s. Cooper-Hewitt Museum, 1973-56-1; gift of the Peale Museum, Baltimore

Figure 17-10 The mark of "M. H. Birge and Sons Co. USA," a firm that has been manufacturing wallpaper in Buffalo, New York, since 1878, is printed along the margin of this Art Nouveau pattern. It is an unused remnant of a paper hung about 1902–4 in Easton's Castle, an ambitious house built in 1889–90 in Aberdeen, South Dakota. The 19½-inch-wide sample is machine-printed in green, red, and tan on a yellow ground. The discovery of this pattern in South Dakota emphasizes the role of wallpaper as a disseminator of styles throughout America. Cooper-Hewitt Museum, 1972-38-25; gift of Dr. and Mrs. Samuel J. Holman

> maintains, instead of effacing, the architectural lines of a room. . . . it is readily damaged, soon fades, and cannot be cleaned; while from the decorative point of view there can be no comparison between the flat meanderings of wall-paper pattern and the strong architectural lines of any scheme of panelling, however simple. Sometimes, of course, the use of wall-paper is a matter of convenience, since it saves both time and trouble; but a papered room can never, decoratively or otherwise, be as satisfactory as one in which the walls are treated in some other manner.[4]

The very rich of New York were the people among whom Wharton had spent her life. They were the subjects of many of her novels, including *The House of Mirth* and *The Age of Innocence,* and they were the audience to whom she addressed her keen and tasteful observations on the effective arrangement of fine furnishings. Her taste epitomized not only the reaction of her generation against the dark, overcrowded interiors of their parents' houses, but also an upper-class aspiration to reproduce the light, bright grandeur of palatial European rooms.[5]

While both elegant socialite decorators and zealous reforming architects with socialist affinities were rejecting wallpapers, middle-class Americans continued to like their papered walls. For such people, Candace Wheeler (1827–1923) wrote her *Principles of Home Decoration,* published in 1903. She recommended various other wall treatments as preferable to wallpaper, including paneling and paint, but allowed that, "of course, wallpaper must continue to be the chief means of wall-covering, on account of its cheapness, and because it is the readiest means of sheathing a plaster surface." She described the wall as a "restraining bound" that everyone wanted to disguise or make "masquerade as a luxury."[6]

Mrs. Wheeler was one of many critical taste makers around the turn of this century for whom ingrain papers stood out as the only American-made products of the wallpaper trade retaining any dignity. Although she herself had designed wallpaper patterns in the 1880s *(figure 16-24),* by the time she published her book in 1903 from the current selection she could recommend only plain-colored ingrain papers. She admired the quality of their colors—"soft and charitable"—an effect created by their "variable surface, without reflections." They provided "variation without contrast in wall surface . . . one of the most important" matters in house decoration. And quite practically, they provided "good backgrounds for pictures."[7]

Pattern design for wallpaper, a subject that had briefly been the focus of artistic and architectural attention, had now been relegated to the periphery. It was to remain for many years a subject of little critical interest outside the decorating trades in America.

CONTINUING ENGLISH INFLUENCE IN AMERICA

The eclipse of interest in patterned wallpaper was not quite so abrupt in England. Leading English architects like C. F. A. Voysey (1857–1941) and important designers like Lewis F. Day (1845–1910), Walter Crane, and William Morris continued to make new designs for wallpapers well into the 1890s and in some cases past the turn of the century. The London firms of William Woollams and Company and Jeffrey and Company brought to Chicago's World Columbian Exposition of 1893 exhibitions that included new wallpapers by important English designers *(figure 17-5).* Walter Crane even designed a wallpaper decoration especially for the exposition—"The Columbian Frieze," illustrating allegorical figures and ships representing the four quarters of the world. It was printed by an American firm, Nevius and Haviland of New York.[8] Americans who had been "artistically enlightened" during the reign of the Aesthetes and had not recovered from the experience were often forced to turn to English imports, or to designs by Englishmen, to satisfy their tastes.

Although major English designers continued to provide patterns for wallpaper manufacturers, the character of many of their patterns had changed. Some showed a predilection for figural compositions in which elaborate shading made the human forms appear rounded and quite real. Naturalistic elements were exquisitely rendered in baroque- and Renaissance-inspired patterns. American millionaires hung them in their splendid classical and rococo rooms, despite the harsh words of taste makers for wallpaper in general. Indeed, a preoccupation with sumptuous, luxurious effects dominated the exhibitions of Jeffrey and Woollams at Chicago. These patterns catered to the same love of grandeur so freely revealed in the grand houses designed by Beaux Arts architects (see pages 477–8). The patterns were often laden with flocking and mica, and were rendered in the high relief of Anaglypta and Lincrusta, or in embossed and lacquered leather. Almost all featured quantities of gold embellishments.[9]

Other wallpapers exhibited by British manufacturers at the Columbian Exposition showed the influence of a new style: Art Nouveau. Its curvaceous forms and distinctive stylization were apparent in the wallpaper patterns of C. F. A. Voysey and Walter Crane (although Crane would have disclaimed the connection with a style

458

Color plate 99 (opposite) "Oranges and Lemons Say the Bells of St. Clements" is the title of this nursery paper designed for Jeffrey and Company in 1902 by Miss Dorothy Hilton of the Birmingham School of Art. It is 22 inches wide. English designers had established the taste for wallpapers derived from illustrations for children's storybooks, and from nursery rhymes: Walter Crane's first such design dated from 1875, and in 1893 the British firm of David Walker produced an engraved—"sanitary"—wallpaper version of illustrations of the months by Kate Greenaway (1846–1901). There is a pirated American version of the Greenaway paper in the Cooper-Hewitt collections. Cooper-Hewitt Museum, 1946-97-2

Color plate 100 This machine-printed Art Nouveau pattern bears the manufacture's mark "F. Arthur, 18 Motcomb St., S.W." A duplicate of the English pattern appears in early photographs of the Glessner house in Chicago, designed by H. H. Richardson and built in 1885–6. The sample shown here is 27 inches wide. Cooper-Hewitt Museum, 1936-5-4; gift of Annie May Hegeman

100

101

Color plate 101 Abstracted, flattened, boldly outlined motifs were featured in Walter Crane's "May Tree Frieze." It was first exhibited in England's Arts and Crafts Exhibition of 1896, the year of its design. In producing this 22-inch-wide frieze, the London firm of Jeffrey and Company used traditional wooden blocks to hand-print it. It was issued with a fill pattern by Crane called "Meadow Flowers." Cooper-Hewitt Museum. 1955-144-7; gift of Roger Warner

Color plate 102 (overleaf) These unused remnants of paper—a "side-wall" pattern and its matching 19-inch-wide frieze—typify turn-of-the-century commercial wallpapers, which were popular from coast to coast. They were preserved in a house in Greeley, Colorado. Machine-printed, they bear the mark of the S. A. Maxwell Company, a firm founded late in the nineteenth century and still active in Chicago. Cooper-Hewitt Museum, 1972-50-6, -7; gifts of the Greeley Municipal Museum, Greeley, Colorado

102

he considered degenerate). Art Nouveau carried the nineteenth-century interest in creating two-dimensional flat patterns past the turn of the century. The influence of this style *(color plate 100)* left some mark on the American wallpaper trade *(figures 17-8 through 17-10),* although writers in American decorating journals registered only faint enthusiasm for it. The "Great White City," which dazzled so many Americans at the Chicago fair, had solidified a preference for architectural classicism that extended to interior decorations, and it made wallpapers derived from classical styles more acceptable to the great masses of home decorators than the more eccentric forms of Art Nouveau.

Another group of English wallpapers, which did have great popularity in America during this period, was also quite outside the classical tradition. These were simplified landscapes and figures, many of them specifically designed as nursery papers. They were printed in bold outlines filled with flat colors *(color plate 99).* They exploited the charm of storybook illustrations, for which Walter Crane *(figure 17-5)* had created a large market. Their popularity continued well into the 1920s. A New York firm, W. H. S. Lloyd, imported such papers from Sanderson of London, publishing illustrated advertisements for them in several magazines *(figure 17-7a).* The most popular vehicle for these storybook illustrations on wallpapers were friezes. Indeed, during the 1890s and until the First World War, friezes of every sort proved to be the heartiest survivors from the frieze–fill–dado schemes of the 1870s and 1880s.

TURN-OF-THE-CENTURY AMERICAN WALLPAPERS

Although most of the major architects, decorators, and designers had lost interest in creating wallpaper patterns, the basic business of making wallpaper seemed to suffer no economic ill effects. The American manufacturers catered to the popular craving for novelty, as critics called it, and brought out new lines seasonally. Further, *The Wall-Paper News and Interior Decorator* in 1908 reported that eighty percent of the consumption of wall coverings in large cities was "forced," explaining:

> Every fall, in order to rent an apartment, or a house, or even a place of business, the landlord must redecorate the walls. Therefore the fall trade in the wallpaper industry is confined mainly to large cities.

The article also noted that, in contrast, the industry's "spring trade" swelled in country regions where seasonal housecleaning demanded frequent repapering.[10]

All this proved to be extremely profitable. By 1893, the trade had grown large enough to support the publication of a periodical of its own. The first issue of *The Wall-Paper News and Interior Decorator* appeared in that year, the same year John Beverly Robinson lamented the trade's abandonment of the high standards that had been established during the reign of Aestheticism. During this period, scrolls and realistic flowers—evocations of the mid-century rococo revival—dominated the repeating patterns of the day *(color plate 102, figure 17-4).* Some of these rococo patterns were updated with borrowings from Art Nouveau, while others were grafted onto neoclassical patterns derived from late-eighteenth- and early-nineteenth-century designs by the brothers Adam *(figure 17-18).* Often, these neo-Adamesque patterns incorporated elaborately swagged strings of pearls. Abstract and two-dimensional patterns based on American Indian designs were also introduced by several firms during this period, satisfying those who had not abandoned the principles of Owen Jones and other theorists of the mid-nineteenth century. Finally, American manufacturers provided their customers with quantities of ever-popular stripes.

COLORS

A palette of saccharine pastel shades, in which any hue approaching white assumed a creamy look, dominated many of the American patterns. Color-blending was omnipresent, one pastel shading into another in even gradations, especially in the background areas of patterns *(color plate 102, figure 17-2).* When stronger, darker colors were used, they were often overprinted with fine networks of lines, dots, or stripes in black, but sometimes these were in brown and olive shades, which gave them a slightly soiled appearance. Forest green, dark murky blue, and dense red verging on maroon were favorite colors. They were used together, often on tan grounds with gold highlights as well as in monochromatic schemes that blended from the deeper hues to the pastel. Finally, metallic gold, glittering mica, or other powdery, sparkling additives were sprinkled lavishly on the surface of almost every wallpaper produced by the big American factories of this period.

FRIEZES, CROWN HANGINGS, AND OTHER ELABORATE BORDERS

Although the wallpaper dado fell from favor during the 1890s, as we have seen, the wallpaper frieze did survive and, in fact, became even more popular than it had been during the previous decades. Friezes were printed to match virtually all repeating patterns *(color plate 102, figures 17-4, 17-16, 17-17),* and they grew ever wider through the decade, dominating turn-of-the-century walls. In many rooms, wood paneling was extended far above the dado's old boundary of the chair rail, the fill pattern was eliminated altogether, and a wide wallpaper frieze topped this exaggerated dado. In another favored decorating scheme, simple papers—colored ingrain, cartridge, or oatmeal paper or a basic stripe—covered the walls below the wide frieze *(figure 17-3).*

Some of these English-inspired wide friezes were among the most distinctive wallpapers of the period, particularly landscapes and storybook scenes *(color plate 101, figures 17-1 through 17-4 and 17-14).* These were often rendered as boldly outlined, flat-colored motifs to be pasted around the tops of walls, especially for children's rooms.

More elaborate ways of connecting wall patterns with friezes added to the period's emphasis on the upper part of the wall. American manufacturers featured "crown hangings" in many of their advertisements *(figures 17-11, 17-12),* and they survive in a number of rooms from these decades. These were printed pattern elements, usually plant and flower forms such as tree trunks, roses on trellises, wisteria or grape vines, that rose from the base of the walls in widely spaced stripes. These vertical tree trunks or the like fit perfectly into the horizontal motifs in the frieze, which seemed to grow and spread naturally to "crown" the tops of the walls.

Just before World War I, advertisements and articles in decorating magazines began to show interiors in which wallpaper borders had been elaborately manipulated to outline doors, windows, and any other interruptions in the walls, like fireplaces. Borders were also used to form panels in a variety of shapes, including large rectangles, ovals, and squares, which were particularly favored in bedrooms of the period *(figure 17-13).* Some of the magazine illustrations show floral borders with related swag friezes, reminiscent of the borders popular about a hundred years earlier, although the versions of the early twentieth century are rendered in more monochromatic and pastel shades and are more realistically delineated.

Not only did decorators of the period favor the use of matching swags and borders, but they also exploited the possibilities of matching the patterns printed on papers with those on the textiles used in curtains, upholstery, and bedspreads. The printing of papers and textiles using the same wood blocks had been common enough in France a hundred years earlier, but not until this late-nineteenth-century period did their use constitute a decorating fad in America. In July of 1915 the editors of *The*

17-11

Figure 17-11 (opposite) *The Wall-Paper News and Interior Decorator* for November 1907 illustrated this stylish Art Nouveau "Crown Decoration manufactured by C. and J. G. Potter." In such "crown hangings" the traditional dado at the base, the fill, and the frieze at the top were adapted. Although each element was printed as a separate part, it was designed with elements that would form bridges or links to the adjoining piece of the set, so that when hung on walls of any height, the wallpaper decoration would appear to have been custom-designed for the wall. Here, the frieze would have included just a bit of the plain narrow part of the stem, as would have the dado or bottom border. These stems could have been easily and neatly matched at the top and bottom of the fill paper. Cooper-Hewitt Museum library

Figure 17-12 This "Crown Decoration" was called "American Beauty Rose" by its manufacturer, the York Card and Paper Company, who illustrated it in the October 1908 issue of *The Wall-Paper News and Interior Decorator.* Crown hangings with motifs as naturalistically rendered as those shown here were extremely popular during the early years of this century. Cooper-Hewitt Museum library

17-12

17-13

Figure 17-13 The use of naturalistic floral papers as borders to outline panels and ceilings was very stylish in the years leading up to World War I. This papered room with its new Colonial Revival furniture was illustrated by the Ellison Furniture and Carpet Company in the November 1912 issue of *The Wall-Paper News and Interior Decorator.* Cooper-Hewitt Museum library

Craftsman Magazine described "a charming plan not only to repeat in the draperies of the room the color of the walls and their borders, but also to echo the actual designs in curtains, rugs, and even furniture."[11] In *The Craftsman*'s scheme, the decorative motifs from the wallpapers and textiles were to be copied and painted by hand on enameled furniture, especially in bedrooms.

Sometimes the same patterns or matching borders were used on ceilings, but even when they were not so closely coordinated, ceilings were usually papered during this period. Many of the ceiling papers were decorated with faint, nebulous patterning printed in creamy tones or in shining mica on off-white.

WALLPAPER AND THE ARTS AND CRAFTS MOVEMENT

Although the Arts and Crafts Movement had begun in England forty years earlier, during the 1860s, it was not until the turn of the century that the movement took hold in America. Crusading spirits within the movement had urged true artist-craftsmen to design and execute their own adornments for the houses in which they lived. Therefore, for walls in a craftsman's house, the logical course would seem to have been to pick up brush and/or stencil, and to paint ornaments of one's own devising on the structurally revealed, uncovered, "honest" flat walls. Ironically, however, major figures in the English Arts and Crafts Movement—Morris and Crane—rank among the greatest designers of wallpaper. In fact, the householder who was persuaded that the ideas derived from these English theorists were the true guides could cite authoritative precedent for any wall treatment: paper, plain surfaces, or original painted ornaments.

Early in the twentieth century, *The Wall-Paper News* carried articles that illustrated those wallpapers the industry deemed appropriate for use with the straight-lined oak furniture labeled "Craftsman" or "Mission." Recognizing that the dainty pastel florals then in vogue were unsympathetic backgrounds for the sober, heavy forms of this furniture, they showed a variety of strong English designs *(color plate 101),* plus bold stripes and plain-colored papers *(figures 17-1, 17-3)* produced by American manufacturers.

Another source of advice on the papers appropriate for the craftsman's home came from a source within the Arts and Crafts Movement itself, *The Craftsman,* mentioned above. This was the monthly publication through which Gustav Stickley (1857–1942) disseminated to Americans his interpretations of the ideas of John Ruskin, William Morris, and the English Crafts Movement. It was the most widely distributed publication of the American Arts and Crafts Movement. Stickley produced it first in Syracuse, New York, in 1901, then moved it to New York City in 1905, where he continued to use it to popularize ideas of the movement until 1916. Over the years, its pages reveal a gradual dilution of crafts principles in an attempt to meet the demands of cheap production and to offer novelties to the public. The magazine's articles and advertisements, however, provide an invaluable guide to wall treatments popular among those middle-class Americans who built bungalows and cottages as well as larger, simply constructed houses in the Craftsman style, and who bought the straight-lined furniture produced by Stickley and his imitators.

Predictably, Stickley's magazine gave some space to articles focusing on one-of-a-kind wall decorations executed by individual artist-craftsmen. A typical article in the issue for April 1910, entitled "A House with Interesting Mural Decorations," described and illustrated a bungalow in a Kansas City suburb. The son of its owners had executed "impressionistic" landscapes, above plaster wall spaces that had been stippled, using a sponge, in gray and golden-yellow distemper colors. Stenciling was another wall decorating technique favored for Craftsman homes, both as a do-it-yourself technique and as one that could be professionally executed while still catering to the unique tastes and designs of those who commissioned the stenciling. Advertisements frequently appeared in *The Craftsman* for stencil patterns and for the supplies needed to execute them.

In spite of the enticements to decorate in these individually expressive and artistic ways, wallpaper was not banished from the middle-class Craftsman-inspired home. Followers of Gustav Stickley used ingrains and plain tints and even patterned paper in many of their houses. In the first illustration in the first issue of *The Craftsman,* a photograph captioned "Suggestions for a dining room" showed patterned paper filling the space between a high wood-paneled dado and the ceiling. There were also advertisements for wallpaper in the magazine. These reveal an aspiration to appeal to those with "Art Principles." They also reveal, however, that the only way to assure this appeal was to call into play the seductions of novelty and ingenuity. An advertisement for the Allen Higgins wallpaper company in the issue for October 1908 epitomizes the confused state of the wallpaper trade's design standards and its attempts to please all the antithetical tastes of the day:

NOVEL
WALL PAPERS

> We are making odd and distinctive things, to match Craftsman Furniture and general furnishings . . . papers with fabrics to match . . . Rich tapestries, Self Tones on Special Grounds—Special Friezes—Pretty, Dainty Florals in Natural colors—Chambrays and Art Linens (latest oddities for Chambers)—Tiles —Orientals—Monk's Cloth Backgrounds.
> Vogue Papers offer Splendid and unusual opportunities for artistic home decorating.[12]

The advertiser's identification of the "artistic" with a search for novelty and oddness is both amusing and revealing, as is the evidence it gives of the advertiser's perception of the special readership of Stickley's magazine.

Patterns conforming to the nineteenth-century reformists' standards of flatness and abstraction continued to be offered in *The Craftsman* by some advertisers. One, W. H. S. Lloyd of New York, showed the English "Art wall hangings" and friezes of London's Sanderson and Sons. Papers by Walter Crane and William Morris were also illustrated in "Craftsman Interiors" featured by the magazine. But these "principled" imports were expensive and beyond the means of most Craftsman bungalow dwellers.

When Stickley opened an overly ambitious and ill-fated department store in New York City for the sale of his "Craftsman-Approved" products, the wall coverings department included papers in equal measure with burlaps and fabrics. Indeed, cloth wall hangings were the most frequently advertised wall treatments in the magazine. "Tapestrolea . . . a fabric of loosely woven strands which gives the desired soft, friendly texture" in colors with the "fresh unobtrusive shades of nature,"[13] and "Fab-ri-Ko-na . . . woven wall coverings in rich shades" were regularly advertised between 1907 and 1916. There were also numbers of ads for "Sanitas . . . oilcloth . . . the washable wall covering . . . in hundreds of dull finished reproductions of paper and fabrics for living rooms . . . in plain and glazed tile effects for bathrooms, kitchens, and pantries."[14] The fact that advertisements for these eye-fooling products continued to appear in *The Craftsman* indicates that the desire for novel, tough, easily cared-for wall coverings was stronger than were moral considerations about honest wall surfaces and the flatness of pattern design.

THE NEW "DECORATOR'S TASTE"

By World War I, a reaction to the decorative styles of the 1870s and 1880s and to the derivations from those styles in Craftsman interiors of the turn of the century

COLORGRAVURE FRIEZE, "THE AUTO-AEROPLANE," SIZE 60 INCHES LONG, 14 INCHES WIDE
NO. 121 A; GRADE, 60 CENTS A SHEET

COLORGRAVURE FRIEZE "THE AUTO-AEROPLANE" — SIZE 60 INCHES LONG, 14 INCHES WIDE
NO. 121 B; GRADE, 60 CENTS A SHEET

COLORGRAVURE FRIEZE "THE AUTO-AEROPLANE" — SIZE 60 INCHES LONG, 14 INCHES WIDE
NO. 121 C; GRADE, 60 CENTS A SHEET

17-14

Figure 17-14 These "Colorgravure" friezes suggest but one of the dozens of novel printing and production techniques introduced to the wallpaper trade around the turn of the century. With their jaunty airplanes and speeding automobiles, these friezes must have looked modern indeed in 1912. They appeared that year in the October issue of *The Wall-Paper News and Interior Decorator,* in an advertisement for the Robert Griffin Company of New York. Cooper-Hewitt Museum library

had set in. A new decorator's spirit was popularized most effectively by Elsie de Wolfe (1865–1950), whose book *The House in Good Taste* was published in 1913. She, like Edith Wharton and many others of her generation, had tired of the relatively dark and dense effects created by the woodwork, paneling, and strongly colored patterns on walls and at windows so popular during the late nineteenth century. Her first impulse was to bring lightness into interiors, and to add some gaiety, missing from the sober designs derived from English theorists.

In the chapter titled "The Treatment of Walls," de Wolfe describes her preference for "plain and dignified painted walls, broken into graceful panels by the use of narrow moldings." Walls should be painted in light tones: cream color was the choice for her own home. When it comes to discussing wallpapers, she contradicts herself constantly, without ever seeming to realize it. She cites the cant of Jones and Eastlake that the wall should be treated as a flat surface decorated only with conventionalized patterns, yet she goes on to illustrate rooms papered in chintz patterns with naturalistic birds and flowers, which became her trademark. Nevertheless, she wrote "Pictorial flowers and lifelike figures have no place upon" the wall. On the other hand, a landscape paper might be nice in a hallway.

At one point she only begrudgingly admits that wallpaper was necessary where the walls were cracked, and concedes that "properly selected wallpapers are not to be despised." However, her mood changes elsewhere, and she is beguiled by "a pale yellowish cream wallpaper" that she finds "very charming with woodwork of white." And with even more rapture she notes "how enchanting" the new black and white papers are, singling out the designs of no less a figure than one of Austria's most avant-garde architects, Josef Hoffman (1870–1955), whom she described as "the great Austrian decorator."

In practice de Wolfe had "a delightful time" decorating with self-conscious chic, conveniently forgetting the rules according to her whim. One is left with an impression that de Wolfe had read the English books and felt obliged to demonstrate her familiarity with what had become accepted principles, but when it came to actually choosing papers for her clients, the most famous of which was the Colony Club of New York, her taste was for the bright, gay products of the French, with their exquisitely lifelike motifs.

On one point she was firm: "those dreadful friezes . . . perpetuated by certain wall paper designers are very bad form and should never be used. Indeed, the very principle of the ordinary paper frieze is bad; it darkens the upper wall unpleasantly."[15]

17-15

Figure 17-15 (opposite) The faces of Charles Dana Gibson's unmistakable "Gibson Girls" gaze from this "Bachelor's Wall Paper" of 1902. It represents yet another novelty introduced to the trade. It was copyrighted by the Life Publishing Company and was advertised with illustrations in the newspapers of many American cities. In this 22-inch-wide sample, machine-printed by the Birge Company of Buffalo, New York, the faces are printed in black over light blue patches of color, giving a mottled effect over a white ground. The ladies' hair is blue. Cooper-Hewitt Museum, 1971-58-3; gift of the Philadelphia Museum of Art

Figure 17-16 Warren, Fuller and Company of New York, the same firm that had advertised their Tiffany wallpapers a decade earlier in the booklet *"What Shall We Do With Our Walls?,"* issued a catalogue in 1894 entitled "Indoors." This illustration from that catalogue shows their all-American version of a "side-wall" pattern, a wide frieze, and a ceiling paper. Victoria and Albert Museum library

17-16

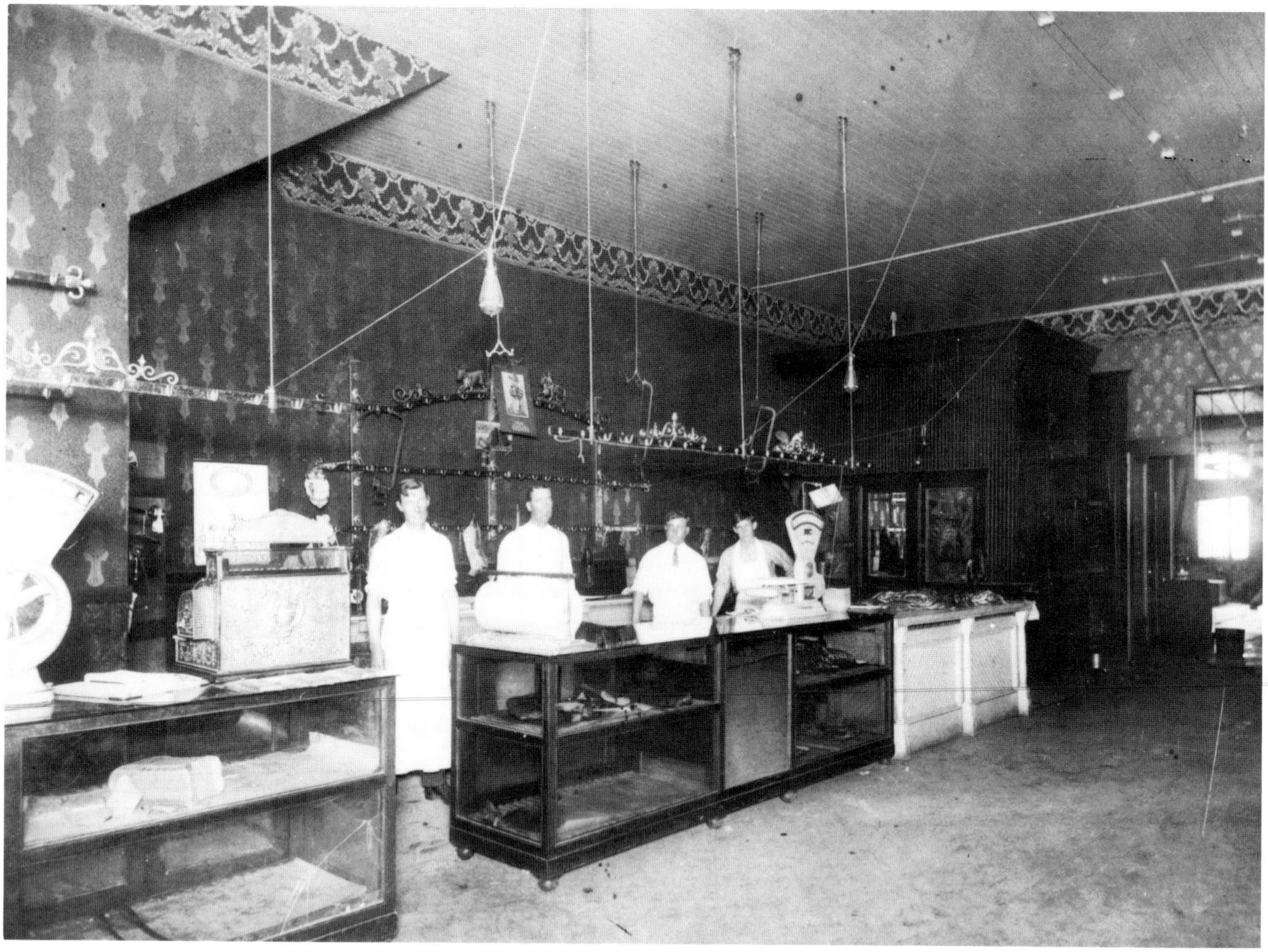

17-17

Figure 17-17 In 1915, wallpapers hung with matching friezes were nearly ubiquitous. Here, the decorations in a meat market in Medford, Oregon, suggest how the mass production of the booming wallpaper industry had affected the look of American interiors of every type. Courtesy of the Southern Oregon Historical Society

17-18

George Leland Hunter (1867–1927) was a writer of more sober and well-researched articles on furnishings, fabrics, and wallpapers and the history of their stylistic development. These were widely published during the same period, appearing in *Country Life in America, Suburban Life, House and Garden,* and *American Homes and Gardens.* In 1913 his articles on wallpaper were collected in a book entitled *Home Furnishing.* They reveal an appreciation of patterns based on design principles that the nineteenth century had cast in opposing and irreconcilable roles. His writings do not exaggerate the differences between conventionalized or abstract patterns and realistic scenes, nor do they suggest that the principles engendering the different kinds of wallpaper were mutually exclusive, as do the inconsistent statements of Elsie de Wolfe. Rather, Hunter reveals an appreciation of a wide spectrum of designs. He could praise and illustrate wallpapers by William Morris and Walter Crane on the same pages with French patterns displaying the most realistic of flowers, birds, and human figures. Hunter commented of Morris: "He knew what so many forget or never learn, that in the creation of masterpieces of art the hand is more important than the head, and the execution than the design."[16] It is debatable that this was an adequate assessment of Morris's accomplishment in these designs, but it clearly reflects the priorities not only of a historian of design during this period, but also of the stylish decorators and housekeepers who were his contemporaries. They were all much more interested in wallpapers that exhibited exquisitely drawn lines, luxurious printing, georgeous color effects, and other signs of skilled, expensive workmanship than they were in any theories lying behind the designs. Visual qualities and the mark of the hand skilled in execution, able to imitate nature with proficiency, were valued above evidence of the workings of the human mind and spirit, which had been so prized by John Ruskin.

As noted above, by the turn of the century avant-garde architects had lost interest in wallpaper design, and many had ceased to use wallpaper at all. However, another group of American architects still found wallpapers legitimate decorations. These

Figure 17-18 (opposite, right) During the late nineteenth century, revivals of late-eighteenth century neoclassical styles for wallpapers often looked like this. A certain attenuation of forms and a crowding of elements within a pattern, combined with colorings in saccharine pastel combinations — here yellow and cream on blue — give an effect very different from the Adamesque models on which they were based. This 19-inch-wide sample was accompanied by a wide frieze with these same vases and pearl swags printed over a ground shaded from blue to white. Cooper-Hewitt Museum, 1941-103-7; gift of Mr. Thomas Molloy

architects followed the teachings of the École des Beaux Arts and participated in the revival of classical architectural styles. Many in this group designed colonial revival houses with interior treatments following models of the past. For middle-class interiors, they encouraged the reproduction of earlier wallpaper styles. Decorators of these houses used re-creations of eighteenth-century patterns in period rooms, but in addition they installed many anachronistic derivations from mid-nineteenth-century models that had been marketed by their manufacturers as "Colonial."

Architectural practitioners in the revival styles shared the interest of decorators like Elsie de Wolfe in achieving *effects* suggestive of the past in rooms adapted to contemporary use and comfort. They were not attempting to make archaeologically correct interiors suitable for museum display. The Beaux Arts houses reflect the same kind of thinking expressed by de Wolfe when she turned her attention to the use of reproduction, as opposed to genuine antique, furniture: she was quite pleased to recommend good reproductions, asking quizzically: "The effect is the thing you are after, isn't it?"[17]

In the opposite camp, many contemporary architects, schooled in the ideas derived from Pugin and Ruskin, considered these "effects" superficial, pretentious, and frivolous, if not dishonest. They came to associate wallpaper not only with the oppressive density of patterning that had burdened the recent past, but also with the chic decorators whose goals were no more serious than the creation of effects these architects deemed appropriate to the theater, but not to daily life. Many of them wrote off wallpaper as a mere decorator's tool, antithetical to good architecture. Ideas of the 1870s and 1880s, grounded in and expressive of moralistic architectural theories, had died a natural enough death in such an atmosphere.

Figure 17-19 A large colorful lithograph, this poster was distributed attached to a calendar for 1903 by the company that then claimed to be the largest wallpaper manufacturers in the world. Meticulously detailed are not only the factory for printing patterns, but also the paper mill, shown in the inset at the upper left, as well as a splendid array of athletic activity in the far background. Author's collection

SYRACUSE
PAPER & PULP CO.
Syracuse, N.Y. U.S.A.
PULP MILL
SKANEATELES, N.Y.
ALL GRADES OF
WALL PAPER
LARGEST WALL PAPER FACTORY IN THE WORLD
SYRACUSE PAPER & PULP CO.
WAREHOUSE
FACTORY
OFFICE
USE SYRACUSE WALL PAPER—THE BEST MADE
1906 JANUARY 1906
1 2 3 4 5 6
7 8 9 10 11 12 13
14 15 16 17 18 19 20
21 22 23 24 25 26 27
28 29 30 31
1906 FEBRUARY 1906
1 2 3
4 5 6 7 8 9 10
11 12 13 14 15 16 17
18 19 20 21 22 23 24
25 26 27 28
1906 MARCH 1906
1 2 3
4 5 6 7 8 9 10
11 12 13 14 15 16 17
18 19 20 21 22 23 24
25 26 27 28 29 30 31
1906 APRIL 1906
1 2 3 4 5 6 7
8 9 10 11 12 13 14
15 16 17 18 19 20 21
22 23 24 25 26 27 28
29 30
1906 MAY 1906
1 2 3 4 5
6 7 8 9 10 11 12
13 14 15 16 17 18 19
20 21 22 23 24 25 26
27 28 29 30 31
1906 JUNE 1906
1 2
3 4 5 6 7 8 9
10 11 12 13 14 15 16
17 18 19 20 21 22 23
24 25 26 27 28 29 30
1906 JULY 1906
1 2 3 4 5 6 7
8 9 10 11 12 13 14
15 16 17 18 19 20 21
22 23 24 25 26 27 28
29 30 31
1906 AUGUST 1906
1 2 3 4
5 6 7 8 9 10 11
12 13 14 15 16 17 18
19 20 21 22 23 24 25
26 27 28 29 30 31
1906 SEPTEMBER 1906
1
2 3 4 5 6 7 8
9 10 11 12 13 14 15
16 17 18 19 20 21 22
23/30 24 25 26 27 28 29
1906 OCTOBER 1906
1 2 3 4 5 6
7 8 9 10 11 12 13
14 15 16 17 18 19 20
21 22 23 24 25 26 27
28 29 30 31
1906 NOVEMBER 1906
1 2 3
4 5 6 7 8 9 10
11 12 13 14 15 16 17
18 19 20 21 22 23 24
25 26 27 28 29 30
1906 DECEMBER 1906
1
2 3 4 5 6 7 8
9 10 11 12 13 14 15
16 17 18 19 20 21 22
23/30 24/31 25 26 27 28 29
Why not have a Big Wall Paper Factory at your Back
ONE THAT GIVES YOU WALL PAPER FROM THE TREE TO THE WALL
BUY OUR LINE—SAVE THE MIDDLEMAN'S PROFIT

CONCLUSION

When pioneers of the Modern Movement took sides against decorators and against architects of the Beaux Arts School, the two factions squared off in fundamental disagreement about the value of pattern and ornament within buildings. Because the Modernists struck wallpaper from the agenda of topics worthy of architectural concern, influential twentieth-century theorists have largely ignored it.

Sigfried Giedion (1888–1968) was one of the most widely read among them. In *Space, Time and Architecture,* published in 1941, he excluded ornament, including wallpaper, from those buildings that he and others of his generation were willing to accept as examples of the new style*less,* "pure" architecture of the twentieth century. Giedion indeed regarded that architecture as the sole heir of history. But "Modern" architecture, for which the forms and theories were solidified during the 1920s, is now recognized as having been simply another style—The International Style—with its own hermetic criteria and distinctive visual mannerisms. Nevertheless, in its heyday, during the middle of the twentieth century, most architects saw it as the apotheosis of the rational processes by which all forms were derived from function, were expressive of structural necessities, and were untainted by decorations. A preoccupation with function as the basic generator of forms dominated architectural theory as well as a great many mid-twentieth-century studies of nineteenth-century design. These studies downplayed all nineteenth-century designs except those that demonstrated the domination of functional concerns. This view influenced studies of the nineteenth-century well into the 1960s. Today, as contemporary architects, designers, scholars, and critics are reassessing the accomplishments of the nineteenth century and recognizing the positive role of the French École des Beaux Arts in exploiting the expressive and empathetic qualities of ornament within the whole fabric of the human environment, the century's interest in wallpaper design is no longer being seen as a ridiculous aberration. This generation now appreciates the importance of getting back to the whole truth of nineteenth-century design and theory rather than emphasizing only one prejudiced point of view. More than that, architects of our own period are again trying to find formal and iconographic bases for ornament, which is exactly what the nineteenth-century designers were rather desperately trying to do.

This book has emphasized the nineteenth century because it was wallpaper's great period—the period when wallpaper was produced in unprecedented and unsurpassed variety and quantity and when it was the subject of theoretical analysis. That period, broadly considered, began in the late eighteenth century as the industrial revolution gathered force, moved into its most full-blown phase beginning in the 1840s, and continued past the turn of the twentieth century.

The nineteenth century was also wallpaper's great period of *use* in America. Because it was present in the poorest houses and the richest, and because it grew out of and was shaped by both massive democratic forces and the powers of capitalism, wallpaper emerges as a key product for assessing the arts of the period. As major elements creating the architectural environments in which most people lived, wallpapers constitute an important mass of art historical documentation. This importance reconfirms the futility of the old distinctions between the fine and the decorative arts.

For example, recognition of the nineteenth century's infatuation with wallpaper suggests a great deal about the period's basic desire to create a safe, ordered, and integrated environment. This analysis reveals points of connection with the whole sweep of art history. Tracing such concerns back to their beginnings in the West leads to Rome. A similar concern for environmental qualities dominated Roman art, just as the desire to create a safe and integrated environment dominated Roman life as a whole. Environmental art—art that created and controlled well-defined spaces—was the great accomplishment of the Romans, an accomplishment that stands in sharp contrast to the sculptural art of the Greeks, in which individuals are depicted asserting their individual wills against the universe, in ways that can be assessed as anarchic.[1] The nineteenth century's sensibilities aligned them and their arts not with the Greeks, but with the Romans. The middle class, which formed the century, was determined, like the Romans, to create a secure environment for itself.

Reflecting that alignment, in 1880 the booklet with which this book began, "*What Shall We Do with our Walls?*" by Clarence Cook, quite aptly went straight to the Romans for classical precedent. On page one Cook stated:

> I think there never will be a better way found for treating the walls of rooms than the old way—of which Pompeii shows us so many examples—of coloring the plaster when it is fresh, with harmonious ground tints, relieved with painted decoration of lines, geometric or flowering patterns, garlands of flowers, dancing nymphs and fluttering cupids, with, not seldom, complete pictures even—their subjects drawn from the mythology of the people.[2]

Thinking past the polemicizing rhetoric of nineteenth-century theorists who insisted on stratifying the "fine" above the "decorative" arts, and on applying distinct rules for designs within each category, Clarence Cook revealed in this passage his appreciation of the classical Italian mastery both of repeating flat patterns and of illusionistic murals for decorating walls. He went on to lament the fact that in his own day:

> People who have learned the practice of Art, who have had art forced into them in training schools and schools of design, do not so overflow with art-productiveness as did these Southern Italians, who lavished design upon everything that they made, from pots and pans, weighing-scales and lamps, to their houses and temples.[3]

Even in a slight work like a booklet promoting wallpapers, Cook followed his mentor John Ruskin in decrying the compartmentalization of life in the nineteenth century, which had resulted in separating art out from the business of living. For Cook in 1880, "with us, all is, thus far, perfunctory and mechanical."

His lament echoes the sentiment expressed by Ruskin in a passage published twenty-one years earlier. Ruskin's reaction to the threat of mechanization and industry in the public world of 1859 had taken the form of retreat into a domestic refuge. However, his defensive strategy in the face of the changing world of the nineteenth century reads as a positive exhortation:

> We are about to enter upon a period of our world's history in which domestic life, aided by the arts of peace, will slowly, but at last entirely, supersede public life and the arts of war. For our own England, she will not, I believe, be blasted throughout with furnaces; nor will she be encumbered with palaces. I trust she will keep her green fields, her cottages, and her homes of middle life: but these ought to be, and I trust will be, enriched with a useful, truthful, substantial form of art.[4]

Ironically, the fact that Ruskin encouraged his contemporaries to take art home contributed to that fracturing of the cultural structure that so pervasively characterized the age. In contrast to the Romans, who applied coherent artistic and architectural standards to their homes, their public baths, their marketplaces, their amphitheaters, and their temples, nineteenth-century English and American theorists all but gave up hope for controlling and ordering vast elements—like factories—within their environments. They encouraged their hearers and readers to concentrate on the only portions of the environment that seemed controllable—their homes. In apparent sympathy with the published theory, Americans in fact devoted unprecedented thought and effort to interior decoration. This helps to explain why the study of physical objects, including the furnishings with which people surrounded themselves during this period, be they chairs, chests, or wallpapers, leads so directly to larger concerns of the era, and so quickly suggests theoretical interpretation. Emerson was more than a century ahead of the new school of scholars who now study what they call "material culture" when he observed in 1859:

> Is it not plain, that not in senates, or courts, or chambers of commerce but in the dwelling-house must the true character and hope of the time be consulted? These facts are, to be sure, harder to read. It is easier to count the census, or compute the square extent of a territory, to criticise its polity, books, art than to come to the persons and dwellings of men and read their character and hope in their way of life. Yet we are always hovering round this better divination. In one form or another we are always returning to it.[5]

APPENDIX A COLORS COMMONLY USED IN EIGHTEENTH- AND NINETEENTH- CENTURY WALLPAPER MANUFACTURE

"Of the colors proper to be used for paper hangings . . . for common designs done with water only . . ."

As published by Robert Dossie, *The Handmaid to the Arts* (London, 1758). The information quoted below is from the 1796 edition, pp. 305–11.

"For *red,* lake, vermilion, rose pink, and red ochre.
For *blue,* Prussian blue, verditer, and indigo.
For *yellow,* the yellow berry wash, Dutch pink, and yellow ochre.
For *green,* verdigrise, or a mixture of the blue colours with the yellow colours, particularly with the yellow berry wash.
For *orange,* vermilion, or red lead, with Dutch pink.
For *purple,* a wash made of logwood, or a mixture of the lake, or rose pink, with deep coloured Prussian blue, or with indigo.
For *black,* ivory black, and, in some nicer cases, lamp black.
For *white,* whiting; and for the heightenings, white lead."

In addition, Dossie advised that carmine might "occasionally" be used "where hangings of more delicate designs and greater value are to be painted . . . but it must be laid on with the pencil, and employed sparingly, otherwise it would too much enhance the expense." In addition to carmine, lake and very bright Prussian blue were cited as "dearer colours." The same colors could be used in varnish, according to Dossie, but must "be had of the makers . . . for this purpose . . . dry," whereas for use in the water-based medium, colors should be purchased in the moist state. Verdigrise and crystals of verdigrise "are with advantage used in varnish, though not proper to be commixt with water . . . turmeric . . . gives a very good yellow . . . in varnish . . . but . . . only on varnished grounds." Dossie also recommended the use of Indian lake, "improperly called safflower," with varnish to make "a much brighter pink ground than any at present".

Rosamond D. Harley, in an excellent recent book, *Artists' Pigments, c. 1600–1835* (New York: American Elsevier Publishing Company, 1970), has provided detailed information about the nomenclature, chemical makeup, and history of the colors listed by Dossie, and by other writers on wallpaper. The following notes on Dossie's color vocabulary are derived from Harley's book, a very useful tool for anyone seeking to interpret color words in early documents or analyzing wallpaper samples of an early period.

Lake: An organic red, made from the females and eggs of insects known as *Coccus lacca,* indigenous to Asia and India. A high-priced color.

Vermilion: Red mercuric sulphide, which occurs naturally in England, though it was also imported. Extremely brilliant. HgS

Rose pink: An organic red, a less brilliant pigment of inferior durability made by mordanting brasil dye on chalk.

Red ochre: Iron oxide. Occurs widely in England. Fe_2O_3

Prussian blue: Potassium ferric ferrocyanide. The first of the artificial pigments with a known history, discovered about 1710. Ms. Harley further describes it as "reasonably priced." $KFe[Fe(CN)_6]$

Verditer: Copper carbonates, manufactured as the byproducts of silver refining in blues and greens. Tending toward green on exposure, blue verditer was more expensive than green verditer. $Cu(OH)_2 \cdot CuCO_3$

Indigo: An organic blue derived from the leaves of the plant genus *Indigotera.* The color was imported from India.

Dutch pink: An organic yellow, obtained from a dye, probably a berry wash. Characteristically very transparent.

Yellow ochre: Iron oxides. A range of oxides, varying from a dull yellow to red and brown. Opaque. Mainly $Fe_2O_3 \cdot H_2O$ or Fe_2O_3.

Verdigris: Copper acetate. The oldest of the manufactured copper greens. It has a tendency toward blue. Verdigris was cheap and easy to obtain in the eighteenth century. $Cu(CH_3COO)_2 \cdot 2Cu(OH)_2$.

Redlead: Orange-red pigment produced by prolonged heating of lead. It was inexpensive, and manufactured on an industrial scale. Pb_3O_4.

Whiting: Chalk—Calcium carbonate. $CaCO_3$

White lead: Lead carbonate. Manufactured on a fairly large scale in England. $2PbCO_3 \cdot Pb(OH)_2$.

Carmine: An organic red made from dried bodies of *coccus cacti,* cochineal insects. A truer red than lac lake from India, which has a purplish cast. Carmine fades rapidly on exposure to light.

Turmeric: An organic yellow, a dye derived from the tuberous roots of varieties of *Curcama,* which grows in India and southeast Asia.

"The colors used by the paper-hangers are the following," according to Andrew Ure's *Dictionary of Arts, Manufactures, and Mines* (London, 1839). A later edition (New York: D. Appleton, 1863) was used in transcribing this list:

Whites These are either white-lead, good whitening, or a mixture of the two.

Yellows These are frequently vegetable extracts; as those of weld, or of Avignon or Persian berries, and are made by boiling the substance with water. Chrome yellow is also frequently used, as well as the *terra di Sienna* and yellow ochre.

Reds are almost exclusively decoctions of Brazil wood.

Blues are either Prussian blue, or blue verditer.

Greens are Scheele's green, a combination of arsenious acid, and oxyde of copper; the green of Schweinfurth, or green verditer; as also a mixture of blues and yellows.

Violets are produced by a mixture of blue and red in various proportions, or they may be obtained directly by mixing a decoction of logwood with alum.

Browns, blacks, and grays. Umber furnishes the brown tints. Blacks are either common ivory or Frankfort black; and grays are formed by mixtures of Prussian blue and Spanish white.

Ure's list provides documentation that new, cheaper manufactured colors had been adopted to make paper hangings in the years since Dossie had described paper stainers' colors. The most significant innovations on Ure's list were chrome yellow, Scheele's green, and Schweinfurt green, colors with known dates of discovery.

Chrome yellow: Derived from chrome ore. $PbCrO_4$. The poisonous color varies from light yellow to strong orange-yellow and has a tendency to discolor on exposure. Discovered in 1797, it was fully described in 1809, and became widely available only after quantities of chrome ore were discovered in the United States in 1820. After that time it became cheap.

Scheele's green: Copper arsenite, a manufactured copper green, was discovered by a Swedish chemist, Scheele, in 1775. Instructions for its manufacture were not published until 1778. In 1812, a process for its manufacture was patented in England.

Schweinfurt green: Copper aceto-arsenite, was first produced commercially in Schweinfurt, Germany, in 1814; the first publication of a method for making the color followed in 1822. A more durable and intense color than other copper greens, it also has a tendency to blacken.

"*Manufacture of paper hangings: the colors employed,*" as listed in the unsigned article published in the periodical *The Decorator* London, 1, no. 8 (September 1864).

"The whites used are French chalk, good whitening; and in some works, white lead is mixed with the latter. The yellows are chrome yellow, *terra de sienna,* yellow ochre, and when vegetable extracts are used, Persian berries. The reds are afforded by decoctions of woods, such as Brazil wood, &c. The blues are artificial ultramarine, Prussian blue, or blue verdila. Some colors are produced by mixtures, such as greens from blues and yellows, and Scheele's green is also used. Violets, browns, blacks, and greys are procured from various vegetable and mineral sources, and from mixtures. All colours are rendered adhesive and consistent by being worked up with gelatinous size or a weak solution of glue."

This note taken from a London trade publication of 1864 documents the use in the wallpaper trade of another datable, relatively new color, artificial ultramarine.

Artificial ultramarine: An inorganic blue of soda, silica, alumina, and sulphur. $Na_{8-10}Al_6Si_6O_{24}S_{2-4}$. A French manufacturer said he had manufactured the color in 1826, but did not reveal his methods until 1828, after which it was commercially produced.

APPENDIX B CHECKLIST OF AMERICAN MANUFACTURERS IN BUSINESS BEFORE 1845

The following checklist of American paper stainers and wallpaper manufacturers in business before 1845 (the period when all production of wallpaper was done by hand) is arranged by city. As gleaned from bills, orders, and invoices, as well as from city directories and newspaper advertisements, it cannot give a complete picture for several reasons. Full runs of such documentation are not always available. Craftsmen did not always advertise or list themselves in directories. Directories often misspell their names. Nor has every source of information from every part of America been exhausted. The list summarizes documentation assembled by the author to date, but is far from definitive. It is hoped that publication of this book will bring forth further documentation to be added to the research files at the Cooper-Hewitt Museum.

The range of dates given for individual craftsmen and firms reflects documented manufacturing activity—the earliest documentation of activity and the latest available. The businesses may well have begun at earlier dates or continued past the last date given here. In a few cases, the range of dates incorporates assumptions. For example, William Doyle of Boston is documented as a paper stainer in 1794 and again in 1796, but not in 1795. It is more than likely that he also worked at the trade in 1795. When craftsmen worked independently, as well as in partnership, they are given more than one listing. For example, Adrian Janes of Hartford worked alone in 1821, as well as from 1831 to 1838, and is so listed. His partnerships, which date from 1822 to 1828 and from 1838 to 1844, are also listed separately. In addition, it should be noted that a number of craftsmen moved from town to town. Thomas S. Webb, for example, worked in Albany, Hartford, and Providence sequentially.

Boston

Boriken, Edward	1810
Bumstead, Josiah	1796–1820
———, Josiah (F.) & Son	1820–28
———	1828–32
Under variant company names the Bumstead family continued to manufacture wallpaper until the end of the nineteenth century.	
Clough, Ebenezer	1795–1817
Cook, John C. & Charles	1831–6
———, Charles	1837–41
Doyle, William	1794–6
Edes, William	1805–6
———, Margaret	1807
Ford, Thomas	1820
Grant, Charles	1821–8
Grant, Moses	1789–1805
———, Moses & Son	1806–10
———, Moses, Jr. & Co.	1811–28
Hovey, Joseph	1786–94
Hurlbert, Jesse P.	1826–8
Ingraham, Francis, Jr.	1800
Marshall, William	1822–9
May, William	1791–8
May & Tolman	1842
Merriam & Brothers	1835–45
Mitchell, Alexander	1805
Pearson, Henry	1821
Prentiss, Appleton	1791–7
Prentiss and May (Appleton Prentiss and William May)	1789–91
Redfern, William	1816–28
Redman, Thomas	1798–1822
Sawyer & Allen	pre-1846
Spear, Charles E.	1823–8
Thayer, Joseph	1826–8
Welsh, John	1786–9

Hartford

Bolles, Isaac & Co.	1812
———, Isaac & J. (Isaac and Jeremiah)	1812–13
Janes, Adrian	1821; 1831–8
Janes & Bolles (Adrian Janes and Edwin Bolles)	1822–8
Janes & Robbins (Adrian Janes and Rowland A. Robbins)	1838–44
Jones, Daniel	1824–5
Jones & Putnam (Daniel Jones and George Putnam)	1821–3
Mills & Danforth (Zecheriah Mills and Edward Danforth; George Putnam joined in 1815.)	1813–15
Mills & Webb (Zecheriah Mills and Thomas S. Webb)	1793
Mills, Danforth & Co.	1816
Mills, Zecheriah	1793–1813
Putnam, George	1819–21; 1824
Putnam & Roff (George Putnam and Amos B. Roff)	1823–4

Rathbone, George S. & Co. (Partnership with Daniels S. Gladding terminated in 1809.)	1808–9; 1813
Webster, James	1803–4
Woodbridge & Putnam (Henry W. Woodbridge and George Putnam)	1816–19
Woodbridge, James R.	1816

New York

Barnes, R. C.	1844–6
Bates, Stephen	1810
Brown, Joshua	1840
Brown, Nathan	1844–56
Carter, Mathews	1840
Case, Wesley	1840–6
Chaveau & Lacarrière	1805
Christy & Constant (Thomas Christy and Samuel S. Constant) This firm—with name variations—continued in business until the 1880s. In some years it is listed as a paper manufacturer; in others, as a paper hanging company.	1844–59
Christy & Robinson	1840–1
Colles, John	1787–91; 1792–1809
———, Richard	1798
———, Richard and John	1791–2
Cooper, Richard	1840
Cottie, Grant	1792–1805
Crygier, Cornelius and John	1797–1801
Day, Thomas, Jr.	1840–6
Doncourt & White	1840–1
Failing, Joseph	1830
Faye, Thomas	1835–61
Fitzgerald, John	1820–5
Harwood, Archibald	1838–40
Hazen, J. H.	1840
Hickey, John	1756
Jones, John	1825–46
Jones, Thomas	1844–6; 1851–6
Jones, J. & T. & Smith (John Jones, Jr., Thomas Jones, and Henry A. Smith)	1850–1
Laforge, Samuel	1820
Leeson, Daniel	1780–3
Lesage, Julius	1844–6
Mooney, William	1790
Ovington, William	1802
Peacock, George	1825–30
Peuscher, George	1840
Pratt & Hardenbergh	1850
Pratt, J. H. & J. M.	1844–7
Prince, R.	1844
Roberts, David	1830
Rugar, John	1765
Sackett, William H.	1844
Sells, John	1805
Shaw, Abner	1830
Stammers, William	1840–4
Strong, Orange	1810–20
Sutphen & Breed	1855–6
Sutphen & Partridge	1844–6
Thériat, Augustus R.	1844
Thériat, E. & C.	1846
Valentine, Joseph	1820
Whiting, Francis H. N.	1844–6
Winans, Isaac	1830–7

Philadelphia

Ashmead, Benjamin	1793
Ashmead, John	1785–97
Ashmead, Thomas	1791
Austin, S.	1806
Beaty, John	1830–45
Belrose & Son	1845–6
Belrose, Son & Blanchard	1844–5
Caldeleugh & Thomas (Robert Caldeleugh and Daniel Thomas) Variant spellings are Caldcleugh, Caldclugh, and Caldclurg (McClelland).	1806
Carnes, Burrill and Edward	1790–4
Chardon, Anthony & Co.	1794–1826
Cook, John	1809–11
Dickinson, Ann	1788
Dickinson, Joseph	1784–8
Fleeson, Plunket Fleeson, an upholsterer, sold hangings "Manufactured in Philadelphia." As yet there is no documentation that he manufactured them, as Nancy McClelland assumed. He was established as early as 1739, but his first advertisement for paper hangings "Manufactured in Philadelphia" appeared in 1769.	1769–late 1770s
Howell & Brothers	1835–19(?)

Howell, John B	1817–1925
Hurley, Thomas	1802
Law, Samuel	1790–1810
Le Collay and Chardon	1789–90
Longstreth, Charles & Sons	1845
Orth & Smith	1811
Poyntell, William	1790–1800
Pugh, Isaac	1828–46
Ryves and Ashmead	1785–6
Ryves and Fletcher	1774–5
Ryves, Ashmead & Poyntell (H. Ryves, J. Ashmead, and W. Poyntell)	1787
Ryves, Edward	1776
Slaughter, Charles C.	1842–5
Slaughter, Francis	1837–40
Smith, Charles	1809
Thomas and Caldeleugh It is likely that this firm is the same as that of Caldeleugh & Thomas, which advertised its "New Manufactory" in 1806.	1798
Van Meter, John	1837–58
Virchaux & Co. (Henry T. Virchaux)	1813–14

Other

Albany, New York	
Howell, John	1813–16
Steel, Lemuel	1822–48
Webb, Thomas S.	1793–5
Baltimore, Maryland	
(Smith, Asa) Although McClelland lists Smith as a manufacturer no documentation has been found to confirm that a person of this name manufactured wallpaper in Baltimore.	
Thomas and Caldeleugh	1798–1809
Williams, Abraham	1803–4
Lee, Massachusetts	
Laflin, W. & W. & Co.	1825–33
New Bedford, Massachusetts	
James, William	1807
Perkins, John	1809–46
Perkins, Smith & Co.	1853–?
New Brunswick, New Jersey	
Hardenbergh, John P.	1844–9
Howell, M. A.	1840–1869
Newark, New Jersey	
White, William P. & Co.	1844–6
Poughkeepsie, New York	
Christy, Thomas & Co.	1835–55
Providence, Rhode Island	
James, Samuel	1803
Newell, Robert	
Schaub, Peter	
Thurber, Samuel, Jr. Newell, Schaub, and Thurber worked together.	1800
Webb, Thomas S.	1800–7
Springfield, New Jersey	
Mackay and Dixey	1790
Steubenville, Ohio	
Cole, James	1820–2
Troy, New York	
Orr, A. & W.	1835–91
Worcester, Massachusetts	
Barry, Joseph From 1859 until 1877 the firm was called Bigelow, Hayden & Co. and was located at Roxbury, Massachusetts.	1840–59

APPENDIX C WALLPAPER REFERENCE COLLECTIONS

1. Bibliothèque Forney
   Hôtel de Sens
   1, rue du Figuier
   Paris, France 75000
2. Bibliothèque Nationale, Department des Estampes
   58, rue Richelieu
   Paris, France 75000
3. Cooper-Hewitt Museum
   Smithsonian Institution
   2 East 91st Street
   New York, New York 10028
4. Deutsches Tapetenmuseum
   (formerly Schloss Wilhelmshöhe)
   Brüder-Grimm-Platz 5
   Kassel, W. Germany 3500
5. Metropolitan Museum of Art
   Print Room
   82nd Street and Fifth Avenue
   New York, New York 10028
6. Musée des Arts Décoratifs
   Palais du Louvre
   Pavillon de Marsan
   107 rue de Rivoli
   Paris, France 75000
7. Musée de L'Impression sur L'Étoffes
   rue des Bonnes-Gens
   Mulhouse, France 68100
8. Museum of Art; Rhode Island School of Design
   224 Benefit Street
   Providence, Rhode Island 02903
9. National Park Service
   a) Boston: North Atlantic Regional Office
      National Park Service
      15 State Street
      Boston, Massachusetts 02109
   b) Philadelphia: Independence National Historical Park
      313 Walnut Street
      Philadelphia, Pennsylvania 19106
10. Society for the Preservation of New England Antiquities
    144 Cambridge Street
    Boston, Massachusetts 02114
11. Stowe-Day Foundation
    77 Forest Street
    Hartford, Connecticut 06105
12. Victoria and Albert Museum
    Print Department
    Exhibition and Cromwell Roads
    London SW 1, England
13. Whitworth Art Gallery
    University of Manchester
    Whitworth Park, Oxford Road
    Manchester, England

APPENDIX D SOURCES

For information about wall coverings, manufacturers currently reproducing old patterns—their addresses and the range of designs offered by each—and information about the preservation and restoration of old wallpapers, the following organizations and journals are suggested as sources:

American Preservation: The Magazine for Historic and Neighborhood Preservation

Bracy House
620 East Sixth Street
Little Rock
Arkansas 72202

American Association for State and Local History

(History News)
1400 Eighth Avenue South
Nashville
Tennessee 37203

Association for Preservation Technology

(Newsletter)
c/o Ann A. Falkner (1980)
Box 2487, Station D, Ottawa
Ontario K1P 5W6
Canada

Building Conservation
Northwood Publications, Ltd.
10–16 Elm Street
London WC 1X
England

The Conservation Center for Art and Historic Artifacts

(Newsletter)
260 South Broad Street
Philadelphia
Pennsylvania 19102

The Conservator: The Annual Journal of the United Kingdom Group

c/o Nigel Williams (1980)
Department of Conservation and Technical Services
The British Museum
London, WC1 B3DG
England

The Decorative Arts Chapter of the Society of Architectural Historians

(Newsletter)

c/o Deborah D. Waters, Secretary (1980)
Winterthur Museum
Winterthur
Delaware 19735

National Trust for Historic Preservation

(Historic Preservation)
(Preservation News)

Occasional publications about sources of materials for preserving and restoring old houses

1785 Massachusetts Avenue, N.W.
Washington, D.C. 20036

Old House Journal
69A Seventh Avenue
Brooklyn
New York 11217

Resources Council, Inc.
979 Third Avenue
New York
New York 10022

Society of Architectural Historians

(Newsletter)
Room 716
1700 Walnut Street
Philadelphia
Pennsylvania 19103

The Victorian Society in America

(Nineteenth Century)
(Newsletter)
East Washington Square
Philadelphia
Pennsylvania 19106

Wallcoverings Information Bureau
66 Morris Avenue
Springfield
New Jersey 07081

Wallcoverings Magazine
(especially annual directory issue)

Publishing Dynamics, Inc.
2 Seleck Street
Stamford
Connecticut 06902

FOOTNOTES

Many of the newspaper advertisements cited in this book are quoted from later transcriptions of advertisements that originally appeared in eighteenth- and early-nineteenth-century newspapers. Titles of the newspapers vary, and are given in these footnotes as they appear in the following compilations of advertisements:

George Francis Dow, *The Arts and Crafts in New England, 1704–1775* (Topsfield, Mass.: The Wayside Press, 1927);

Rita Susswein Gottesman, *The Arts and Crafts in New York,* vol. 1, 1726–76, vol. 2, 1777–99, vol. 3, 1800–04 (New York: The New-York Historical Society, 1938, 1954, 1965);

Alfred Coxe Prime, *The Arts and Crafts in Philadelphia, Maryland, and South Carolina,* Series 1, 1721–85, Series 2, 1786–1800 (Topsfield, Mass: The Wayside Press for the Walpole Society, 1929, 1932). Hereinafter, these volumes will be referred to as Dow, Gottesman, and Prime.

Davida Deutsch transcribed a large number of advertisements that she graciously gave to the author. In addition, Lee Roberts has been good enough to contribute many other transcriptions. These unpublished transcriptions will be briefly acknowledged below as Deutsch or Roberts. In Chapters 9 and 10, all of the references to advertisements in Washington, D. C., newspapers are based on transcriptions that Anne Golovin gave to the author. Many additional transcriptions of miscellaneous advertisements have been contributed by Nancy Goyne Evans and David Kiehl. The author thanks all of these people for so generously sharing their findings.

The author has also transcribed a number of advertisements directly from old newspapers, and when this is the case, no citation of a secondary source is given.

INTRODUCTION

1 Samuel and Joseph C. Newsom, *Picturesque Californian Homes, no. 2* (San Francisco: S. and J. C. Newsom, 1885), p. 5. David Gebhard was good enough to bring this reference to the author's attention.

2 Linda Nochlin, *Realism* (New York: Penguin Books, 1971), pp. 224–5.

CHAPTER 1– THE EARLIEST AMERICAN WALL HANGINGS

1 Twentieth-century scholarship in the fields of American architecture and decorative art makes clear that the dominance of English styles continued from the seventeenth century right through the nineteenth century and faltered only with the rise of French-inspired Beaux Arts Classicism around the turn of the twentieth century. Major works on architecture providing sound bases for study in these fields include, for architecture (in chronological order for the periods covered):

Morrison, Hugh. *Early American Architecture.* New York: Oxford University Press, 1952.

Pierson, William H. *American Buildings and Their Architects: The Colonial and Neo-Classical Styles.* Garden City, N.Y.: Doubleday, 1970.

Scully, Vincent J. *The Shingle Style and the Stick Style.* 1955; rev. edn. 1971. New Haven and London: Yale University Press,

Murray, Richard N.; Pilgrim, Dianne H.; and Wilson, Richard Guy. *The American Renaissance, 1876–1917.* (Catalogue of an Exhibition, Brooklyn Museum, Oct. 13–Dec. 30, 1979) New York: The Brooklyn Museum, 1979. This study includes decorative art as well as architecture.

For American furniture and other decorative arts (in chronological order for the periods covered):

Cooper, Wendy. *In Praise of America: The American Decorative Arts, 1650–1830: Forty Years of Discovery Since the 1929 Girl Scouts Loan Exhibition.* (Catalogue of an Exhibition, National Gallery of Art, 1980.) New York: Alfred A. Knopf, 1980.

Downs, Joseph. *American Furniture, Queen Anne and Chippendale Periods in the Henry Francis du Pont Winterthur Museum.* New York: The Macmillan Company, 1952.

Montgomery, Charles, and Kane, Patricia, eds. *American Art, 1750–1800: Towards Independence.* (Catalogue of an Exhibition, Yale University Art Gallery and Victoria and Albert Museum, 1976.) Boston: New York Graphic Society, 1976.

Montgomery, Charles. *American Furniture: The Federal Period.* New York: The Viking Press, 1966.

Tracy, Berry. *Classical America, 1815–1845.* (Catalogue of an Exhibition, Newark Museum, Apr. 26–Sept. 2, 1963.) Newark, N.J.: Newark Museum Assoc., 1963.

Metropolitan Museum of Art. *19th Century America: Furniture and Other Decorative Arts.* (Catalogue of an Exhibition, Apr. 16–Sept. 7, 1970.) Edited by Berry B. Tracy. New York: New York Graphic Society, 1970.

Clark, Robert Judson, ed. *The Arts and Crafts Movement in America.* (Catalogue of an Exhibition, Princeton University Art Gallery, 1972.) Princeton, N.J.: Princeton University Press, 1972. This study includes architecture as well as decorative arts.

For guidance to bibliographies on brass, ceramics, glass, pewter, silver, and textiles in America see Montgomery, Charles. *A List of Books and Articles for the Study of the Arts in Early America.* Winterthur, Del.: The Henry Francis du Pont Winterthur Museum, 1970.

2 Peter Thornton in *Seventeenth-Century Interior Decoration in England, France and Holland* (New Haven and London: Yale University Press, published for the Paul Mellon Centre for Studies in British Art, 1978) emphasizes the more elaborate wall treatments, describing paneled walls on pp. 71–4. For a comprehensive account of British wallpaper history see Alan V. Sugden and John L. Edmondson, *A History of English Wallpaper 1509–1914* (New

York: Charles Scribner's Sons, 1925; London: B. T. Batsford, 1926). Also useful are: C.C. Oman, *Catalogue of Wall Papers, Victoria and Albert Museum* (London: Board of Education, 1929), and a more recent essay by Clare Crick, "The Origins and Development of Wallpaper," introduction to *Historic Wallpapers in the Whitworth Art Gallery* (Manchester: Whitworth Art Gallery, 1972).

3 Peter Laslett, *The World We Have Lost,* 2nd edn. (New York: Charles Scribner's Sons, 1973), pp. 27, 40, 44. In 1688 Gregory King reckoned there were 200 noble families including about 1,000 people; 800 families of baronets, 600 of knights: 3,000 families of esquires, and 12,000 of gentlemen.

4 Oman, *Catalogue of Wall Papers, Victoria and Albert Museum,* p. 3 and *n.,* p. 73.

5 Abbott Lowell Cummings, introduction to *Rural Household Inventories: 1675–1775* (Boston: The Society for the Preservation of New England Antiquities, 1964), p. xxxiv.

6 Richard Beale Davis, ed., *William Fitzhugh and His Chesapeake World 1676–1701: The Fitzhugh Letters and Other Documents* (Chapel Hill: University of North Carolina Press for the Virginia Historical Society, 1963), pp. 142, 175.

7 Advertisement of Benjamin Church, *Boston Newsletter,* Aug. 22, 1745, as quoted by Walter Kendall Watkins, "The Early Use and Manufacture of Paper Hangings in Boston," *Old Time New England,* 12, no. 3 (Jan. 1922): 111–12. Watkins also notes two other examples of early-seventeenth-century use of tapestry in Boston.

8 Advertisement of Stephen Callow, *The New-York Gazette Revived in the Weekly Post-Boy,* Nov. 6, 1749 (Gottesman 1: 134).

9 Advertisement of John Webster, *Pennsylvania Journal,* Aug. 20, 1767 (Prime 1: 214).

10 As published by Benno Foreman in *Winterthur Newsletter,* Jan. 1970, quoting "Pepperrell Manuscripts," *New England Historical and Genealogical Register,* 49 (Apr. 1865): 147.

11 Cummings, introduction to *Rural Household Inventories: 1675–1775,* p. xxxiv.

12 As quoted by Horace L. Hotchkiss, Jr., "Wallpapers Used in America, 1700–1850," in Helen Comstock, ed., *The Concise Encyclopedia of American Antiques,* 2 vols. (New York: Hawthorn Books, 1958), 2:488.

13 Advertisement of Roper Dawson, *The New-York Gazette or the Weekly Post-Boy,* June 3, 1762 (Gottesman 1: 123).

14 Quoted by Watkins, "Early Use of Paper-Hangings," p. 109.

15 James Birket, *Some Cursory Remarks Made by James Birket in His Voyage to North America 1750–51* (New Haven: Yale University Press, 1916), p. 28.

16 In addition, importation from Holland is documented by at least one eighteenth-century advertisement, that of Cornelius Crygier of New York, which appeared in *The Diary or Loudon's Register,* Sept. 21, 1792 (Gottesman 2, 144).

17 Advertisement of John Mason, upholsterer, *Pennsylvania Journal,* Philadelphia, July 19, 1770 (Prime 1: 208–9).

18 Birket, *Some Cursory Remarks,* p. 8.

19 See footnote 2, Chapter 1.

CHAPTER 2– HOW WALLPAPER WAS MADE

1 Robert Dossie, *The Handmaid to the Arts* (London, 1758). The author was able to use not this first known edition, but another London edition published by A. Millar, W. Law, and R. Carter in 1796, in 2 volumes, now at the Victoria and Albert Museum Library. All quotations from Dossie that follow in this chapter are from this 1796 edition, and all are to be found within an appendix entitled "On the Manufacture of Paper Hangings," vol. 2, pp. 304–16. Dossie's work is quoted so extensively here in Chapter 2 that individual citations are not noted subsequently. All the quotations that follow can be found within this thirteen-page appendix.

2 Nancy McClelland, *Historic Wall-Papers from Their Inception to the Introduction of Machinery* (Philadelphia and London: J. B. Lippincott Company, 1924), p. 30. Walter Kendall Watkins, "The Early Use and Manufacture of Paper Hangings in Boston," *Old Time New England,* 12, no. 3 (Jan. 1922): 110. E. A. Entwisle, *A Literary History of Wallpaper* (London: Batsford, 1960), p. 82, quoting Luke Herbert, *Engineers' and Mechanics' Encyclopaedia,* 1836.

3 Watkins, "Early Use of Paper Hangings," pp. 109, 110.

4 As late as 1796 Josiah Bumstead of Boston advertised in *Columbian Centinel,* Oct. 1, 1796: "Wanted; a number of reams of hanging paper." (Deutsch) The use of the word "reams"—a ream was made up of about five hundred sheets of paper—suggests that at that late date in the eighteenth century it remained the job of the paper stainer rather than that of the paper maker to join sheets to form rolls. Although paper mills had been established here as early as 1690, Americans still had to import large quantities of foreign paper. It seems probable, therefore, that European hanging paper stock could have been imported in its raw state and decorated by American paper stainers. The earliest record known to the author specifying American production of paper specifically for hangings is a 1791 diary reference to paper mills at Watertown, Massachusetts. These mills were reported to be "employed in the making of paper for the Blocks and Stamps used for Hangings, & c." See Watkins, "Early Use of Paper Hangings," p. 119.

5 *The Compleat Appraiser, The 4th Edition. Greatly Improved* (London, 1770), p. 3. Florence Montgomery was good enough to bring this source to the author's attention.

6 Advertisement of Appleton Prentiss, *Independent Chronicle,* Boston, Apr. 20, 1791. As quoted by Watkins, "Early Use of Paper Hangings," p. 117. Advertisement of Moses Grant, *Independent Chronicle,* Boston, Apr. 2, 1789. Quoted by Watkins, "Early Use of Paper Hangings," p. 115.

Advertisement of Appleton Prentiss and William May, *The Mas-*

sachusetts Centinel, Boston, Sept. 23, 1789 (Deutsch). Later in the nineteenth century, 8 yards was to become the American standard for a roll of wallpaper.

7 For a more detailed discussion of the dating of the earliest use of machine-made papers by paper stainers, see Chapter 14, p. 302.

8 Andrew Ure, *A Dictionary of Arts, Manufactures, and Mines in Two Volumes* (New York: D. Appleton, 1863), p. 327. First edition London, 1839.

9 On painting techniques and materials for Chinese wallpapers, see: Alan V. Sugden and John L. Edmondson, *A History of English Wallpaper 1509–1914* (New York: Charles Scribner's Sons, 1925; London: B. T. Batsford, 1926), p. 102. C.C. Oman, *Catalogue of Wall Papers, Victoria and Albert Museum* (London: Board of Education, 1929), p. 69. McClelland, *Historic Wall-Papers,* p. 90. "Chinese Export Wallpapers," an unpublished ms. written for a graduate course at Columbia University by Elizabeth McLane in the study files of the Cooper-Hewitt Museum.

10 Evidence that American paper stainers used stencils is provided in the "Invoice, of sundry Paperhangings, Prints, and Utensills [sic], remaining on hand at the Manufactory No 43 Marlborough Street," Appleton Prentiss, invoice no. 39 in the invoice book of a commission merchant, Mss. 761 (1786–97), Baker Business Library, Harvard University, Cambridge, Mass. Although no precise date is written on the invoice, it appears to date from 1788–9.

11 Wood blocks, or "prints," were the most important of the paper stainer's tools. The listing of tools belonging to Appleton Prentiss of Boston (see footnote 10, this chapter), includes "13 Setts Prints" valued at £101.2.0 and "26 Setts Prints" valued at £66.12.0. These were the most numerous, and costliest, utensils on hand at the manufactory. They were much more important than the "25 Stancils" already mentioned, valued at only fifty shillings (see footnote 2, this chapter).

Engraving done on printing presses using metal plates had been utilized early in the history of paper staining to make single sheets of patterning. Some engraving was continued in the craft well into the eighteenth century. A Boston merchant in 1767 offered "beautiful copper-plate furniture Paper for Rooms"—advertisement of Samuel Fletcher, *Boston News-Letter,* Oct. 1, 1767 (Dow, p. 152). British records do include eighteenth-century patents for engraving paper hangings. However, wood block printing was the standard way of ornamenting wallpaper prior to the introduction of cylinder printing machines, and it was continued for finer work long past that time. In fact, there are still a very few craftsmen block-printing fine wallpapers in Alsace, London, and Paris.

12 While Dossie describes a simple leather-covered block on which the color is spread so that it can be picked up on the "print," other writers describe much more complicated devices.

James Smith in *The Mechanic or Compendium of Practical Inventions* (London: Henry Fisher, 1824), p. 168, described "the woollen cloth sieve on which the colour is laid and spread by a boy (called the Tere boy) with a hair-brush. This cloth sieve is laid upon a leather sieve impervious to moisture, and it floats upon some gum-liquor, in a wooden vessel." (David W. Kiehl was good enough to bring this source to the author's attention.)

Ure, *A Dictionary of Arts,* p. 326, calls it a "swimming-tub," explaining: "The tub in which the drum or frame covered with calf-skin is inverted, contains water thickened with pairings of paper from the bookbinder . . ."

13 Ure, *A Dictionary of Arts,* p. 326.

14 McClelland, *Historic Wall-Papers,* p. 59. While Dossie provided English paper stainers with instructions for making their own flock, the French paper stainers, according to Ure, purchased their flock in a white state and dyed it to their own specifications. From the 1780s, there is evidence that American paper stainers were flocking some of their products: the 1789 inventory of a Boston craftsman, John Welsh, included eight bags of colored flocks valued at forty shillings. Watkins, "Early Use of Paper-Hangings," p. 115.

15 Ure, *A Dictionary of Arts,* p. 327.

16 In addition to flocking and fake flocking, other powdery substances were adhered to the surfaces of wallpapers. The "frosted" and "spangled" papers advertised in America during the eighteenth and early nineteenth centuries were probably made from what Dossie described as "that kind of talc called isinglas [mica] which, being reduced to a gross flaky powder, has a great resemblance to thin silver scales or powder." Like flock, spangles were scattered over areas prepared with varnish and were used both for grounds and for applied patterns.

CHAPTER 3– EIGHTEENTH-CENTURY ENGLISH WALLPAPER STYLES

1 See pp. 102–3.

2 Alan Sugden and John L. Edmondson, *A History of English Wallpaper 1509–1914* (New York: Charles Scribner's Sons, 1925; London: B.T. Batsford, 1926), p. 40, quoting an advertisement that appeared in a London newspaper, *The Postman,* Dec. 10, 1702.

3 "Invoice of Sundries sent to America," Manuscript in New York Public Library, Schuyler Papers, Box 2. Schuyler was in England during 1761 and 1762. The invoice was brought to the author's attention by Kristin Gibbons. Perhaps the puzzling word "Dover" is a corruption of "Divers" for diverse—meaning multicolored? D° is an abbreviation for ditto.

4 Sugden and Edmondson, *A History of English Wallpaper,* p. 40*n.,* quoting correspondence of Mrs. Delaney. Other variant spellings of "caffy" include "caffa" and "caphas" and (French) "*caffas.*" Florence Montgomery has suggested to the author that *cafard* and *caffart* seem to mean "counterfeit" or "imitation" in French, reinforcing the probability that *caffas* wallpapers are flocked imitations of damask.

5 "Stoco"—wallpaper ornaments imitating plaster work. "Ni-

cholls"—perhaps relates to the word "Nicol" given in the 1933 edition of *The Oxford English Dictionary* as an obsolete and rare variant of Nickel . . . Nickel-green, annabergite. "Tripolys"—possibly polished papers: *The Oxford English Dictionary* describes Tripoly/Tripoli as a fine earth used as a polishing powder. Or again, perhaps "tripolys" on Van Schuyler's invoice is a gross corruption of the word "trophies."

6 Edna Donnell in "The Van Rensselaer Wall Paper and J. B. Jackson: A Study in Disassociation," *Metropolitan Museum Studies* 1, Part 1 (Feb. 1932): 77–108, resoundingly disproved the attribution of Lee and Van Rensselaer papers to Jackson.

7 A trade card of the paper stainer William Masefield, in the Banks Collection, British Museum (illustrated as figure 43 in Sugden and Edmondson, *A History of English Wallpaper*), includes "Paintings of Landscapes, Festoons and Trophies."

8 Willard S. Randall, *The Proprietary House in Amboy* (Perth Amboy, N. J.: The Proprietary House Association, 1975), pp. 12, 13, 17. Lee Roberts kindly brought this to the author's attention.

9 These included: "Paper Machee for the Ceiling of two Rooms, one of them 18 Feet Square, the other 18 by 16 with Cr. [corner] Chimneys." "Invoice of Sundry Goods to be ship'd by Mr. Washington of London for the use of G. Washington 1757 April 15," in John C. Fitzpatrick, ed., *The Writings of George Washington from the Original Manuscript Sources 1745–1799,* 39 vols. (Washington D.C.: U.S. Government Printing Office, 1931), 2 (1757–1769): 23.

10 Advertisement of John Blott, *The South Carolina Gazette,* Charleston, S.C., May 11, 1765 (Prime 1: 275). Advertisement of James Reynolds, Carver and Gilder, *The Pennsylvania Chronicle,* Philadelphia, Nov. 7–14, 1768 (Prime 1: 225).

11 Quoted by Esther Singleton, *Social New York under the Georges* (New York: D. Appleton, 1902), p. 43.

12 Sugden and Edmondson, *A History of English Wallpaper,* p. 72.

13 The room is illustrated by Eric Entwisle in *The Book of Wallpaper* (London: Arthur Barker, 1954) as plate 32.

14 Advertisement of Joseph Dickinson, *Pennsylvania Journal,* Philadelphia, Aug. 25, 1784 (Prime 1: 281).

15 The "Gothic" pattern illustrated at the bottom in *color plate 23* bears close comparison to elements in plates VI and IX of Batty and Thomas Langley's *Gothic Architecture Improved by Rules and Proportions,* London, 1742. Reissued 1972, from the London edition of 1747, by Benjamin Blom, Inc., New York.

16 Advertisement of Charles Digges, *Maryland Gazette,* Annapolis, Md., Mar. 5, 1761 (Prime 1: 274). Advertisements of Thomas Lee, both in issues of the *Boston News-Letter,* May 17, 1764 (Dow 1: 152), and Jan. 24, 1765, and in issues of the *Boston Post-Boy and Advertiser,* July 30, 1764, and Sept. 3, 1764.

17 Advertisement of Zecheriah Mills, *Connecticut Courant,* Hartford, Conn., Jan. 25, 1796.

18 Evidence that Revere sold imported wallpapers is provided by advertisements in the *Independent Chronicle,* Boston, Jan. 1, 1784, through Sept. 30, 1784 (Deutsch).

19 James Fenimore Cooper, *The Pioneers* (New York: Grosset and Dunlap, 1832), pp. 56–7. Albert J. von Frank brought this passage to the author's attention.

20 C. C. Oman, *Catalogue of Wall Papers, Victoria and Albert Museum* (London: Board of Education, 1929), p. 37. Oman cites the specific act of law imposing death for offenses against the regulations as 46 George III c. 112. In this catalogue, Oman further discusses tax stamps on pp. 47 and 48. More detailed treatment of excise marks appears in Ada Longfield Leask, "Excise Marks on English and Irish Wallpapers," *Antiques,* 88, no. 2 (Aug. 1965): 216–19. And in Ada Longfield Leask, "Some Later Excise Marks on English Wall Papers," *The Paper Maker* (Wilmington, Del.: Hercules Powder Co., Inc.), 35, no. 2 (1966). Since the monogram GR was the most frequently used mark, and since the letters and numbers accompanying the monogram have not been decoded, precise dating of most English excise marks is not yet possible.

21 Florence Montgomery, *Printed Textiles: English and American Cottons and Linens 1700–1850* (New York: Viking Press: A Winterthur Book, 1970), pp. 36–46.

22 Skipwith Papers, Earl G. Swem Library, The College of William and Mary in Virginia. William B. Hill brought these papers to the author's attention. In this survey, space does not permit the consideration of the Irish wallpapers exported to America. Irish styles followed the English. A system levying excise payments paralleled the English system, with some excise stamps on the papers incorporating an Irish harp rather than the British crown. Notices of the arrival of paper hangings from Ireland appeared in Boston and Philadelphia newspapers of the 1780s. The Irish paper stainers and their products have been described in published articles by Ada Longfield Leask, including: "Old Wallpapers in Ireland," *Journal of the Royal Society of Antiquities in Ireland,* 81 (1951); "Stucco and Papier Mache in Ireland," *Journal of the Royal Society of Antiquities in Ireland,* 78 (1948); "Wallpaper and Legislation," *Journal of The Royal Society of Antiquities in Ireland,* 92 (1968).

23 Advertisement of Warrington and Keene, *The Virginia Gazette and Independent Chronicle,* Richmond, Va., June 28, 1788.

24 Eric A. Entwisle, *A Literary History of Wallpaper* (London: B. T. Batsford, 1960), p. 56, quoting what he cites only as *Hamwood Papers of the Ladies of Llangollens . . .* Edited by Mrs. G. H. Bell, 1930. Writing under date July 13, 1788.

25 See Betty Ring, "Memorial Embroideries by American Schoolgirls," *Antiques,* 100, no. 4 (Oct. 1971): 571, fig. 2.

CHAPTER 4– EIGHTEENTH-CENTURY FRENCH WALLPAPER STYLES

1 Joel Munsell, *A Chronology of Paper and Paper-Making* (Albany: J. Munsell, 1857), p. 19.

2 Henri Clouzot and Charles Follot in their book *Histoire du Papier Peint en France* (Paris: Éditions d'Art Charles Moreau, 1935) traced through the eighteenth century the developments in France that led to that country's preeminence in wallpaper production. The history of the craft was further studied in a very useful exhibition catalogue, *Trois Siècles de Papiers Peints* (Paris: Musée des Arts Décoratifs, 1967), with an essay by Jean-Pierre Seguin. Nancy McClelland, *Historic Wall-Papers from Their Inception to the Introduction of Machinery* (Philadelphia and London: J. B. Lippincott Company, 1924), also deals with the subject at some length, and gives a lively account of Réveillon and his factory under siege in 1789.

3 The following list of artists whose designs were used by Réveillon is compiled from Clouzot and Follot, *Histoire du Papier Peint,* p. 66, and from McClelland, *Historic Wall-Papers,* pp. 415–30. Full names and dates are not given in either source but when available have been taken from Benezit, *Dictionnaire des Peintres.* Clouzot and Follot comment on the difficulty of identifying Réveillon's designers, and mention that Réveillon may simply have used designs engraved after these masters rather than actually hiring them to produce wallpaper designs:

Félix Boisselieur, otherwise unknown except through records that list him as a designer for Réveillon.

Juste-Nathan Boucher, called François, le jeune, 1736–82.

(François ?) Cietti, if François, active by 1758. The designer named Cietti who worked for Réveillon is recorded as having died in 1794.

Desrais, 1752–89.

Jean-Baptiste Fay, worked for Réveillon 1775–89.

Jean-Baptiste Huet, 1745–1811.

Lafosse, perhaps Jean-Charles Delafosse, 1734–89 (?)

Étienne de Lavallée-Poussin, 1733–93.

Paget, otherwise unmentioned except in records noting that he directed design for Réveillon.

Jean Louis Prieur, le jeune, 1759–93.

4 The papers were purchased through New York and Boston dealers. Bills of 1794 and 1795 enumerating purchases for the Phelps–Hatheway house from Moses Grant of Boston and Cornelius Crygier of New York survive at the New York State Library. Arthur W. Leibundguth graciously provided photocopies of these bills for the author's use.

5 The "old man" would have been Jean Jacques Arthur, active 1789–c.1794 in association with François Robert. Jefferson's letter from which I have quoted was published by Julian P. Boyd, ed., *The Papers of Thomas Jefferson* (Princeton, N.J.: Princeton University Press, 1961), 16 (1789–1790): 322.

6 Wallpapers at the John Brown house are more fully described by Wendy Cooper in "The Purchase of Furniture and Furnishings by John Brown, Providence Merchant, Part II, 1788–1803," *Antiques,* 102, no. 4 (Apr. 1973): 734–43. The color illustration in *color plate 12* is used through the courtesy of Wendy Cooper.

7 Henry Wansley, *An Excursion to the United States of North America in the Summer of 1794,* 2nd edn., rev. (Salisbury, England: J. Easton, 1798), p. 123.

8 The design is illustrated by Clouzot and Follot, *Histoire du Papier Peint,* p. 69.

9 Advertisement of Thomas and Caldeleugh, *Federal Gazette and Baltimore Daily Advertiser,* Apr. 12, 1798 (Deutsch).

10 William Strickland, *Journal of a Tour in the United States of America, 1794–1795,* ed. Rev. J. E. Strickland (New York: The New-York Historical Society, 1971), p. 117. The author is indebted to Nancy Goyne Evans for this reference. Additional examples of French wallpapers of the late eighteenth century hung in America are illustrated in McClelland, *Historic Wall-Papers,* especially papers in the Quincy house, Quincy, Mass., pp. 235, 243, and papers in the Imlay house in Allentown, N.J., p. 257.

11 Advertisement of Francis Delorme, *General Advertiser,* Philadelphia, Nov. 18, 1790 (Prime 2: 219). Delorme was also spelled "De L'orme."

12 Advertisement of Burrill and Edward Carnes, *Pennsylvania Packet,* Philadelphia, Apr. 27, 1793 (Prime 2:275)

CHAPTER 5– ORIENTAL WALLPAPERS

1 Elizabeth McLane, "Chinese Export Wallpapers," unpublished ms. written for a graduate course at Columbia University, 1972, p. 19. A copy is in the study files of the Cooper-Hewitt Museum.

Alan V. Sugden and John L. Edmondson, *A History of English Wallpaper 1509–1914* (New York: Charles Scribner's Sons, 1925; London: B. T. Batsford, 1926), p. 97, assert that "A set invariably comprised twenty-five sheets, four feet wide and twelve feet long." Nancy McClelland, *Historic Wall-Papers from their Inception to the Introduction of Machinery* (Philadelphia and London: J. B. Lippincott Company, 1924), p. 90, calls the paper the Chinese used "rice paper." Sugden and Edmondson (p. 102) add that the "Chinese hand-made paper was generally pasted on stout cartridge paper affixed to coarse linen or canvas for convenience of hanging."

2 McLane, "Chinese Export Wallpapers," pp. 11–12.

3 Waldemar H. Fries identified twenty-five birds pasted on the wallpaper at Temple Newsam as cut from ten plates in the first volume of Audubon's *The Birds of America,* according to Brenda Greysmith, *Wallpaper* (New York: Macmillan, 1976), p. 70.

4 Advertisement of James Rivington, *The New York Daily Gazette,* Aug. 11, 1790.

5 Advertisement of Thomas Hurley, upholsterer, *Pennsylvania Packet,* Philadelphia, May 9, 1786.

6 Lady Jean Skipwith, Prestwould, Clark County, Va., to James Maury, Aug. 27, 1795, Skipwith Papers (ViW(SPS)), Earl G. Swem Library, The College of William and Mary in Virginia.

William B. Hill generously brought these papers to the author's attention.

7 Robert Dossie, *The Handmaid to the Arts,* 2 vols. (London: A. Millar, W. Law, and R. Carter, 1796; first known edn., 1758), 2:311. Advertisement of Joseph Dickinson, *Pennsylvania Packet,* Philadelphia, Sept. 7, 1785 (Prime 1: 281).

8 "Invoices of Sundry Goods Ship'd by Rich[d] Washington . . . [to] George Washington," Nov. 2, 1757, in Invoices and Letters, 1755–1766, in Original Manuscripts in the Washington Papers at the Library of Congress, Washington, D.C., p. 7. This and all subsequent references to wallpaper in the papers of George Washington were brought to the author's attention by Christine Meadows.

9 Advertisement of Robert Moore, cabinet maker, *Dunlap's Maryland Gazette or the Baltimore General Advertiser,* July 3, 1775 (Prime 1:278).

10 "Invoice, of sundry Paperhangings, Prints, and Utensills [sic], remaining on hand at the Manufactory No 43 Marlborough Street," Appleton Prentiss, Invoice no. 39 in the invoice book of a commission merchant, Mss. 761 (1786–97), Baker Business Library, Harvard University, Cambridge, Mass. Although no precise date is written on the invoice, it appears to date from 1788–9.

11 Thomas Hancock of Boston to John Rowe, London, Jan. 23, 1737. Portions of this letter were published by Walter Kendall Watkins in 1922 in "The Early Use and Manufacture of Paper-Hangings in Boston," *Old Time New England,* 12, no. 3 (Jan. 1922): 110. Sugden and Edmondson published a more complete version of the letter in 1925 in *A History of English Wallpaper,* pp. 49–50, and accompanied it with information about Dunbar of Aldermanbury.

12 Advertisement of John Arthur, *The New-York Gazette and the Weekly Mercury,* Nov. 22, 1773 (Gottesman 1: 316).

13 Advertisement of Henry W. and Lewis Phillips, *New-York Evening Post,* Feb. 17, 1804 (Gottesman 3: 161).

14 As quoted by Rodris Roth in "Interior Decoration of City Houses in Baltimore: The Federal Period," *Winterthur Portfolio 5* (1969), p. 67.

15 "Invoice of Goods, to be shipped Nath. Lyt. Savage, by the first ship bound to York. May 8, 1768," in the papers of Nathaniel Littleton Savage, transcriptions in the Library of the Museum of Early Southern Decorative Arts, Salem, N.C., from the originals either in the county records of Northampton County, Va., or in the Savage papers at the Huntington Library, San Marino, Ca. Susan Stitt generously gave this reference to the author.

16 George Washington to Robert Morris, Oct. 2, 1787. Fitzpatrick, John Clement, ed., *The Writings of George Washington from the Original Manuscript Sources, 1745–1799,* 39 vols. (Washington D.C.: United States Government Printing Office, 1931–44), 29 (Sept. 1786–June 1788, printed Jan. 1939): 283.

17 "Mercantile Value of the Fine Arts. No. V.—Paper-Hanging," *The Art-Union: A Monthly Journal of the Fine Arts and the Arts Decorative and Ornamental* (6, no. 70) (London, July 1, 1844): 182.

CHAPTER 6– EIGHTEENTH CENTURY AMERICAN WALLPAPERS

1 Gottesman 1: 237; Dow: 152; J. Leander Bishop, *A History of American Manufactures from 1608–1860* (Philadelphia: Edward Young, 1868), 1: 209, cites *Dodsley's Annual Register,* 8:55, and 9:62.

Advertisement of Plunket Fleeson, *Pennsylvania Gazette,* Philadelphia, Oct. 19, 1769. Nancy McClelland appears to have been in error when she gave a date of 1739 as marking the beginning of wallpaper manufacturing in America. (Nancy McClelland, *Historic Wall-Papers from Their Inception to the Introduction of Machinery* (Philadelphia and London: J.B. Lippincott Company, 1924), p. 241. Apparently she assumed that when Fleeson went into the upholstery business in 1739 he must have begun to manufacture paper hangings. In fact, this does not seem to have been the case. Fleeson's advertisements, which are lengthy and quite specific, and have been studied in detail, do not indicate that he ever manufactured wallpapers. Only in 1769 does he begin to advertise that he sells papers "manufactured in America," although he does not specify who actually made them. See also Milo M. Naeve, "Plunket Fleeson, Eighteenth-Century Upholsterer," *Winterthur Newsletter,* Jan. 28, 1958, pp. 3–4.

2 *The New-York Packet,* Aug. 5, 1788.

3 *The Daily Advertiser,* New York, Oct. 26, 1789. The parade is also described in the *Massachusetts Centinel,* Boston, Oct. 28, 1789.

4 Justin Winsor, *The Memorial History of Boston,* 4 vols. (Boston: Osgood, 1883), 4:78, n. 2.

5 As quoted by Nancy McClelland, *Historic Wall-Papers,* p. 238.

6 A report on the legislation prohibiting imports to Massachusetts, including paper hangings, appeared in the *Massachusetts Centinel,* Boston, Dec. 2, 1786.

7 John W. Maxson, Jr., "American Papermakers in the Great Tariff Debate," *The Paper Maker,* 31 (Wilmington, Del.: The Hercules Powder Co., Inc., 1962): 1.

8 Advertisement of Edward Ryves, *Pennsylvania Journal,* Philadelphia, Oct. 18, 1785 (Prime 1:280).

Advertisements of Joseph Dickinson, *Pennsylvania Packet,* Philadelphia, Oct. 18, 1785 (Prime 1:281–2), and Apr. 29, 1786 (Prime 2:277).

Notice of the arrival in Baltimore of Colay (Le Collay), Chardon, and Orinard, *Daily Advertiser,* New York, May 22, 1789, and advertisement of Le Collay and Chardon, *Federal Gazette,* Philadelphia, July 22, 1789 (Prime 2:281). The New York newspaper revealed: "It is uncertain in what place these gentlemen will fix themselves, but whatever it shall be, their success is most earnestly wished, and amounts to a certainty."

9 Advertisement "WANTED for the New Paper-Hanging Manufactory, A Person who can speak English and French as a Clerk," advertiser not named, *Pennsylvania Packet,* Philadelphia, June 25, 1789 (Prime 2:286).

Advertisement for Engravers in Wood and printers repeated in English and French by an unnamed advertiser, *Daily Advertiser,* New York, Nov. 24, 1794.

Advertisement of Cornelius and John Crygier, *Mercantile Advertiser,* New York, July 1, 1799, and Apr. 17, 1799. Lee Roberts found this advertisement and was good enough to give it to the author.

10 James Smith, *The Mechanic or Compendium of Practical Inventions* (London: Henry Fisher, 1824), p. 168, speaks of "Tereboys." Andrew Ure, *A Dictionary of Arts, Manufactures, and Mines in Two Volumes* (New York: D. Appleton, 1863; first edn. London, 1839), p. 326, uses the French *"tireur."*

11 Advertisement of William James, *New Bedford Mercury,* Aug. 14, 1807.

12 William Poyntell of Philadelphia called for "Print Cutters," adding "Neat Workmen at drafting and cutting, may have employ for the winter," in an advertisement in the *Pennsylvania Packet,* Dec. 4, 1797 (Prime 2:285). Le Collay and Chardon of Philadelphia advertised: "Wanted, Two experienced stampers (or Printers)," in the *Federal Gazette,* Philadelphia, July 22, 1789 (Prime 2: 281).

13 Advertisements of Burrill and Edward Carnes, *Federal Gazette,* Philadelphia, Sept. 20, 1792 (Prime 2:275), and *Pennsylvania Packet,* Philadelphia, Apr. 27, 1793 (Prime 2:275). Advertisement of Anthony Chardon, *Pennsylvania Packet,* Philadelphia, Mar. 18, 1794 (Prime 2:275).

14 The business of William Poyntell is treated at length by Horace Hotchkiss in "Wallpaper from the Shop of William Poyntell," *Winterthur Portfolio 4* (Charlottesville, Va.: University Press of Virginia, 1969), pp. 26–33. The quotation is from advertisement of William Mooney, *The New York Daily Gazette,* July 29, 1790.

Edwin P. Kilroe, *Saint Tammany and the Origin of the Society of Tammany or Columbian Order in the City of New York* (New York: M. B. Brown Printing & Binding Co., 1913), p. 119.

15 Advertisement of Joseph Hovey, *The Massachusetts Centinel,* Boston, May 20, 1789 (Deutsch). Advertisement of William Poyntell, *Federal Gazette,* Philadelphia, Oct. 9, 1792 (Prime 2:284).

16 Advertisement of Joseph Dickinson, *Pennsylvania Packet,* Philadelphia, Jan. 30, 1786 (Prime 2:276). Advertisement of A. S. Norwood, *New York Gazette and General Advertiser,* Feb. 15, 1799 (Gottesman 2:151). Advertisement of Zecheriah Mills, *Connecticut Courant,* Hartford, Feb. 4, 1807, as quoted by Phyllis Kihn, "Zecheriah Mills, Paper-Hanging Manufacturer of Hartford, 1793–1816," *Connecticut Historical Society Bulletin,* 26, no. 1 (Jan. 1961):27.

17 Advertisement of Joseph Hovey, *The Massachusetts Centinel,* Boston, May 7, 1788.

18 Advertisements of Joseph Dickinson, *Pennsylvania Journal,* Philadelphia, Aug. 25, 1784 (Prime 1:281), and *Pennsylvania Packet,* Philadelphia, Mar. 9, 1787 (Prime 2:278). Advertisement of Josiah Bumstead, *Columbian Centinel,* Boston, Oct. 1, 1796.

19 Advertisement of Gerardus Duyckinck, Jr., for papers made by John Colles, *Daily Advertiser,* New York, Apr. 12, 1787 (Deutsch).

20 Advertisement of Joseph Dickinson, *Pennsylvania Packet,* Philadelphia, July 5, 1786 (Prime 2:276).

21 Advertisement of Burrill and Edward Carnes, *Pennsylvania Packet,* Philadelphia, Apr. 27, 1793 (Prime 2:275).

22 As quoted by Walter Kendall Watkins, "The Early Use and Manufacture of Paper-Hangings in Boston," *Old Time New England,* 12, no. 3 (Jan. 1922): 117.

23 "Invoice, of sundry Paperhangings, Prints, and Utensills [sic], remaining on hand at the Manufactory No 43 Marlborough Street," Appleton Prentiss, Invoice no. 39 in the invoice book of a commission merchant, Mss. 761 (1786–97), Baker Business Library, Harvard University, Cambridge, Mass. Although no precise date is written on the invoice, it appears to date from 1788–9.

24 Advertisement for papers made by Prentiss and May, paper stainers of Boston, to be sold at Greenleaf's Printing Office in New York, *New York Journal and Patriotic Register,* July 20, 1790 (Gottesman 2:155).

CHAPTER 7– THE PAPER HANGER'S STOCK IN TRADE

1 Advertisement of William May, *The Massachusetts Centinel,* Boston, Feb. 4, 1789 (Deutsch).

2 Advertisement of Cornelius and John Crygier, *Mercantile Advertiser,* New York, July 1, 1799. Advertisement dated Apr. 17, 1799.

3 Advertisement of William Doyle, *Columbian Centinel,* Boston, Apr. 30, 1796.

4 Advertisement of William May, *The Massachusetts Centinel,* Boston, Feb. 4, 1789 (Deutsch).

5 Advertisement of Appleton Prentiss, *Columbian Centinel,* Boston, Mar. 10, 1792.

6 Advertisement of Joseph Hovey, *Independent Chronicle,* Boston, July 15, 1790, as quoted by Walter Kendall Watkins, "The Early Use and Manufacture of Paper-Hangings in Boston," *Old Time New England,* 12, no. 3 (Jan. 1922): 117.

7 Advertisement for "Poyntell's American Manufactory of Paper Hangings," *Dunlap's American Daily Advertiser,* Philadelphia, Mar. 2, 1793.

8 Advertisement of Zecheriah Mills & Co., *Connecticut Courant,* Hartford, Jan. 25, 1796, as quoted by Phyllis Kihn, "Zecheriah Mills, Paper-Hanging Manufacturer of Hartford, 1793–1816," *Connecticut Historical Society Bulletin,* 26, no. 1 (Jan.

1961): 23–4.

9 Advertisement of William May, *Independent Chronicle,* Boston, Aug. 4, 1791, as quoted by Watkins, "Early Use of Paper-Hangings," p. 117.

10 Advertisement of William Mooney, *The Diary,* New York, Aug. 18, 1796.

11 Advertisement of Thomas and Caldeleugh, *Federal Gazette and Baltimore Advertiser,* July 5, 1798 (Deutsch).

12 Advertisement of Moses Grant, *Columbian Centinel,* Boston, Apr. 15, 1797 (Deutsch).

13 Molding and oakleaf borders advertised by John and Cornelius Crygier, *Mercantile Advertiser,* New York, Apr. 17, 1799 (Roberts). "Carved work" borders advertised by Prentiss and May, *Independent Chronicle,* Boston, July 22, 1790, as given in Watkins, "The Early Use of Paper-Hangings," p. 117. "Dental" borders by Appleton Prentiss, *Independent Chronicle,* Boston, Apr. 20, 1791, as given in Watkins, "The Early Use of Paper-Hangings," p. 117.

14 Invoice of "Messrs Dawes, Stephenson & Co., London, Aug. 3, 1799," itemizing purchases "Bot of James Duppa, No. 34 ? Broad Street," Skipwith Papers, Earl G. Swem Library, The College of William and Mary in Virginia.

15 Invoice, Sept. 27, 1763, in Invoices and Letters, 1755–1766, in Original Manuscripts in the Washington Papers at the Library of Congress, Washington D.C., p. 111.

16 As quoted in Alan V. Sugden and John L. Edmondson, *A History of English Wallpaper 1509–1914* (New York: Charles Scribner's Sons, 1925; London: B. T. Batsford, 1926), p. 51.

17 Cited by Horace L. Hotchkiss, Jr., "Wallpapers Used in America 1700–1850," *The Concise Encyclopedia of American Antiques,* ed. Helen Comstock, 2 vols. (New York: Hawthorn Books, 1958), 2: 486.

18 Advertisement of Daniel Leeson, *Royal Gazette,* New York, Nov. 11, 1780 (Gottesman 2:154).

19 Advertisement of William Mooney, *New York Gazette: and the Weekly Mercury,* July 31, 1780 (Gottesman 2:148).

20 From a list of instructions of Thomas Jefferson, New York, to William Short, Paris, Apr. 6, 1790, as quoted by Julian P. Boyd, ed., *The Papers of Thomas Jefferson* (Princeton, N.J.: Princeton University Press, 1961), 16 (1789–1790): 322.

21 George Washington to Clement Biddle, Jan. 17, 1784. John C. Fitzpatrick, ed., *The Writings of George Washington from the Original Manuscript Sources, 1745–1799,* 39 vols. (Washington, D.C.: United States Government Printing Office, 1931–44), 27 (June 1783–Nov. 1784, printed Sept. 1938):305. George Washington to Clement Biddle, June 30, 1784. Fitzpatrick, ed., *The Writings of George Washington,* 27:430. See footnote 9, Chapter 8.

22 "Plain papers of different colors," advertisement of Francis Delorme, *Charleston City Gazette and Advertiser,* July 6, 1793 (Prime 2:276). "Every kind of Fashionable plain paper," advertisement of William Doyle, *Columbian Centinel,* Boston, Apr. 30, 1796.

Advertisement of William Poyntell, *Dunlap's American Daily Advertiser,* Philadelphia, Mar. 2, 1793. The Mackay and Dixey advertisement is quoted, apparently in full, but without reference to its source, by Nancy McClelland, *Historic Wall-Papers from their Inception to the Introduction of Machinery* (Philadelphia and London: J. B. Lippincott Company, 1924), p. 265.

23 Henry Remsen, New York, to Thomas Jefferson, Nov. 19, 1792, in the Jefferson Papers, Library of Congress, as published in "Clues and Footnotes," *Antiques,* 96, no. 6 (Dec. 1969): 808.

24 "Would have to mine," meaning "which I also would like to have for my walls," Watkins, "Early Use of Paper-Hangings," p. 110.

25 Advertisement of Plunket Fleeson, *Pennsylvania Gazette,* Philadelphia, Oct. 19, 1769 (Prime 1:276).

26 These phrases taken from advertisements of Joseph A. Fleming, *The Daily Advertiser,* New York, June 11, 1792 (Deutsch), and of John I. Post, *Mercantile Advertiser,* New York, June 24, 1799 (Gottesman 2:152).

27 Advertisement of Appleton Prentiss, *Columbian Centinel,* Boston, Mar. 10, 1792 (Deutsch).

28 "Chintz festoon borders," advertisement of "Paper-Hanging Warehouse, Corner of Black Horse Alley, in Second-street," *Pennsylvania Packet,* Philadelphia, Oct. 16, 1789 (Prime 2:286). "Feather Festoon" and "Plain Festoon," advertisement of Cornelius and John Crygier, *The Mercantile Advertiser,* New York, July 1, 1799 (Roberts). Advertisement of William May, *Independent Chronicle,* Boston, Aug. 4, 1791, as quoted by Watkins, "Early Use of Paper Hangings," p. 117. Advertisement of William Doyle, *Columbian Centinel,* Boston, Apr. 30, 1796.

29 Robert Dossie, *The Handmaid to the Arts,* 2 vols. (London: A Millar, W. Law, and R. Carter, 1796; first known edn., 1758), 2:315.

30 Advertisement of Prentiss and May, *Independent Chronicle,* Boston, July 22, 1790, as quoted by Watkins, "Early Use of Paper Hangings," p. 117. See also footnote 16, chapter 2.

31 "Gilt" papier-mâché borders, advertisement of Plunket Fleeson, *Pennsylvania Gazette,* Philadelphia, Oct. 19, 1769 (Prime 1:276).

"Patch," advertisement of Appleton Prentiss, *Columbian Centinel,* Boston, Mar. 10, 1792 (Deutsch).

"Lemon" and "race," advertisement of Cornelius and John Crygier, *Mercantile Advertiser,* New York, July 1, 1799 (Roberts).

"Feather" and "rose," advertisement of Cornelius and John Crygier, *New-York Gazette and General Advertiser,* Apr. 21, 1797 (Gottesman 2:144).

"Garland," advertisement for "#40 Cornhill," *Columbian Centinel,* Boston, Aug. 11, 1792 (Deutsch).

"Best fruit Border," manuscript invoice "Mr. Hathe. [Hatheway] bot of Moses Grant, Bost. 22 Oct 1794," in Phelps Papers, New York State Library, Albany, N.Y. Arthur W. Leibundguth graciously furnished a photocopy of the invoice for the author's use.

"Fantail" and "canopy," 1789 inventory of the estate of John Welsh, Jr, paper stainer of Boston, as quoted in Watkins, "Early Use of Paper-Hangings," p. 115.

32 *The Compleat Appraiser, The 4th Edition. Greatly Improved* (London, 1770), p. 4. The compiler of *The Compleat Appraiser,* identified only as "an Eminent Broker," commented in introducing the passage quoted: "To set down all the *particular* Names that *Wall-Paper* is distinguished by would be endless." The author is indebted to Florence Montgomery for this reference.

Advertisement of William Bucktrout, *Williamsburg Gazette,* May 9, 1771.

33 The items are: "96 Yds blue embost Paper, w^{te} Mosaic" and "144 Y^{ds} Crimson embost paper, Wte Mosaic," Nov. 2, 1757. "Invoice of Sundry Goods Ship'd by Richd Washington . . . on the proper Acct. and risque of the Honble Geo. Washington . . . Nov. 2, 1757," in Invoices and Letters, the Washington Papers, Library of Congress, p. 7.

34 See, for instance, advertisement of N. M. Ryder, *City Crier and Country Advertiser,* Boston, Mar. 2, 1848, where "common" papers are distinguished from "grounded" and "satin-faced" papers by price ranges that differ widely.

35 Advertisement of Moses Grant, *Columbian Centinel,* Boston, Apr. 15, 1797 (Deutsch). Advertisement of Josiah Bumstead, *Columbian Centinel,* Boston, May 6, 1797 (Deutsch). Advertisement of Josiah Bumstead, for his Boston store, in *The Providence Gazette,* Apr. 11, 1795 (Deutsch).

36 Advertisement of William Mooney, *The Diary,* New York, Aug. 18, 1796 (Roberts).

37 For assigning dates to the lesser patterns that turn up so frequently in American houses of the eighteenth century, there are particularly useful volumes containing small-scale patterns, with documentation for their dates and makers, at the Bibliothèque Nationale in Paris. Registry books for designs that were patented during the very last years of the eighteenth century include quantities of small-scale cheap patterns as well as high-style examples.

CHAPTER 8– COLOR SCHEMES AND THE USE OF WALLPAPERS

1 Lady Jean Skipwith, Prestwould, Clark County, Va., to James Maury, Aug. 27, 1795, Skipwith Papers (ViW(SPS)), Earl G. Swem Library, The College of William and Mary in Virginia.

2 Unsigned article on trade of France with the United States, *Morning Chronicle,* New York, Mar. 10, 1803 (Gottesman 3:293).

3 "Two to seventeen colours," advertisement of Prentiss and May, *Independent Chronicle,* Boston, July 22, 1790, as quoted by Walter Kendall Watkins, "The Early Use and Manufacture of Paper-Hangings in Boston," *Old Time New England,* 12, no. 3 (Jan. 1922): 117.

"Two to twenty-six colours," advertisement of Burrill and Edward Carnes, *Pennsylvania Packet,* Philadelphia, Apr. 27, 1793 (Prime 2:275).

"Light and dark figured," advertisement of Cornelius and John Crygier, *New-York Gazette and General Advertiser,* Apr. 21, 1797 (Gottesman 2:145).

Advertisement of William Poyntell, *Pennsylvania Packet,* Philadelphia, Mar. 21, 1791 (Prime 2:283).

4 Advertisement of William Poyntell, *Aurora,* Philadelphia, Nov. 3, 1797 (Prime 2:285).

5 George Washington to Richard Washington, Sept. 1757, John C. Fitzpatrick, ed., *The Writings of George Washington from the Original Manuscript Sources, 1745–1799,* 39 vols. (Washington D.C.: United States Government Printing Office, 1931–44), 2 (1757–69, printed June 1931): 138.

6 Advertisement John Colles, *Daily Advertiser,* New York, Apr. 11, 1787 (Gottesman 2:153).

7 Advertisement in the form of a handbill for the Blue Paper Warehouse, 1609–1702, in the Bagford Collection, British Museum. Reproduced in Alan V. Sugden and John L. Edmondson, *A History of English Wallpaper 1509–1914* (New York: Charles Scribner's Sons, 1925; London: B. T. Batsford, 1926), plate 33.

8 Walter Kendall Watkins, "The Early Use and Manufacture of Paper-Hangings in Boston," *Old Time New England,* 12, no. 3 (Jan 1922): 110.

9 George Washington to Clement Biddle, June 30, 1784. Fitzpatrick, ed., *Writings of George Washington,* 27:430.

10 These drawings are supposed to have been commissioned for Diderot's *Encyclopaedia* but were never published as prints. They have been repeatedly reproduced in publications about wallpaper. See Brenda Greysmith, *Wallpaper* (New York: Macmillan, 1976), plate 29; and for the full series of 7 plates, which include 85 figures, see Heinrich Olligs, ed., *Tapeten, Ihre Geschichte bis zur Gegenwart,* 3 vols. (Braunschweig: Klinkhardt and Biermann, 1969), 3: following 16.

11 "An inventory of the Personal Estate of His Excellency Lord Botetourt began to be Taken the 24th of Octo^{r} 1770," Botetourt Manuscripts, Virginia State Library, Richmond, Va. Oznabrigs, Ozenbriggs, and Osnaburg were all names for coarse linen fabric of a type associated with Osnabrück, Germany, but also made in England and elsewhere.

12 Advertisement of Thomas Hurley, *Pennsylvania Packet,* Philadelphia, Oct. 5, 1786 (Prime 2:280).

13 Manuscript note accompanying invoice of "Messrs Dawes, Stephenson & Co., London, Augt 3, 1799," itemizing purchases "Bot of James Duppa, No 34 *?* Broad Street," Skipwith Papers, Earl G. Swem Library, The College of William and Mary in Virginia.

14 As given in Sugden and Edmondson, *A History of English Wallpaper,* p. 51.

Wendy Cooper has brought to the author's attention a particularly detailed eighteenth-century American reference to the use of lining paper under a wallpaper. It occurs in a letter: Anne Elizabeth

Clark Kane, possibly New York or New London, Friday the 24th, late 1790s or early 1800s, to Harriet Clark, as given by Joseph K. Ott in "John Innes Clark and His Family—Beautiful People in Providence," *Rhode Island History,* 32, no. 4 (Nov. 1973): 129.

"With the paper I have sent some suitable lining—that is such as should be first well pasted all over the wall of the room and the paper put over it—this is to prevent any stains from the wall soaking thro to discolour the paper or injure its appearance and is always done in this country where a room is intended to be well papered. Great care should be taken to lay the first paper fair and smooth . . ."

15 Advertisement of Richard Bird, *The South Carolina Gazette,* Charleston, Sept. 11, 1762 (Prime 1:274).

16 Advertisement of Edward Ryves, *Pennsylvania Mercury,* Philadelphia, Apr. 21, 1786 (Prime 2:285).

17 Deborah Franklin to Benjamin Franklin, October 6–13 ?, 1765, *The Papers of Benjamin Franklin,* ed. Leonard W. Labarre and others (New Haven and London, 1968), 12:294–8, as quoted by Howard H. Peckham in "Clues and Footnotes," *Antiques,* 97, no. 6 (June 1970): 936.

18 John David Schoepf, *Travels in the Confederation, 1783–1784,* trans. and ed. J. Morrison (Philadelphia, 1911), 1:60. As quoted by Frederick S. Koontz in "Clues and Footnotes," *Antiques,* 98, no. 6 (Dec. 1970):948.

19 Advertisement of Cornelius and John Crygier, *Mercantile Advertiser,* New York, July 1, 1799 (Roberts).

20 Advertisements for papers for staircases include those of William Kidd, *Virginia Gazette,* Williamsburg, May 19, 1774, and of William Poyntell, *Pennsylvania Gazette,* Philadelphia, Sept. 3, 1783.

21 Thomas Jefferson, New York, to William Short, Paris, Apr. 6, 1790, enclosure, in *The Papers of Thomas Jefferson,* ed. Julian P. Boyd (Princeton: Princeton University Press, 1961), 16 (1789–1790):318.

22 "Invoice of Sundry Goods Ship'd by Rich[d] Washington . . . on the proper Acct. and risque of the Hon[ble] George Washington . . . ," Nov. 2, 1757. Invoices and Letters, 1755–66 in Original Manuscripts at The Library of Congress, Washington, D.C., p. 7.

23 Advertisement of John Blott, *The South Carolina Gazette,* Charleston, May 11, 1765 (Prime 1:275).

24 Advertisement of Roper Dawson, *The New-York Gazette or the Weekly Post-Boy,* June 3, 1762 (Gottesman 1:123).

25 Advertisement of Francis Delorme, *City Gazette and Daily Advertiser,* Charleston, Jan. 13, 1796 (Prime 2:221).

26 Chimney boards were not an American invention. Sugden and Edmondson, *A History of English Wallpaper,* p. 81, cite a bill of 1769 for work done by Thomas Chippendale at Nostel Priory, including "Covering Chimney boards with blue verditer paper and putting borders round ditto."

27 Advertisement of John Blott, *The South Carolina Gazette,* Charleston, June 9, 1766 (Prime 1:275).

28 Advertisement of John Mason, *Pennsylvania Chronicle,* Philadelphia, Oct. 28, 1771 (Prime 1:209).

29 Advertisement of William Poyntell, *Gazette of the United States,* Philadelphia, Mar. 17, 1796 (Prime 2:284). The *Oxford English Dictionary* helps to explain the advertiser's rather loose word usage in describing the area over the fireplace opening as "breast works": "Breast" is "applied to various surfaces, or parts of things analagous in shape, position, etc. to the human breast: the forefront, face, swelling or supporting surface." Poyntell seems to have been improvising on the more usual "Chimney-breast," which the *O.E.D.* defines as "that projecting part of the wall which is between the chimney-flue and the room."

30 Advertisement by an unnamed merchant at No. 40 Cornhill, *Columbian Centinel,* Boston, Aug. 11, 1792 (Deutsch).

31 Advertisement of Joseph Hovey, *Independent Chronicle,* Boston, July 15, 1790, as quoted by Watkins, "Early Use of Paper-Hangings," p. 117.

32 Advertisement of an unnamed merchant at No. 40 Cornhill, *Columbian Centinel,* Boston, Feb. 24, 1796 (Deutsch).

33 "Maha[y]" is an abbreviation for "Mahogony"—"Invoice of Sundry Goods Ship'd by Robt Carry & Comy . . . on account and risque of Colo George Washington . . . ," Aug. 1759, in Invoices and Letters, the Washington Papers, Library of Congress, p. 17.

34 Advertisement of Joshua Blanchard, *Boston News-Letter,* May 14, 1767 (Dow, 152).

35 Advertisement of John Blott, *The South Carolina Gazette,* Charleston, Jan. 31, 1771 (Prime 1:200).

36 Advertisement of (first name not given) Booden, *The South Carolina Gazette,* Charleston, Dec. 16, 1756 (Prime 1:275).

37 Advertisement of Francis Delorme, *Charleston City Gazette and Advertiser,* July 6, 1793 (Prime 2:276).

38 Abbott Lowell Cummings, ed., *Rural Household Inventories, establishing the Names, Uses and Furnishings of Rooms in the Colonial New England Home, 1675–1775* (Boston: The Society for the Preservation of New England Antiquities, 1964), p. 213.

39 Advertisement of John Blott, *The South Carolina Gazette,* Charleston, Jan. 19, 1765 (Prime 1:275).

40 "A great variety of cheap paper hangings" just imported from London were advertised by Lewis Deblois, *Boston Gazette,* Jan. 5, 1761 (Dow, 151).

41 Advertisements of William Poyntell, *Pennsylvania Packet,* Philadelphia, Apr. 13, 1783 (Prime 1:279); *Pennsylvania Journal,* Philadelphia, Aug. 16, 1783 (Prime 1:279).

42 Advertisement of Joseph Dickinson, *Pennsylvania Packet,* Philadelphia, Apr. 29, 1786 (Prime 2: 277).

43 Advertisement for papers made by Prentiss and May of Boston, sold at Greenleaf's Printing Office, New York. *New-York Journal, and Patriotic Register,* Feb. 17, 1791 (Gottesman 2: 156).

44 "List of the Different Kinds of Masons', Bricklayers', and

Plasterers' Work," 1801, McComb Papers, New-York Historical Society, New York City. This reference was brought to the author's attention by Lee Roberts.

45 Trade card of Richard Worley, 1730–60, Landauer Collection, New-York Historical Society, New York City, as published in *Early American Trade Cards from the Collection of Bella C. Landauer,* with critical notes by Adele Jenny (New York: William Edwin Rudge, 1927), plate 1.

46 Advertisement of Joshua Blanchard, *Boston News-Letter,* Jan. 24, 1765 (Dow, 152).

47 Advertisement of William Vans, *The Essex Gazette,* Salem, Mass., Apr. 23, 1771 (Deutsch).

48 Advertisement of Daniel Leeson, *New-York Gazette, and the Weekly Mercury,* Mar. 3, 1783 (Gottesman 2: 154).

49 Advertisement of Joseph Dickinson, *Pennsylvania Packet,* Philadelphia, Sept. 7, 1785 (Prime 1: 281).

50 Advertisement of Joseph Dickinson, *Pennsylvania Packet,* Philadelphia, July 5, 1786 (Prime 2: 276).

51 Advertisement of William Mooney for papers made by Mackay and Dixey at Springfield, N.J., *The New York Daily Gazette,* Aug. 13, 1790.

52 Advertisement of T. Greenleaf for papers made by Appleton Prentiss at Boston, *New-York Journal, and Patriotic Register,* Feb. 17, 1791 (Gottesman 2: 155).

53 Advertisement of Josiah Bumstead, *Columbian Centinel,* Boston, May 6, 1797 (Deutsch).

54 Charles F. Montgomery, *American Furniture: The Federal Period* (New York: Viking, 1966), p. 25.

55 Advertisements of William Poyntell, *Pennsylvania Packet,* Philadelphia, Apr. 12, 1783 (Prime 1: 279), and *Pennsylvania Journal,* Philadelphia, Aug. 16, 1783 (Prime 1: 279).

56 Advertisement of Joseph Dickinson, *Pennsylvania Packet,* Philadelphia, July 14, 1787 (Prime 2: 278).

57 Advertisement of Joseph Dickinson, *Independent Gazetteer,* Philadelphia, Dec. 20, 1786 (Prime 2: 277).

58 News item in *Independent Gazetteer,* Philadelphia, July 17, 1790, reporting on the manufactory of Burral (sic) Carnes under the firm of Le Collay and Chardon (Prime 2: 286).

INTRODUCTION

1 John Claudius Loudon (1783–1843), *An Encyclopaedia of Cottage, Farm and Villa Architecture and Furniture,* Book I, *Cottage Furnishings* (London: Longman, Rees, Orme, Brown, Green & Longman, rev. edn. 1836; earliest edn. located, 1833), p. 279.

2 A. J. Downing, *The Architecture of Country Houses* (New York: Dover, 1969; reprint of 1850 edn.), p. 23.

3 Marion Foster Washburne, "About Walls and Wallpapers," *The Decorator and Furnisher,* 10, no. 4 (July 1887): 124.

4 Gottfried Semper (1803–79, German architect) made the statement in *Wissenschaft, Industrie und Kunst* (Brunswick, Germany, 1852) as given in translation by Nikolaus Pevsner, "High Victorian Design," *Studies in Art, Architecture and Design,* vol. 2, *Victorian and After* (New York: Walker and Co., 1968), p. 47.

5 Charles L[ocke] Eastlake, *Hints on Household Taste in Furniture, Upholstery, and Other Details* (Boston: J. R. Osgood and Co., 1872; 1st London edn., 1868), p. 115.

6 Jules Desfossé and Hippolyte Karth, "The Manufacture of Paper Hangings in France," *The Furniture Gazette,* London, Mar. 14, 1874, p. 255.

CHAPTER 9– FRENCH SCENIC WALLPAPERS

All subsequent references to advertisements in newspapers published in Washington, D. C., were transcribed by Anne Golovin, who generously made available to the author xerox copies of her transcriptions.

1 Continuous paper was first used by Zuber in 1820, according to company archives, but well into the 1830s handmade sheets were used for most scenics.

2 Eric A. Entwisle, *French Scenic Wallpapers 1800–1860* (Leigh-on-Sea, England: F. Lewis, 1972), p. 49.

3 Eugène A. Fauconnier ("Chef de fabrication" for Desfossé et Karth for twenty-two years), in ms. "Notes de E. A. Fauconnier sur les 'Desfossé' et Anciens du Papiers Peint," 1935/1936 in the research files of the Cooper-Hewitt Museum, details which blocks from several manufacturers were bought by Desfossé et Karth, and at what intervals they were used to print new sets of given scenic subjects.

4 Fauconnier, in "Notes," gives the number of colors used to print many of the scenics he lists as Desfossé et Karth products. The numbers vary: 1,099 colors for *"l'Eden"*; 400 colors for *"Le Jardin d'Armide"*; 57 colors for *"Les Quatres Saisons"* with figures *en grisaille.*

5 Sidney George Fisher, "The Diaries of Sidney George Fisher, 1844–1849," *Pennsylvania Magazine of History and Biography,* 86 (1962): 66.

6 Musée des Arts Décoratifs, *Trois Siècles de Papiers Peints* (Paris: Musée des Arts Décoratifs, 1967), p. 16, no. 26. The two overdoors described in this exhibition catalogue measure 67 × 97.5 cm. and 68.5 × 96.5 cm.

7 Advertisement of Francis Delorme, *Charleston City Gazette and Daily Advertiser,* Sept. 16, 1794 (Prime 2:220); advertisement of William Poyntell, *Federal Gazette,* Philadelphia, Mar. 11, 1795 (Prime 2:284).

8 By 1809, the son of François Robert had sold the business to Joseph Guillot. Musée des Arts Décoratifs, *Trois Siècles de Papiers Peints,* p. 16.

9 Nancy McClelland, *Historic Wall-Papers from their Inception to the Introduction of Machinery* (Philadelphia and London: J.B. Lippincott Company, 1924), p. 294.

10 Jean Zuber, Père, *Réminiscences et Souvenirs* (Mulhouse:

Veuve Bader & Cie., 1895). The author thanks Francis de Ransart of Zuber et Cie. both for calling this volume to her attention as well as his gift of a photo reproduction of it, now in the library of the Cooper-Hewitt Museum.

11 Zuber, *Réminiscences et Souvenirs,* pp. 55–7. The magnitude of producing these tableaux seems to have driven Dollfus, Zuber's head workman, literally crazy. In his *Réminiscences,* p. 58, Zuber recounts the event in 1808:

> *Notre mélangeur Dollfus, qui avait commencé l'impression du tableau l'Hindoustan, avait été si troublé et si surexcité par ce nouveau travail qu'il devint fou en quelques jours; je fus obligé de le remplacer avec beaucoup de peines et de fatigues; j'y arrivai avec l'aide de Mongin et pendant toute une année je fus attelé à cette pénible besogne, car il nous fut impossible de former ou de trouver un autre mélangeur.*

12 *Décors Panoramiques Zuber* (Rixheim, Alsace: Zuber et Cie., [1975?]). This catalogue makes use of the extensive research in the Zuber archives by Jean Pierre Seguin, Odile Kammerer, and others. Dates assigned to Zuber scenics in this chapter rely heavily on this work.

13 C. F. Mansfield offered the "Lady of the Lake" view in an advertisement in the *Public Ledger,* Philadelphia, Mar. 25, 1836. Most likely this view is a Zuber paper.

14 Musée des Arts Décoratifs, *Trois Siècles de Papier Peints,* p. 68, no. 244.

15 The advertisement of C. F. Mansfield cited in footnote 13 called this view the "Battle of Navarino."

16 Charles Edward Lester, ed., *Glances at the Metropolis* (New York: Isaac D. Guyer, 1854), p. 37.

17 Two of the designers of scenics and *décors* who worked for several rival manufacturers include a craftsman who is identified in the French records only as Wagner and one Charles Muller. Wagner worked for the firms of Desfossé et Karth, Lapeyre, and Mader, while Muller worked for the firms of Delicourt and Desfossé et Karth. See Musée des Arts Décoratifs, *Trois Siècles de Papier Peints,* pp. 39, 44, 53, and 58 and pp. 36, 39–40 respectively regarding these two designers.

18 Victor Poterlet, active from 1830 until the 1870s, also designed for his own firm, as well as for most of the major French wallpaper manufacturers, including Dauptain, Delicourt, Desfossé et Karth, and Dumas. See Musée des Arts Décoratifs, *Trois Siècles de Papier Peints,* pp. 34–6, 39, 48, 50, 59–60.

19 Musée des Arts Décoratifs, *Trois Siècles de Papier Peints,* p. 44. Clare Crick, "Two Wallpapers by Dufour et Compagnie," *Pharos '78* (Museum of Fine Arts, St. Petersburg, Fla.), 15, no. 1 (June 1978): 5.

20 As quoted by Nancy McClelland, *Historic Wall-Papers,* p. 403.

21 As quoted by Entwisle, *French Scenic Wallpapers,* p. 34.

22 Crick, "Two Wallpapers by Dufour et Compagnie," p. 3. Rue de Beauveau is now Rue Beccaria. The wallpaper firm of Follot now occupies the site.

23 Musée des Arts Décoratifs, *Trois Siècles de Papier Peints,* p. 44.

24 Musée des Arts Décoratifs, *Trois Siècles de Papier Peints,* p. 46, no. 129, cites an example belonging to the Musée des Arts Décoratifs *"en grisaille rehaussé de bleu et d'ocre rose."*

25 Musée des Arts Décoratifs, *Trois Siècles de Papier Peints,* p. 44. Fauconnier, in "Notes," lists the sets of wood blocks acquired by Desfossé et Karth from Mader and his successors, including the firms of Dufour, Dauptain, Brière, Clerc et Margeridon, Kob et Pick, Jouanny, as well as lesser manufacturers. Therefore, it is very difficult to determine who actually printed wallpaper made from these blocks.

26 Heinrich Olligs, ed., *Tapeten Ihre Geschichte bis zur Gegenwart,* vol. 2 (Braunschweig: Klinkhardt & Biermann, 1970), pp. 325–7. Henri Clouzot, *Tableaux-Tentures de Dufour & Leroy* (Paris: Librairie des Arts Décoratifs, [1930?]). In this chapter, the dates assigned to Dufour wallpapers were based on materials in *Trois Siècles de Papiers Peints* and *Tapeten Ihre Geschichte bis zur Gegenwart,* which had relied heavily on *Tableaux-Tentures* and H. Clouzot and Ch. Follot, *Histoire du Papier Peint en France* (Paris: Éditions D'Art Charles Moreau, 1935). While writing *Histoire du Papier Peint,* Clouzot and Follot had access to the Dufour account books for 1824–31, which have since disappeared.

27 Advertisement of James H. Foster, *New England Palladium,* Boston, Dec. 2, 1817; advertisement of Samuel Robinson, *Daily National Intelligencer,* Washington, D.C., May 15, 1826.

28 See pages 195, 203.

29 C. C. Oman, *Catalogue of Wall-Papers, Victoria and Albert Museum* (London: Board of Education, 1929), p. 65. Since *"Vues de Londres"* depicts both London Bridge, built in 1831, and the old Westminster Bridge, superseded by the new Westminster Bridge in 1846, it probably was issued about 1840.

30 Advertisement of C. F. Mansfield, *Public Ledger,* Philadelphia, Mar. 25, 1836, offered this set for sale. Andrew Ure, *A Dictionary of Arts, Manufactures, and Mines,* 2 vols. (New York: D. Appleton, 1863; 1st edn. London, 1839), 2: 326, says that "the history of Psyche and Cupid, by M. Dufour, has been considered a masterpiece in this art, rivalling the productions of the pencil in the gradation, softness, and brilliancy of the tints."

31 An advertisement of Samuel Robinson, *Daily National Intelligencer,* Washington, D.C., Oct. 22, 1819, suggests a new date of publication; heretofore, this paper has been dated 1824.

32 Heretofore, this pattern has been dated 1823 and 1825 respectively by Olligs, ed., *Tapeten Ihre Geschichte,* p. 231, and Musée des Arts Décoratifs, *Trois Siècles de Papier Peints,* p. 47. However, several advertisements of Samuel Robinson, *Daily National Intelligencer,* Washington, D.C., Oct. 22, 1819; *Daily Na-*

tional Intelligencer, May 22, 1820; and *Daily National Intelligencer,* Sept. 15, 1821, offered this pattern. It was also offered by S. P. Franklin, *Daily National Intelligencer,* Washington, D.C., Feb. 3, 1823.

33 During the Trojan War, Antenor was the head of the peace party. As a result, his life was spared when the city fell. Afterwards, he wandered until he founded Cyrene or Patavium, known today as Padua.

34 Musée des Arts Décoratifs, *Trois Siècles de Papier Peints,* p. 48.

35 "The Island Home of Paul and Virginia," *Appleton's* Magazine, Dec. 25, 1869, pp. 584–5.

36 This view has been recorded in the following houses: Andrews–Spofford house, Salem, Mass.; Morse house, New Berlin, N.H.; Prestwould, near Clarksville, Va.; Woodlawn, Lexington, Ky.; and the Bonaparte house, Philadelphia, Pa.

37 "Le Parc Français" has been recorded in three houses: Cabell house, Lynchburg, Va. (demolished); Prestwould, near Clarksville, Va.; and the King Caesar house, Duxbury, Mass.

38 This scenic is documented in four houses: the Hanover, N.H., house of a Professor Young (the scenic now at the Boston Museum of Fine Arts); Nicholas Ward Boylston house, Princeton, N.J.; Homewood Manor, Princeton, Mass.; and the State Street house of a Dr. Phillips, Windsor, Vt.

39 In the 1920s, this paper was recorded at the Boston address, 40 Beacon Street, then the Daniel P. Parker house.

40 Musée des Arts Décoratifs, *Trois Siècles de Papier Peints,* p. 36. This firm, while agents for Zuber, was also associated—between 1838 and 1859—with Delicourt, likewise a Zuber agent.

41 Fauconnier, "Notes," p. 30.

42 Musée des Arts Décoratifs, *Trois Siècles de Papier Peints,* p. 39.

43 Fauconnier, "Notes," p. 30.

44 This scenic is documented in four houses: J. Fenimore Cooper house, Cooperstown, N.Y.; Daniel P. Parker house, Boston, Mass.; "house of Mrs. Harkness," N.J., as noted in the research files of the Cooper-Hewitt Museum by Grace Lincoln Temple during the 1920s; house of Mrs. William McLean, Germantown, Pa.

45 McClelland, *Historic Wall-Papers,* p. 298, notes this paper in the Bailey house in Bath, Me., and an unidentified Salem, Mass., house. Ruth Hairston Early, *Campbell Chronicles and Family Sketches Embracing the History of Campbell County, Virginia 1782–1926* (Lynchburg: J.P. Bell Company, 1927), p. 240, notes that "Don Quixote illustrations [were] used in the house of Captain Labby . . . on Main Street . . . Sancho Panzo tossed in a blanket. . . . Many private homes [in Lynchburg] were newly decorated and furnished in . . . honor [of Andrew Jackson's April 1817 visit]. . . . Among others Dr. John Cabell, an ardent admirer of Jackson selected papering for his home on Main Street. . . . "

46 McClelland, *Historic Wall-Papers,* p. 157.

47 Ibid., p. 158.

48 "Livre de Vente depuis Janvier 1815," ms. vol., Archives Zuber (Rixheim, Alsace, France), pp. 304–5; 354–8.

49 "Livre de Vente depuis Janvier 1815," pp. 792–4.

50 "Livre de Vente depuis le 28 Fevrier 1821 jusqua' à 11 Mars 1842," ms. vol., Archives Zuber, pp. 480, 481, 847.

51 This accounting is for the Philadelphia firm of Blanchard and Hardy for the years 1828–1831 as recorded in the "Grand Livre 1829–1834," ms. vol., Archives Zuber, vol. E, p. 291.

52 Advertisement of S. P. Franklin, *Daily National Intelligencer,* Washington, D.C., Feb. 3, 1823; Mar. 1, 1823.

53 Advertisement of L. Fournier, *Louisiana Courier,* New Orleans, Mar. 28, 1808. Advertisement of Downing & Grant, *Kentucky Gazette,* Lexington, Kentucky, Dec. 2, 1816. Nancy Goyne Evans was good enough to bring this advertisement to the author's attention.

54 Examples of the Captain Cook Paper are recorded in: Peabody, Mass.: Ham house (McClelland, p. 366); Augusta, Maine: Ruel-Williams house (McClelland, p. 366); "In a house near Hoosic Falls, N.Y." (McClelland, p. 366); Salem, Mass: house belonging in the 1920s to Mrs. Charles A. Brown (McClelland, p. 366); Mecklenburg County, N.C., Oak Lawn—see *figure 10-9.*

55 See McClelland, *Historic Wall-Papers,* p. 388, for a listing of thirteen of these hung during the nineteenth century. Additional examples have been recorded in:

Caroline County, Va.: Gay Mont (the paper was destroyed by fire). At Gay Mont it has been replaced by a duplicate set that once hung in Stanton, Va., in a house that belonged to Mrs. R. H. Catlett when McClelland recorded it in 1924.

Waterville, N.Y.: Charlemagne Tower house, paper now in collections of Munson-Williams-Proctor Institute, Utica, N.Y.

Warren, Long Island, N.Y.: Never hung, but purchased from Dufour by Elisha Dyer for use in his house. Now in collections of Winterthur Museum, Wilmington, Del.

Richmond, Va.: Cabell house, paper now in collection of the Association for the Preservation of Virginia Antiquities.

56 McClelland, *Historic Wall-Papers,* p. 381. The paper at Prestwould identified by McClelland as "Venetian Scenes" is *not* Venetian Scenes.

57 Examples of this paper have been recorded in Dorchester, Mass.: Nahum Capen house; Unionville, Ohio: Shandy Hall.

58 Bill of Francis Pares, New York, Nov. 30, 1835, to Mr. Van Gorder for "Suberbs of Rome, 2 setts Colored View . . . $120." Photocopy of the bill and paper in collection of Charles E. Proctor, Warren, Ohio. Long Island, N.Y.: Willow Hills, Joseph Phelon house, paper installed during 1830s.

59 McClelland, *Historic Wall-Papers,* p. 380.

60 Illustrated in McClelland, *Historic Wall-Papers,* pp. 308–13.

61 The letter of Apr. 16, 1819, is preserved by the descendants

of John H. Bernard. One of them, James S. Patton, graciously sent a transcription to the author. His transcription was included in a letter of Feb. 27, 1972, to the author, and has been used here.

62 Charles E. Lester, ed., *Glances at the Metropolis* (New York: Guyer, 1854), p. 37.

63 Harper's Ferry National Historical Park, photo files, No. NHF 1947. The fragments show "The Worship of Vesta" in shades of ochre.

64 The bill is at the Baker Business Library, Harvard University, Case 13, Wendell Family Papers, Jacob Wendell, Bills, 774: Household Supplies.

65 These may be wholesale prices. Quoted from transcriptions from Gilman's account book graciously sent to the author by a descendant of Ephraim Gilman (1778–1852), Mrs. Elsie R. H. Roberts, in a letter June 11, 1973.

66 The prices are taken from advertisements of:

James H. Foster, *New England Palladium,* Boston, Oct. 15, 1813: 25¢–$2.50/roll.

John Perkins, *New Bedford Mercury,* Aug. 1, 1817, 4,500 rolls, 42¢.

Samuel Robinson, *Daily National Intelligencer,* Washington, D.C., Nov. 17, 1821, common-priced paper hangings 37¢ apiece.

C. F. Mansfield, *Public Ledger,* Philadelphia, Mar. 25, 1836: American and French paper hangings from 25¢ apiece to more costly.

67 Mss. bills in the Jackson Papers, Hermitage, Tenn., Jan. 2, 1836, and Oct. 25, 1836. It is interesting that some portions of the Hermitage Telemachus *(figure 9-24)* are printed on continuous, machine-made paper, while others are on handmade paper. Perhaps the factory would have charged less for sets of the paper printed on the cheaper stock made by machines.

68 However, the inn burned before the paper arrived and it was later hung in a private house in Warren, Ohio. A set of the same paper was originally purchased for a hotel in Buffalo, N.Y., but eventually hung by Col. Robert Harper, proprietor of this hotel, in his home, Shandy Hall, Unionville, Ohio. Jarius B. Barnes was good enough to give this information to the author in a letter of Feb. 7, 1974.

69 Harriet Martineau, *Retrospect of Western Travel in 2 Volumes* (London: Sanders and Otley; New York: Harper and Brothers, 1838), 1:83–4. The author thanks Christopher Monkhouse for this reference.

70 Benjamin Silliman, *The World of Science, Art, and Industry, Illustrated from Examples in the International Exhibition* (New York: Putnam and Co., 1854), pp. 202–3.

71 Clarence Cook, *"What Shall We Do With Our Walls?"* (New York: Warren Fuller and Co.), p. 16.

72 Christopher Dresser, *Development of Ornamental Art in the International Exhibition* (New York and London: Garland Publishing, 1978; reprint of Day and Son publication, 1862), p. 44.

73 Advertisement of S. P. Franklin, *Daily National Intelligencer,* Washington, D.C., Oct. 20, 1825.

74 Billhead of W. H. Sacket, New York, with manuscript bill of sale to A. C. Laurence, Sept. 6, 1857.

75 Bill of Francis Regnault [Richmond, Va.], to Mr. H(umbert) Skipwith [Prestwould], Clarksville [Va.], Dec. 24, 1831. Skipwith Papers, Box 6, Earl G. Swem Library, The College of William and Mary in Virginia. The author is grateful to Nancy Goyne Evans for bringing to her attention this most interesting of all the nineteenth-century bills for hanging wallpaper known to the author.

CHAPTER 10– FRENCH WALLPAPER ORNAMENTS OF THE EARLY AND MID-NINETEENTH CENTURY

1 Advertisement of Thomas and Caldeleugh, *Federal Gazette and Baltimore Advertiser,* Apr. 12, 1798 (Deutsch).

2 Advertisement of Woodbridge and Putnam, *Connecticut Courant,* Hartford, Aug. 6, 1816.

3 1840s billhead of Josiah F. Bumstead, in the collection of Robert L. Raley: bill of sale to J. A. Beckwith (the printed "4" in the date is crossed through and replaced with a "5").

4 1853 handbill of J.F. Bumstead & Co. describing "Spring Stock of 1853," Joseph Downs Manuscript Collection, Henry Francis du Pont Winterthur Museum.

5 See Robinson's advertisements in *Daily National Intelligencer,* Washington, D.C., for Sept. 10, 1819, including French "Decorative figures, for rooms, recesses, & ceilings," and advertisement for Oct. 2, 1820, including French "ornaments for ceilings."

6 Advertisement of George Platt, *New York Evening Post,* Jan 2, 1845.

7 Photograph and information about Oaklawn, built 1818, provided by Mrs. William L. Stratton. See footnote 54, Chapter 9.

8 Sébastien Le Normand illustrated this ceiling paper in his *Nouveau Manuel Complet du Fabricant de Papiers Peints* (Paris: la librarie Encyclopédique de Roret, 1856; earlier edns. known from 1830, 1832, etc.). Le Normand cites Dufour et Leroy as the leading French manufacturers. Chapter 4, plate 2, illustrates the ceiling shown as *color plate 47,* calling it *"La Toilette de Vénus."*

9 The ceiling paper at Swansbury was described in letters of June 25 and Nov. 5, 1973, from Nan Jay Barchowsky to the author, in which she enclosed pictures and samples. Now in the study files of the Cooper-Hewitt Museum, they show a front stair hall decorated during the nineteenth century with a stone pattern in shades of gray and with a center on the ceiling of the second floor.

10 Original design at Bibliothèque Forney, Paris, Reserve 745.522.2, 1800–1850—"Dessins Originaux et Models de Papiers Peints 1800–40, Tome I," the design marked "52" and "42051."

11 For an illustration of this dado, see Heinrich Olligs, ed., *Tapeten Ihre Geschichte bis zur Gegenwart,* vol. 2 (Braunschweig: Verlag Klinkhardt & Biermann, 1970), p. 74, pl. 352; see also correspondence with Dr. Ben H. Caldwell, Nashville, Tenn., research

files, Cooper-Hewitt Museum.

12 In the account books of the Zuber Archives entries for Walter Crook appear from 1829 through 1832.

13 John Claudius Loudon, *An Encyclopaedia of Cottage, Farm and Villa Architecture and Furniture,* Book 1, *Cottage Furnishings* (London: Longman, Rees, Orme, Brown, Green & Longman, 1836), p. 278.

14 A. J. Downing, *The Architecture of Country Houses* (New York: Dover, 1969; reprint of 1850 edn.), p. 369.

15 James Arrowsmith, *The Paper-Hanger's Companion: A Treatise on Paper-Hanging in which the practical operations of the trade are systematically laid down* (Philadelphia: Henry Carey Baird & Co., 1887; English edns. as early as 1840 are known), preface and pp. 61–2.

16 Advertisement of Peter D. Turcote, *New York Post,* May 18, 1816.

17 Advertisement of S. P. Franklin, *Daily National Intelligencer,* Washington, D.C., Apr. 19, 1820.

18 Advertisement of an unknown wallpaper seller, who gave his address as 40 Cornhill, Boston, *Columbian Centinel,* Boston, July 9, 1800.

19 Advertisement of Samuel Robinson, *Daily National Intelligencer,* Washington, D.C., Sept 15, 1821.

20 See footnotes 48 through 51, Chapter 9, for page references in the Zuber Archives.

21 In 1816, Geisse & Korkhaus bought from Zuber wallpapers designated by the following production numbers that coincide with production numbers assigned to items in the Zuber catalogue now at the Metropolitan Museum of Art: 221 (2 copies); 257 (24 copies); 394 (24 copies); 412 (2 copies); 659 (2 copies); 660 (4 copies).

In 1817 Virchaux & Borrekins bought the following: 221 (1 copy); 257 (2 copies); 394 (2 copies); 412 (1 copy); 1,067 (12 copies); 1,070 (12 copies).

In 1822, Miles A. Burke bought the following: 257 (4 copies); 394 (2 copies); 412 (2 copies); 659 (4 copies); 660 (6 copies).

22 "Mercantile Value of the Fine Arts: No. V—Paper Hanging," *The Art-Union* (London), vol 6. (July 1, 1884): 179, 180.

23 Advertisement of Howell and Brothers, *The Mercantile Register or Business Man's Guide . . . Principal Business Establishments* (Philadelphia: H. Orr, 1846), p. 176. Advertisement of George B. Michael, *St. Louis Intelligencer,* Dec. 19, 1855. The author is indebted to Lynn Springer for transcriptions of this and other St. Louis advertisements.

24 Advertisement of Samuel Robinson, *Daily National Intelligencer,* Washington, D.C., Sept. 10, 1819.

25 See footnote 18 this chapter.

26 Advertisement for M. Werckmeister & Co., *New York Evening Post,* May 8, 1810 (Roberts).

27 Advertisement of Richardson's Paper-Hanging Warehouse, Albany City Directory, 1845–6 (advertisement on cover).

28 Advertisement of G. B. Michael, St. Louis Business Directory, 1847, p. 173.

29 Bill from Charles Grant, Boston, to Ephraim Brown, North Bloomfield, Ohio, July 19, 1821: "1 Fireboard pr. 9/1.50"; "1 Fireboard pr. 6/1.00." Grant was a customer of Zuber. He bought 998.50f. worth of wallpaper from Zuber in 1828–9.

30 Bill from Robert Golder, Philadelphia, Apr. 2, 1835, to Andrew Jackson for wallpaper purchase totaling $503, with a 5 percent discount, reduced to $477. Research files, The Hermitage, Nashville, Tenn.

31 Advertisement of C. F. Mansfield, *Public Ledger,* Philadelphia, Mar. 25, 1836.

CHAPTER 11– FRENCH WALLPAPERS AND THE DEVELOPMENT OF THE AMERICAN CRAFT

1 References to importation from Germany appear fairly regularly, but are not given emphasis, in American nineteenth-century advertisements. Examples are: Advertisements of George B. Michael in *Tri-Weekly St.-Louis Intelligencer,* June 15, 1855, and in *Kennedy's St. Louis City Directory,* 1857; Advertisement of R. Gledhill, *Scribner's Trade Directory,* New York, Nov. 1875–Apr. 1876, vol. 2; Advertisement of Warren, Fuller & Co., in the 1882 catalogue following the volume of the National Academy of Design, *Catalogues of Annual Exhibitions 1881–1890,* at the New-York Historical Society.

2 J. Milbert, *Picturesque Itinerary of the Hudson River and the Peripheral Parts of North America,* trans. Constance D. Sherman (Ridgewood, N.J., 1968), p. xxv. Jay Cantor brought this passage to the author's attention.

3 James Mease, M.D., *The Picture of Philadelphia* (Philadelphia: B. and T. Kite, 1811), p. 75.

4 See John W. Maxson, Jr., "American Papermakers in the Great Tariff Debate," *The Paper Maker,* 31, no. 1 (1962).

5 John Morison Duncan, *Travels Through Parts of the United States and Canada in 1818 and 1819,* 2 vols. (New York: W. B. Gilley; New Haven: Howe and Spalding, 1823), 1:259.

6 *The Proceedings of a Convention of the Friends of National Industry Assembled in the City of New York, Nov. 29, 1819* (New York: C. S. Van Winckle, 1819), [n.p.].

7 *Digest of Accounts of Manufacturing Establishments in the United States and of their Manufactures, Made Under the Direction of the Secretary of State, in Pursuance of a Resolution of Congress of 30th March, 1822* (Washington, D.C.: Gales and Seaton, 1823), pp. 5, 16.

8 *U.S. Congress, Senate, Senate Documents, 22nd Congress, 1st session, in Public Documents Printed by order of the Senate of the United States,* 3 vols. (Washington, D.C.: Duff Green, 1832), vol. 3, document 177.

9 Edwin T. Freedley, *Philadelphia and its Manufactures: A Hand-Book Exhibiting the Development, Variety, and Statistics of*

the Manufacturing Industry of Philadelphia in 1857 (Philadelphia: Edward Young, 1858), p. 371.

10 Albert Gallatin, *Report of the Secretary of the Treasury on American Manufacturers* (Brooklyn: Thomas Kirk, 1810), pp. 11, 13.

11 Tench Coxe, *Digest of the Census, Published by Congress, May, 1813,* as quoted by J. Leander Bishop, in *A History of American Manufacturers,* 3 vols., 3rd edn. (Philadelphia: Edward Young, 1868), 2:152.

12 Louis B. McLane, Secretary of the Treasury, *Documents Relative to the Manufactures in the United States, Collected and Transmitted to the House of Representatives . . . By the Secretary of the Treasury,* in U.S. Congress, House of Representatives, *House Executive Documents,* 2 vols. (Washington, D.C.: Duff Green, 1833), pp. 144–5, 184–5, 456.

13 Advertisement of Zecheriah Mills, *Connecticut Courant,* Hartford, Apr. 18, 1810, as quoted by Phyllis Kihn, "Zecheriah Mills: Paper-Hanging Manufacturer of Hartford, 1793–1816," *Connecticut Historical Society Bulletin,* 26, no. 1 (Jan. 1961):28.

14 Victor S. Clark, *History of Manufactures in the United States 1607–1860* (Washington, D.C.: Carnegie Institute, 1916), pp. 391–6. Betsy Blackmar guided the author to this information about average wages.

15 Advertisements of Ann Dickinson, *The Pennsylvania Packet,* Philadelphia, Feb. 1788; August 20, 1788.

16 Advertisement of Margaret Hovey, *Massachusetts Mercury,* Boston, June 24, 1800. An account of Joseph Hovey's business is given by Walter Kendall Watkins, "The Early Use and Manufacture of Paper Hangings in Boston," *Old-Time New England,* 12, no. 3 (Jan. 1922):115–19.

17 Howell & Co.'s importation of a wallpaper printing machine in 1884 is mentioned in Henry Burn, "American Wall-Papers," in *One Hundred Years of American Commerce,* ed. Chauncey M. Depew (New York: D. O. Haynes & Co., 1895), p. 506, and in Nancy McClelland, *Historic Wall-Papers from Their Inception to the Introduction of Machinery* (Philadelphia and London: J.B. Lippincott Company, 1924), p. 274.

CHAPTER 12– REPEATING PATTERNS OF THE EARLY NINETEENTH CENTURY

1 For examples of a range of designs within this category, see Musée des Arts Décoratifs, *Trois Siècles de Papier Peint* (Paris: Union Centrale des Arts Décoratifs, 1967), illustrations for catalogue, nos. 115, 117, 120, 127, 364.

2 Advertisement of John Perkins, *New Bedford Mercury,* Apr. 13, 1812.

3 Photograph in the study files, Cooper-Hewitt Museum. Bill of Thomas and Caldeleugh, Stationers and Paper Hanging Manufacturers, Baltimore, Oct. 20, 1808, original in Adena State Memorial Archives, Chillicothe, Ohio.

4 Photograph of fragments and information about the Wheeler house contributed by E. Frank Stephenson, Jr., to study files, Cooper-Hewitt Museum.

5 Advertisement of Josiah Bumstead, *Columbian Centinel,* Boston, May 23, 1821.

6 Advertisement of Moses Grant, *Columbian Centinel,* Boston, May 10, 1800. Advertisement of Josiah Bumstead, *Massachusetts Mercury,* Boston, Apr. 18, 1800.

7 Advertisement of John Perkins, *New Bedford Mercury,* Aug. 29, 1828.

8 Advertisement of A. Hyam of Baltimore, *Daily National Intelligencer,* Washington, D.C., Apr. 17, 1826.

9 Advertisement of James H. Foster, *Columbian Centinel,* Boston, July 27, 1825.

10 Advertisement for house at 77 Willow Street, Brooklyn, *New York Commercial Advertiser,* Mar. 12, 1835.

11 Advertisement of Mills and Danforth, *Connecticut Courant,* Hartford, May 3, 1814.

12 Advertisement of Moses Grant, Aug. 31, 1813, as quoted by Nancy McClelland, *Historic Wall-Papers* (Philadelphia & London: J. B. Lippincott Company, 1924), p. 270.

13 Advertisement of Mills and Danforth, *Connecticut Courant,* Hartford, Apr. 13, 1813, as quoted by Phyllis Kihn, "Zecheriah Mills, Paper-Hanging Manufacturer of Hartford, 1793–1816," *Connecticut Historical Society Bulletin,* no. 26, 1 (Jan. 1961): 29. Zecheriah Mills was trained in the Boston paper-staining craft and advertised when he started in business in 1793 that he was receiving his "materials" from Boston to set up the manufactory. Advertisement of Mills and Webb, *Connecticut Courant,* in Kihn, "Zecheriah Mills," p. 21.

14 These archives include those of Zuber et Cie., and records at the Bibliothèque Forney, and the Musée des Arts Décoratifs.

15 *"Irisé"* is a standard French adjective meaning "rainbow-colored, iridescent," as is *"ombré,"* meaning "shaded, tinted." *"Fondu"* is the word used by Andrew Ure in *A Dictionary of Arts, Manufactures, and Mines,* 2 vols. (New York: D. Appleton, 1863; 1st edn. London, 1839), 2:326, where he describes the production of "the *fondu* or rainbow style of paper hangings." *"Fondu"* is probably derived from the French verb *"fondre,"* to melt or dissolve —the colors melt/dissolve into one another in rainbow papers.

J. H. Thayer, in an advertisement in the *Independent Chronicle and Boston Patriot,* Apr. 21, 1827, offered sixteen cases of paper hangings from the "most approved Paris manufacturer" including *Iris* papers, doubtless an Anglicization of *irisé.*

16 Advertisement of Samuel Robinson, *Daily National Intelligencer,* Washington, D.C., May 15, 1826.

17 Matthew Digby Wyatt, *Industrial Arts of the Nineteenth Century,* 2 vols. (London: Day & Son, 1853), vol. 2, opposite plate 95. Wyatt credits, incorrectly, the Swiss with the invention of "rainbowing" for wallpapers.

18 Ure, *A Dictionary of Arts,* 2:326.

19 (a) A pattern found in the Florida Lee Calhoun Memorial Room in Fort Hill, the John C. Calhoun house built in 1803 and enlarged in 1805, is Zuber pattern #2502, first bound into sample books dated 1828–9, Livre 19, Archives Zuber.

(b) Fragments of a paper from a house in Alexandria, Va., now at the Cooper-Hewitt Museum, accession #1973-1-1, gift of Morrison Heckscher, match Zuber pattern #2506, 1830–1.

(c) Paper from a house in Branden, Vt., sent by Mrs. J. J. Kennedy to the Birge Co., Buffalo, for identification, matches Zuber pattern #2758, 1832–3.

(d) Paper from Daniel Busley house, Newburyport, Mass., Society for the Preservation of New England Antiquities, accession #1969, 225, matches Zuber 2598, 1830–1.

(e) Paper from Camden, Me., in the collection of the Society for the Preservation of New England Antiquities, matches the same Zuber pattern, #2598, 1830–1.

20 Advertisement of A. Hyam, proprietor of a Baltimore "Wholesale & Retail Paper Hangings Ware-House," in the *Daily National Intelligencer,* Washington, D.C., Apr. 17, 1826.

21 Advertisement of John Perkins, *New Bedford Mercury,* Sept. 28, 1810.

22 This date of pre-1835 is based on the fact that the gray brick pattern is printed on handmade paper of the type seldom used after this date—perhaps a late cutoff date of c. 1840 would be even safer in dating this pattern.

23 John Claudius Loudon, *An Encyclopaedia of Cottage, Farm and Villa Architecture and Furniture* (London: Longman, Rees, Orme, Brown, Green & Longman, 1836), p. 583.

24 Lee Roberts undertook research on the Jonas Wood house for the City of New York, when the house was moved from its original site at 314 Washington Street and restored by the Housing & Development Administration of the City of New York in 1972. She traced in city directories and in tax records successive occupants of the house and generously made her findings available to the author. See catalogue information accompanying the Cooper-Hewitt Museum accession 1973-8-1 to -8. The house was moved to the corner of Greenwich and Harrington Streets in the Washington Street Urban Renewal Area.

25 The Lemeul Steel paper at the Cooper-Hewitt Museum bears accession # 74-112-8a, -8b; gift of Harrison Cultra.

26 Margaretta Brown, the wife of Senator John Brown, of Frankfort, Kentucky, June 21, 1836, New York, to Mary Watts Brown (Mrs. Orlando Brown), Frankfort. In Archives, Colonial Dames of America in the Commonwealth of Kentucky, Frankfort, Kentucky.

CHAPTER 13– BANDBOXES

1 Samuel Johnson, *A Dictionary of the English Language* (London: 1755).

2 Handbill in the collections of the Cooper-Hewitt Museum, New York, accession no. 1951-42-1; purchased in memory of Eleanor G. Hewitt.

3 Advertisement of Benjamin Joyce, *Daily Advertiser,* New York, July 3, 1795.

4 Inventory of the Estate of Silvain Bijotat, New York, Sept. 28, 1824. Courtesy of the Henry Francis du Pont Winterthur Museum, Winterthur, Delaware. Joseph Downs Manuscript Collection, No. 54-67-21.

5 When the edges of decorative papers overlap newspapers lining a bandbox, the dates on the newspapers do provide a cutoff date for the wallpaper. A newspaper's place of publication can also suggest the area in which a bandbox was made or used. To learn whether the date or location of a newspaper lining reveals anything about the wallpaper in sight on a bandbox, the box must be carefully examined in an attempt to establish which layer of paper overlaps which. In addition, an equally careful attempt should be made to determine whether the patterning in sight is the original decoration pasted on the box or a later recovering.

6 Lilian Baker Carlisle, *Hat Boxes and Bandboxes at Shelburne* Museum (Shelburne, Vt.: The Shelburne Museum, 1960), p. 182.

7 Bandbox bearing label of Henry Cushing, in the collection of the Henry Francis du Pont Winterthur Museum, Winterthur, Del., accession no. 64.2041 a.b. Bandbox bearing label of Joseph S. Tillinghast in the collection of the Boston Museum of Fine Arts, accession no. 54.1783.

8 Carlisle, *Hat Boxes and Bandboxes,* p. 136.

9 Carlisle, *Hat Boxes and Bandboxes,* p. 188.

10 Carlisle in her *Hat Boxes and Bandboxes* writes about the popularity of bandboxes among women working in the New England textile factories, p. vi.

11 Elizabeth Leslie, *Miss Leslie's New Receipts for Cooking,* (Philadelphia: T. B. Peterson and Brothers, 1854), p. 315. The author is indebted to Christian Rohlfing for this reference.

CHAPTER 14– MAKING WALLPAPER BY MACHINE

1 This summary of early-nineteenth-century developments in paper-making machinery relies heavily on Lyman Horace Weeks, *A History of Paper-Manufacturing in the United States, 1690–1915* (New York: Lockwood Trade Journal Co., 1916), pp. 173–6.

2 Jean-Pierre Seguin, "Des Siècles de Papiers Peints," in Musée des Arts Décoratifs, *Trois Siècles de Papiers Peints (Paris: Union Centrale des Arts Décoratifs, 1967), p. 114, and Alan V. Sugden and John L. Edmondson, A History of English Wallpaper, 1509–1914* (New York: Charles Scribner's Sons, 1925; London: B. T. Batsford, 1926), p. 124.

3 Weeks, *A History of Paper-Manufacturing,* p. 199.

4 Handbill of Harris & Winans (65 Canal Street, near Broadway, where New York city directories show that the firm was

located only between 1835 and 1836), in the collections of the Print Study Room, Metropolitan Museum of Art, N.Y., acc. # 54.-90.705.

5 Advertisement of Joseph Freeman, *New Bedford Gazette,* Sept. 4, 1834.

6 "House Furnishing and Decoration: Wall Paper," *The American Builder and Journal of Art,* Chicago, Dec. 1869, p. 231.

7 Harry Weston, "A Chronology of Paper Making in the United States," *The Paper Maker,* no. 2 (1944): 9.

8 Edwin T. Freedley, *Philadelphia & Its Manufacturers, a hand-book . . . of the Manufacturing Industry of Philadelphia in 1857* (Philadelphia: Edward Young, 1858), p. 371.

9 Richard Herring, *Paper & Paper Making, Ancient & Modern* (London: Longman, Brown, Green & Longmans, 1856), p. 118.

10 "House Furnishing and Decoration: Wall Paper," *The American Builder and Journal of Art,* May 1869, 231.

11 James S. Munroe, United States Patent #204, 446, June 19, 1878.

12 Matthew Digby Wyatt, *Industrial Arts of the Nineteenth Century at the Great Exhibition, 1851,* 2 vols. (London: Day & Son, 1853), vol. 2, opposite plate 100.

13 Peter Force, United States Patent #3, 573X, Aug. 22, 1822. Class 101, printing subclass 187. In the patent specifications, Peter Force suggests that motive power other than hand-turning on the crank could be substituted.

14 Sugden and Edmondson, *A History of English Wallpaper,* pp. 125–7.

15 Seguin, "Des Siècles de Papier Peints," pp. 114–16.

16 Henry Burn, "American Wall-Papers," in *One-Hundred Years of American Commerce,* ed. Chauncey M. Depew (New York: D.O. Haynes & Co., 1895), p. 505, and Nancy McClelland, *Historic Wall-Papers from Their Inception to the Introduction of Machinery* (Philadelphia and London: J.B. Lippincott Company, 1924), p. 274.

17 J. Leander Bishop, *A History of American Manufacturers,* 3 vols. (Philadelphia: Edward Young & Co., 1868), 3:305.

18 Horace Greeley, ed., *The Great Industries of the United States* (Hartford, Chicago, and Cincinnati: J. B. Burr, Hyde & Co., 1872), p. 463.

19 "Wall-Paper," *The Decorator and Furnisher,* 16, no. 6 (Sept. 1890):192.

20 William R. Bradshaw, *Wallpaper: Its History, Use and Decorative Importance* (New York: The Joseph P. McHugh Company, 1891), p. 10.

21 Clare Crick, "The Origin & Development of Wallpaper," *Historic Wallpapers in the Whitworth Art Gallery* (Manchester, England: Whitworth Art Gallery, 1972), p. 13.

22 A patent in which the photographic process for making wallpaper is described only in the vaguest of terms was registered in the name of Archibald W. Paull, United States Patent # 140, 533, July 1, 1873.

23 "Household Artists," *The Manufacturer and Builder,* 1, no. 9 (New York, Sept. 1869):279.

24 "American Industries. No. 50.: The Wall Paper Manufacturer," *Scientific American,* 43, no. 4 (July 24, 1880): 53.

25 Bishop, *A History of American Manufacturers,* 2:453.

26 United States Bureau of the Census, Twelfth Census of the United States Taken in the Year 1900, Manufacturers, Part I, Vol. 7 (Washington D.C.: Government Printing Office, 1902), p. 12.

27 "Wall-Paper," p. 192.

28 "The American Trade in Paper-Hangings," *The Furniture Gazette,* London, May 10, 1879, p. 323.

29 Advertisement of Alfred Peats, *The Delineator, A Journal of Fashion, Culture & the Fine Arts* (London and New York: Butternick Publishing Co., Ltd.), Oct. 1895, p. v.

CHAPTER 15– 1840 TO 1870: THE IMPRINT OF MACHINES, THE ELABORATION OF STYLES

1 These arsenical colors were named in mid-nineteenth-century accounts of the manufacture of wallpaper as colors standardly used in the trade. See Appendix A. The history of the discovery and commercial introduction of these colors is recounted by Rosamond D. Harley in *Artists' Pigments c. 1600–1835* (New York: American Elsevier Company, Inc.). "Arsenic in Wall-Papers," *The Furniture Gazette,* London, July 25, 1874, p. 767, and Aug. 8, 1874, p. 821.

2 Advertisement of George B. Michael, *Daily Missouri Republican,* St. Louis, Oct. 15, 1850.

3 A. J. Downing, *The Architecture of Country Houses* (New York: Dover, 1969; reprint of 1850 edn.), p. 398.

4 Jules Desfossé and Hippolyte Karth, "The Manufacture of Paper Hangings in France," *The Furniture Gazette,* London, Mar. 14, 1874, p. 230.

5 "American Wall Paper," *The Furniture Gazette,* London, Oct. 18, 1873, p. 436.

6 *The Journal of Design,* 2, no. 8 (London, Oct. 1849).

7 Nikolaus Pevsner, "High Victorian Design," *Studies in Art, Architecture and Design,* Vol. 2, *Victorian and After* (New York: Walker and Co., 1968), p. 46.

8 Josiah F. Bumstead billhead of the 1840s in the collection of Robert L. Raley. Photocopy, study files Cooper-Hewitt Museum. Bill of sale to J. A. Beckwith, 1850. The printed "4" in the date is replaced by a handwritten "5."

9 Advertisement of Perkins, Smith and Co., *Daily Mercury,* New Bedford, May 6, 1853.

10 Downing, *The Architecture of Country Houses,* pp. 369, 370.

11 H.S.B., "The Domestic Economy of Architecture, Parlor Walls and Wall Ornaments," *The American Builder and Journal of Art,* Chicago, Mar. 1869, p. 69.

12 Actual samples of Spurr's "wood hangings," used as trade

cards, are preserved in the Warshaw collection of the Smithsonian Institution, at the Baker Business Library of Harvard University, and at the Historical Society of Pennsylvania.

13 "Trellis Work Paper" with imitation flock background, manufactured by Horne, 41 Gracechurch Street. *The Journal of Design,* 1, no. 4 (June 1849):118.

14 Pevsner, "High Victorian Design," p. 54.

15 A. W. N. Pugin, *The True Principles of Pointed or Christian Architecture* (New York: St. Martin's Press, London Academy, 1973; reprint of 1st edn., London: John Weale, 1841), p. 34.

16 Charles L[ocke] Eastlake, *Hints on Household Taste in Furniture, Upholstery, and Other Details* (Boston: James R. Osgood and Company, 1872), p. 122.

17 Pugin, *The True Principles,* p. 29.

18 Ibid.

19 Advertisement of Murdock and Crocker, *Palladium and Republican,* New Haven, Mar. 2, 1833.

20 Downing, *The Architecture of Country Houses,* p. 369.

21 Ibid., p. 398.

22 John Claudius Loudon, *An Encyclopaedia of Cottage, Farm and Villa Architecture and Furniture,* Book 1, *Cottage Furnishings* (London: Longman, Rees, Orme, Brown, Green and Longman, 1836), p. 279, item 582.

23 Downing, *The Architecture of Country Houses,* p. 369.

24 Pevsner, "High Victorian Design," p. 48.

25 Downing, *The Architecture of Country Houses,* p. 398.

26 Ibid., p. 398.

27 Advertisement of George B. Michael, *Daily Missouri Republican,* St. Louis, Oct. 15, 1850.

28 Advertisement of William Booth, Richmond City Directory (1845):"Paperhangings consisting of gilt, velvet, tapestry, fresco, plain and common Papers with borders to match."

29 *Catalogue of the New York Exhibition of the Industry of all Nations* (New York: George P. Putnam, 1853). Award to Jules Desfossé for "Large and fine Tableau Decoration Paper."

30 A set of the paper was photographed in place in a house in Portland, Maine, by Ed Polk Douglas, who also reports he has seen another in a second New England house.

31 H.S.B., "The Domestic Economy of Architecture," p. 69.

32 Slides of this paper made by the author are in study files, Cooper-Hewitt Museum.

33 Jean-Pierre Seguin, "Des Siècles de Papier Peints," in Musée Des Arts Décoratifs, *Trois Siècles de Papiers Peints* (Paris: Union Centrale des Arts Décoratifs, 1967), p. 50. This gives the name of the Mannheim firm only as "Englehard." *"Établié à Mannheim de 1830 1905, cette manufacture, qui pendant un temps fut liée par contrat à Desfossé, acheta quantité de dessins en France."*

34 Patent of William Baily, United States Patent # 71,120, for making stamp-gilt paper hangings.

35 A. E. Gordon, *New Brunswick and its Industries* (New Brunswick: Times Printing House, 1873), p. 46.

36 H.S.B., "The Domestic Economy of Architecture," p. 69.

37 Evidence for the popularity of these gilt papers is supported by surviving examples in places as remote as Austin County, Texas. Fragments of one of these patterns, featuring a gilded bird, tinged with green, alternating with a bouquet of flowers, were found there during restoration of the Gregor McGregor house, built in 1861. A narrow green flocked border had been used with the paper, and has been reproduced in the restored house, now one of the Winedale Inn properties of the University of Texas.

38 James Watson Williams papers, Oneida Historical Society, Utica, N.Y. Correspondence 235.70. Letter to his wife, Helen Elizabeth, Apr. 4, 1850.

39 Newark City Directory, vol. 25 (1859–60), p. 490. Advertisements of C. Abbott, 196 Broad St., and of Rutan and Birdsall, 70 Broad St.

40 Circular headed "J. F. Bumstead and Co/invite the attention of purchasers to their Spring Stock of 1853 . . . " in the Henry Francis du Pont Winterthur Museum, Joseph Downs manuscript collection, 74×281. The example at Winterthur bears a postmark and is addressed to Messrs. Merriam Chapin and Co., Springfield, Mass., and is inscribed in what appears to be mid-nineteenth-century script, "J F Bumstead and Co., March 28th 53 *Circular.*"

41 Thomas Webster and Mrs. Paine, *The American Family Encylopaedia of Useful Knowledge* (New York: Derby and Jackson, 1859; first edn. Harper and Brothers, 1845), p. 80.

42 Advertisement of William H. Patton in R. A. Smith, *Philadelphia as it is, being a correct Guide* (Philadelphia, 1852), p. 58.

43 Advertisement of S. H. Gregory and Co. in A. D. Jones, *American Portrait Gallery* (New York: Milton Emerson and Co., 1853), p. 63.

44 Advertisement of Galveston House furnishing warehouse, *Galveston Tri-Weekly News,* May 28, 1857.

45 "Mercantile Value of the Fine Arts: No. V: Paper-Hanging," *The Art Union: A Monthly Journal of the Fine Arts and the Arts Decorative and Ornamental,* London, July 1, 1844, p. 179.

46 1840s billhead in the collection of Robert L. Raley, photocopy. Study files, Cooper-Hewitt Museum.

47 Advertisement of Perkins, Smith and Co., *New Bedford Daily Mercury,* Mar. 12, 1853.

48 Catharine E. Beecher and Harriet Beecher Stowe, *The American Woman's Home* (New York: J. B. Ford and Co., 1869), p. 86.

49 Downing, *The Architecture of Country Houses,* p. 404.

50 Ibid. p. 398.

51 *Eighty Years Progress of the United States: A Family Record of American Industry, Energy and Enterprise,* vol. 1 (Hartford, Conn., 1868), p. 298.

52 H.S.B., "The Domestic Economy of Architecture," p. 69.

53 George Wallis, special report in *New York Industrial Exhi-*

bition: General Report of the British Commissioners (London: Harrison and Son, 1851), p. 69.

54 Advertisement of Beaty and Curry, *The Mercantile Register or Business Man's Guide* (Philadelphia, 1846).

55 Advertisement of John Perkins, *New Bedford Mercury,* Sept. 28, 1810.

56 Advertisement of Joseph Freeman, *New Bedford Weekly Courier and Workingman's Press,* Mar. 12, 1834.

57 Advertisement of John Perkins, *New Bedford Mercury,* Apr. 3, 1812.

58 Advertisement of John Perkins, *New Bedford Mercury,* Aug. 29, 1828.

59 Advertisement of Isaac Pugh, *The Mercantile Register or Business Man's Guide* (Philadelphia, 1846).

CHAPTER 16– THE 1870s AND 1880s: A MAJOR CHANGE IN TASTE

1 Walter Smith, *Art Education, Scholastic and Industrial* (Boston: James R. Osgood and Co., 1872), p. 213.

2 Phoebe Stanton, *Pugin* (New York: Studio Viking, 1971), especially pp. 180–4. Representing the views of industrial designers, the writings of Owen Jones and Christopher Dresser are especially useful. See Owen Jones, *Grammar of Ornament* (New York: Van Nostrand Reinhold Co., 1972; reprint of the 1856 edn.); and Christopher Dresser, *The Art of Decorative Design* (London: Day and Son, 1862) and *Principles of Decorative Design* (London, Paris, New York: Cassell Petter and Galpin, 1873).

3 John Ruskin, *The Two Paths* (1859), in *The Complete Works of John Ruskin,* 39 vols., eds. E. T. Cook and Alexander Wedderburn (London: George Allen, 1903), 16:231.

> You will every day hear it absurdly said that room decoration should be by flat patterns—by dead colours—by conventional monotonies, and I know not what. Now, just be assured of this —nobody ever yet used conventional art to decorate with, when he could do anything better, and knew what he did would be safe. Nay, a great painter will always give you the natural art, safe or not. Correggio gets a commission to paint a room on the ground floor of a palace at Parma: any of our people—bred on our fine modern principles—would have covered it with a diaper, or with stripes and flourishes, or mosaic patterns. Not so Correggio: he paints a thick trellis of vine leaves, with oval openings, and lovely children, leaping through them into the room; and lovely children, depend upon it, are rather more desirable decorations than diaper, if you can do them—but they are not quite so easily done. . . . in all cases whatever, the greatest decorative art is wholly unconventional —downright pure good, painting and sculpture, but always fitted for its place.

4 Jones, *Grammar,* p. 2.

5 Jones, *Grammar,* Proposition 8, p. 5; Proposition 13, p. 6.

6 Jones, *Grammar,* Proposition 23, p. 7.

7 "Modern Wallpapers," *The Furniture Gazette,* London, May 13, 1876, p. 201.

8 Constance Cary Harrison, *Woman's Handiwork in Modern Homes* (New York: Charles Scribner's Sons, 1881), p. 136.

9 Henry Russell Hitchcock, *American Architectural Books,* new expanded edn. (New York: Da Capo, 1976), pp. 85–92.

10 Roger B. Stein, *John Ruskin and Aesthetic Thought in America 1840–1900* (Cambridge, Mass.: Harvard University Press, 1967), p. 73.

11 Stein, *John Ruskin and Aesthetic Thought,* p. 46.

12 Ruskin, *The Two Paths,* p. 290. This sentence echoes Ruskin's earlier statement in *The Stones of Venice:* "All noble ornamentation is the expression of man's delight in God's work." (*Complete Works,* 9:70.)

13 Ruskin, *The Two Paths,* p. 253.

14 Peter Floud, "The Wallpaper Designs of William Morris," *The Penrose Annual,* 54 (1960): 41–5.

15 "Daisy" was the first wallpaper produced by Morris and Company in 1862, but not the first designed. In 1862, Morris, with Webb, first designed "Trellis." Medieval sources served as an inspiration to Morris in other aspects of his art as well. Gothic and medieval elements can be found in much of his poetry, stained glass, furniture, embroidery, and other work. Ray Watkinson, *William Morris as Designer* (New York: Reinhold Publishing Corporation, 1967), pp. 10, 28; Phillip Henderson, *William Morris* (New York: McGraw-Hill Book Company, 1967), pp. 14, 208–11; Paul Thompson, *The Work of William Morris* (New York: The Viking Press, 1967), pp. 68–72, 87, *et passim.*

16 "Modern Wallpapers," p. 302.

17 "Culture and Progress: The William Morris Window," *Scribner's Monthly,* 6, no. 2 (June 1873):245.

18 H. Hudson Holly, *Modern Dwellings in Town and Country, Adapted to American Wants and Climate With a Treatise on Furniture and Decoration* (New York: Harper and Brothers, 1878), p. 152.

19 Mary Jane Smith Madigan, "The Influence of Charles Locke Eastlake on American Furniture Manufacture, 1870–90," *Winterthur Portfolio,* 10 (1975), p. 9.

20 M.E.W.S. (Mary Elizabeth Wilson Sherwood), "Some New York Interiors," *The Art Journal* 3, no. 36 (New York, 1877):329.

21 "The Frieze and The Dado," *Carpentry and Building* (New York) 6, no. 6 (June 1884):106–7.

22 Paragraph in section of paid notices, "Scribner's Miscellany," bound at end of volume, *Scribner's Monthly,* 12 (May–Oct. 1876):5.

23 Peter Floud cited this passage in "The Wallpaper Designs

of William Morris," *The Penrose Annual,* vol. 54 (1960).

24 Harriet Prescott Spofford, *Art Decoration Applied to Furniture* (New York: Harper and Brothers, 1877), p. 184.

25 Charles L[ocke] Eastlake, *Hints on Household Taste in Furniture, Upholstery, and Other Details* (Boston: J.R. Osgood and Co. 1872; 1st London edn. 1868), p. xxiv.

26 Eastlake, *Hints on Household Taste,* p. 116.

27 Spofford, *Art Decoration,* p. 147.

28 *Appleton's Journal,* 10 (New York, Sept. 1873):123.

29 Wilmot Pilsbury, "Principles of Decorative Design," *The Furniture Gazette,* London, Aug. 20, 1873, p. 329.

30 Walter Smith, *Art Education,* p. 187.

31 Spofford, *Art Decoration,* pp. 184–5.

32 James M. Wilcox, "Paper Industry, Stationery, Printing, and Book-Making," in *United States Centennial Commission, International Exhibition 1876 Reports and Awards,* vol. 5, p. 76.

33 Ibid, p. 12.

34 Advertisement of J. L. Isaacs, in J. A. Dacus, *A Tour of St. Louis* (St. Louis: J. W. Buel, Western Publishing Co., 1878), p. 311.

35 Harrison, *Woman's Handiwork,* p. 136.

36 Christopher Dresser, "Ceilings, Walls and Hangings," *The Decorator and Furnisher,* 8, no. 1 (Oct. 1888): 26.

37 See footnote 2, this chapter.

38 Eastlake, *Hints on Household Taste,* p. 116.

39 Clarence Cook, *The House Beautiful* (New York; Charles Scribner's Sons, 1881), p. 19.

40 "Wall Decorations—Paper Hangings," *Carpentry and Building: A Monthly Journal,* 1, no. 4 (New York, 1879).

41 Susan N. Carter, "Principles of Decoration. 1. Wall-Papers," *The Art Journal* 6 (New York, 1880): 5.

42 Clarence Cook, *"What Shall We Do With Our Walls?"* (New York: Warren, Fuller and Co., 1880), p. 10.

43 [George W. Sheldon], *Artistic Houses* (New York: Benjamin Blom, Inc., 1971; reprint of 1883 edn.).

44 "Decorative Art in Our Homes: The Aesthetic in Wallpaper Manufacture," *Frank Leslie's Illustrated Newspaper,* 54, no. 1392 (May 27, 1882): 219.

45 Lloyd Lewis and Henry Justin Smith, *Oscar Wilde Discovers America (New York: Benjamin Blom, 1967; reprint of a book first published in 1936 by Harcourt Brace),* p. 55. He is supposed to have made this comment Jan. 6, 1882.

46 "Decorative Art in Our Homes," p. 219.

47 Cook, *"What Shall We Do With Our Walls?,"* p. 34.

48 "The Competition in Wall-Paper Designs," *The American Architect and Building News* (Nov. 16, 1881), p. 251.

49 "Art Notes," *Inland Architect,* Nov. 1883, unnumbered page.

50 Antoinette F. Downing and Vincent J. Scully, *The Architectural Heritage of Newport, Rhode Island* (Cambridge, Mass.: Harvard University Press, 1952), p. 152.

51 Samuel and Joseph C. Newsom, *Picturesque Californian Homes, No. 2* (San Francisco: S. and J. C. Newsom, 1885), p. 5.

52 For instance, Harrison, *Woman's Handiwork in Modern Homes,* pp. 136–7.

53 "The Frieze and the Dado," *Carpentry and Building,* 6, no. 6 (New York, June 1884):106.

54 Clarence Cook, *"What Shall We Do With Our Walls?",* p. 17.

55 Eastlake, *Hints on Household Taste,* p. 124.

56 *The Morris Exhibit at the Foreign Fair* ([no place or publisher given; presumably London, Morris & Co.], 1883), pp. 16–17.

57 Dresser, *Principles,* p. 85.

58 "Modern Wallpapers," p. 302.

59 R. H. Pratt, "Dados," *The Decorator and Furnisher,* 4, no. 4 (New York, July 1884):138.

60 Dresser, *Principles,* p. 85.

61 Henry T. Williams and C. S. Jones, *Beautiful Homes, or Hints in House Furnishing* (New York: Henry T. Williams, 1878), p. 42.

62 "Wallpapers," *Carpentry and Building,* 2, no. 12 (New York, Dec. 1880): 224.

63 "The Use of Paper for Ceiling Decorations," *Carpentry and Building,* 3, no. 10 (New York, Oct. 1881): 208.

64 *The Morris Exhibit at the Foreign Fair* [no place or publisher given; presumably London, Morris & Co.], 1883), pp. 16–18.

65 Christopher Dresser, "Ceilings, Walls and Hangings: some modified views by Dr. Dresser," *The Decorator and Furnisher,* 13, no. 1 (New York, Oct. 1888), p. 26.

66 Eastlake, *Hints on Household Taste,* pp. 119–20.

67 Ibid., p. 120.

68 Nearly every author of a book or article on decorating cited in these footnotes has something to say on the subject. See for instance Spofford, *Art Decoration,* p. 181 ff.

69 Williams and Jones, *Beautiful Homes,* pp. 34–5.

70 "Some New York Interiors," p. 329.

71 Paid notice of Pottier and Stymus, decorators, *Scribner's Monthly,* Vol. 10 (May–October 1875).

72 "Our Philadelphia Letter," *The Decorator and Furnisher,* 1, no. 1 (Oct. 1882): 20.

73 Sample books of French wallpapers of the 1870s and 1880s preserved in the Musée de l'Impression sur Étoffes, Mulhouse, Alsace, France, include many examples conforming to the English rules and incorporating favorite "Aesthetic" motifs like sunflowers and lilies.

74 Shirley Dare, "Modern Furnishing in America," *The Furniture Gazette,* London, Dec. 24, 1877, p. 473.

75 Journals taking note of the Japanese processes include *American Artisan* (Sept. 14, 1870), *The Decorator and Furnisher* (Feb. 1885), *The Journal of Decorative Art* (Apr. 1884).

76 Seams like those necessary in joining animal hides for wall

coverings make the paper look even more like leather.

77 "The Building Interests," *Industrial Chicago,* 6 vols. (Chicago: Goodspeed Publishing Co., 1891–6), 2 (1891): 533.

78 Alan V. Sugden and John L. Edmondson, *A History of English Wallpaper 1509–1914* (London: B. T. Batsford, 1925), pp. 250ff.

79 Designs for Frederick Beck and Co's. Lincrusta Walton were registered in the United States Patent Office under the names of their designers, including:

P. Groeber: July 1, 1884: Design Patent No. 15114; July 29, 1884: 15173; Aug. 12, 1884: 15238–15242; Jan. 28: 19612–19624; Feb. 25, 1890: 19670.

C. C. Hiscoe: Jan. 21, 1890: Design Patent No. 19598–19601, 19603.

Albert Liesel: July 29, 1884, Design Patent No. 15174; Aug. 12, 1884: 15245, 15246.

Augustin Le Prince: June 24, 1884, Design Patent No. 15093, Aug. 12, 1884, 15247.

80 This advertisement and variations on it ran regularly in a magazine published by the manufacturers of Lincrusta, Frederick Beck, entitled *Beck's Journal of Decorative Art.*

81 Alan Sugden and John L. Edmondson, *A History of English Wallpaper 1509–1914,* p. 250ff.

82 See also advertisement for "Munroe's Patent Ingrain Wall Paper," *The Decorator and Furnisher* 2, no. 2 (May 1883): 64: "... perfectly sanitary, no poisonous coloring matter ... waterproof ... can be washed ... very thick and elastic, and will cover imperfections in walls. ... mostly Wool, and has a soft finish, agreeable to the eye, and makes a superior background for pictures." A. S. Jennings in *Wall Paper Decoration* (London, 1907), p. 57, describes "oatmeal ingrain," explaining that it "is made by the addition of white to the colored pulp, which results in a mottled appearance or series of small irregular spots of white scattered throughout the colored ground."

83 The firm of Warren Fuller and Co., founded in 1855 as "Partridge, Pinchot and Warren," was succeeded by "Whiting, Young and Warren" in 1875, was known as J. S. Warren and Co. before 1880, and as "Warren Fuller and Co." in the early 1880s. It was variously and more or less successively called "Warren, Fuller and Lange" (in the mid-1880s), "Warren Lange and Co." (by 1890), and again "Warren, Fuller and Co." between 1895 and 1901.

CHAPTER 17– 1890 TO 1915: REVIVALS AND THE PURSUIT OF NOVELTY

1 John Beverly Robinson, "Architects' Houses," *Architectural Record* III (July 1893–July 1894), p. 206.

2 Frank Lloyd Wright, "The Cardboard House," in *The Kahn Lectures for 1930: The Princeton Monographs in Art and Archaeology.* (Princeton, N.J.: Princeton University Press, 1931), pp. 78–9.

3 Loos's statement "in a magazine article" is quoted by Henry Russell Hitchcock, *Architecture: Nineteenth and Twentieth Centuries,* 3rd edn. (Baltimore, Md.: Penguin Books, 1968), p. 352.

4 Edith Wharton and Ogden Codman, Jr., *The Decoration of Houses* (New York: Charles Scribner's Sons, 1897), p. 44.

5 R. W. B. Lewis, *Edith Wharton, A Biography* (New York: Harper and Row, 1975), pp. 77–9.

6 Candace Wheeler, *Principles of Home Decoration* (New York: Doubleday, Page & Company, 1903), p. 92.

7 Wheeler, *Principles,* pp. 109, 112.

8 Walter Crane, "Columbia Frieze," *Wallpaper News,* 1, no. 2 (Feb. 1893): 22.

9 Many of these are listed in catalogues of their exhibitions published by English wallpaper manufacturers who exhibited at the fair. These catalogues are preserved in the Warshaw Collection, National Museum of History and Technology, Washington, D.C. They are Catalogue of Wm Woollams & Co. for World's Columbian Exposition, Chicago, 1893, and Catalogue of Jeffrey & Co. for World's Columbia Exposition, Chicago, 1893.

10 *The Wallpaper News and Interior Decorator,* Dec. 1908.

11 "The Walls of Your Home and the New Coverings for them," *The Craftsman,* 28, no. 4 (July 1915): 372.

12 Photograph from *The Craftsman,* 1, No. 1 (Oct. 1901): ii; Advertisement for Allen-Higgins Wall Paper Co. from *The Craftsman,* 15, no. 1 (Oct. 1908): xl.

13 Advertisement for Tapestrolea from *The Craftsman,* 18, No. 6 (Sept. 1910): xxxiv; Advertisement for Fab-Ri-Ko-Na, *The Craftsman,* 13, No. 6 (Mar. 1908): xii; Advertisement for Sanitas from *The Craftsman,* 18, No. 1 (Apr. 1910): 125.

14 *The Craftsman,* 13, no. 6 (Mar. 1908). Advertisement for "Fab-ri-ko-na," p. xii. Advertisement for "Sanitas," p. xvi.

15 Elsie de Wolfe, "The Treatment of Walls," *The House in Good Taste* (New York: The Century Co., 1915), pp. 52–68. The following pages are listed according to the order of the quotations in the text: pp. 59, 54, 62, 66, 66, 66, 60.

16 George Leland Hunter, *Home Furnishing* (New York: John Lane Company; Toronto: Bell & Cockburn, 1913), pp. 193–4.

17 Elsie de Wolfe, "Reproductions of Antique Furniture and Objects of Art," *The House in Good Taste,* p. 268.

CONCLUSION

1 Vincent Scully, *Environment, Act, and Illusion: Architecture, Sculpture and Painting* (To be published: New York: Whitney Library of Design, 1981).

2 Clarence Cook, *"What Shall We Do With Our Walls?,"* (New York: Warren Fuller & Co., 1880), pp. 1–2.

3 Cook, "What Shall We Do With Our Walls?" p. 4.

4 John Ruskin, *The Two Paths* (1859), in *The Complete Works of John Ruskin, 39 vols.,* eds. E. T. Cook and Alexander Wedderburn (London: George Allen, 1903), 16:342–3.

5 Ralph Waldo Emerson, *Emerson's Complete Works,* vol. 7, *Society and Solitude* (Boston: Houghton, Mifflin and Company; New York: The Riverside Press, 1886), pp. 105–6.

SHORT SELECTED BIBLIOGRAPHY

Eighteenth- and nineteenth-century sources have been cited so often in the footnotes that no listing seems necessary here. For further reading and for additional illustrations the following books and articles published during the twentieth century are recommended:

Aslin, Elizabeth. *The Aesthetic Movement.* New York: Praeger, 1969.

Clark, Fiona. *William Morris: Wallpapers and Chintzes.* London: Academy, 1973.

Clouzot, Henri. *Tableaux-Tentures de Dufour & Leroy.* Paris: Librairie des Arts Décoratifs, n.d. (c. 1930).

Clouzot, Henri, and Follot, Charles. *Histoire du Papier Peint en France.* Paris: Éditions D'Art Charles Moreau, 1935.

Dornsife, Samuel A. "Wallpaper." *The Encyclopaedia of Victoriana,* edited by Harriet Bridgeman and Elizabeth Drury. New York: Macmillan, 1975.

Durant, Stuart. *Victorian Ornamental Design.* New York: St. Martin's Press, 1972.

Entwisle, Eric A. *The Book of Wallpaper.* London: Arthur Barker, 1954.

———.*French Scenic Wallpapers 1800–1860.* Leigh-on-Sea, England: F. Lewis, 1972.

———. *A Literary History of Wallpaper.* London: B. T. Batsford, 1960.

———. *Wallpapers of the Victorian Era.* Leigh-on-Sea, England: F. Lewis, 1964.

Evans, Joan, *Pattern: A Study of Ornament in Western Europe from 1180 to 1900.* Oxford: The Clarendon Press, 1931.

Fowler, John, and Cornforth, John. *English Decoration in the 18th Century.* Princeton, N.J.: The Pyne Press, 1974.

Frangiamore, Catherine Lynn. Wallpapers in Historic Preservation. Washington, D.C.: National Park Service, 1977.

———. "Wallpaper: Technological Innovation and Changes in Design and Use." *Technological Innovation and the Decorative Arts: Winterthur Conference Report, 1973,* pp. 277–305. Charlottesville, Va: University Press of Virginia, 1974.

———. "Wallpapers Used in Nineteenth Century America," *Antiques,* vol. 102, 6 (Dec. 1972): 1042–51.

Gombrich, Ernst Hans Josef. *The Sense of Order: a Study in the Psychology of Decorative Art.* Ithaca, N.Y.: Cornell University Press, 1978.

Greysmith, Brenda. *Wallpaper.* New York: Macmillan, 1976.

Hotchkiss, Horace. "Wallpapers Used in America, 1700–1850." *The Concise Encyclopedia of American Antiques.* 2 vols, edited by Helen Comstock. New York: Hawthorne Books, 1958. Vol. 2, pp. 488ff.

Hunter, George Leland. *Decorative Textiles.* Philadelphia and London: J.B. Lippincott & Co., 1918.

Justema, William. *Pattern: A Historical Panorama.* Boston: New York Graphic Society, 1976.

———. *The Pleasures of Pattern.* New York: Reinhold, 1968.

McClelland, Nancy V. *Historic Wall-Papers from Their Inception to the Introduction of Machinery.* Philadelphia and London: J. B. Lippincott Company, 1924.

Musée des Arts Décoratifs. *Trois Siècles de Papiers Peints.* Paris: Musée des Arts Décoratifs, 1967.

National Trust for Historic Preservation in the United States. *Documented Reproduction Fabrics and Wallpapers.* Washington, D.C.: National Trust, 1972.

Olligs, Heinrich, ed. *Tapeten Ihre Geschichte bis zur Gegenwart.* 3 vols. Braunschweig: Klinkhardt & Biermann, 1970.

Oman, C. C. *Catalogue of Wallpapers, Victoria and Albert Museum.* London: Board of Education, 1929.

Peterson, Harold. *Americans at Home.* New York: Scribners, 1971.

Pevsner, Nikolaus. *Studies in Art, Architecture and Design.* 2 vols. Vol. 2, *Victorian and After.* New York: Walker & Co., 1968.

Roth, Rodris. "Interior Decoration of City Houses in Baltimore: The Federal Period." *Winterthur Portfolio 5,* pp. 59–86. Charlottesville, Va.: The University Press of Virginia for the Henry Francis du Pont Winterthur Museum, 1969.

Sanborn, Kate. *Old Time Wall Papers.* Greenwich, Conn.: The Literary Collector Press, 1905.

Seale, William. *The Tasteful Interlude: American Interiors Through the Camera's Eye, 1860–1917.* New York: Praeger, 1975.

Spencer, Isobel. *Walter Crane.* New York: Macmillan, 1975.

Sugden, Alan V., and Edmondson, J. L. *A History of English Wallpaper, 1509–1914.* New York: Charles Scribner's Sons, 1925; London: B. T. Batsford, 1926.

Thornton, Peter. *Seventeenth-Century Interior Decoration in England, France and Holland.* New Haven and London: Yale University Press, published for the Paul Mellon Centre for Studies in British Art, 1978.

Wellman, Rita. *Victoria Royal: The Flowering of a Style.* New York and London: Charles Scribner's Sons, 1939.

Whitworth Art Gallery. *Historic Wallpapers in the Whitworth Art Gallery.* Manchester, England: The Whitworth Art Gallery, 1972.

ACKNOWLEDGMENTS

Work on this book began and ended at the Cooper-Hewitt Museum. Its founders, Eleanor and Sarah Hewitt, had a special interest in wallpaper, and their initial gifts to what was then the Cooper Union Museum (founded in 1897) formed the basis of a collection that has grown to become the largest and most important of its kind in the world. It has acquired that status through the continuing support of a group that had its origins among friends of the Misses Hewitt and expanded to include wallpaper producers and designers throughout the industry. This book owes a great deal to the sponsorship of individual companies and trade organizations like the Wallcovering Information Bureau.

When I came to work at the museum in 1967, the remarkable wallpaper collection first aroused my interest in the subject and I fell heir to the work of generations of curators before me. I began my study during the late 1960s with the knowledgeable advice and encouragement of Christian Rohlfing and Janet Thorpe of the museum staff, of Harvey Smith and William Justema, friends of the museum, and of Abbott Lowell Cummings and Richard Nylander, colleagues who work with the fine collection of wallpapers at the Society for the Preservation of New England Antiquities. The New York State Council on the Arts supported a preliminary research trip to England, France, and Germany.

Lisa Taylor, director of the Cooper-Hewitt Museum, encouraged my research during 1972 and 1973 under a grant from the National Endowment for the Humanities, a grant matched by an industry association then called The Wallcoverings Council. Contributions to the research grant were made by Academy Handprints, Inc.; Birge Wallcoverings, Reed Decorative Products Group-U.S.; Imperial Wallcoverings; Lennon Wallpaper Company; Ronkonkoma Wall Paper Corporation; Specialty Jute Products Corporation, Inc.; Stamford Wall Paper Company; Thomas Strahan Wallcoverings, Decorative Products Division, National Gypsum Company; United Wallcoverings, Division of Collins and Aikman; Wall Trends International; and York Wallcoverings, Inc. I thank these members of the industry both for their support and for their patience during the many succeeding years since they so generously contributed to the project that led to publication of this book.

The research grant supported my study in major libraries and wallpaper collections of this country and of England, France, and Germany. Librarians and curatorial staff members of these institutions who have helped me for days and weeks on end are recognized in lamentably cursory fashion on the list that concludes this acknowledgment. I hope they will forgive the brevity of my thanks, but know that they are sincere.

In the course of research, a nationwide survey requesting information about wallpaper was mailed to museums, historical societies, and a variety of preservation groups. Representatives of more than one thousand institutions as well as individuals responded to that survey, and the information, illustrations, and samples of wallpaper they so generously contributed to the Cooper-Hewitt have been an important source for this book. I am sorry that I cannot personally acknowledge each of them anew, but I thank them all.

The grant also supported my writing of an early draft of the manuscript during 1973, a draft completed and revised during a summer leave of absence in 1975 from my duties as curator of the Atlanta Historical Society. Work on the book continued at a less intense pace during my years in Atlanta, and during two years of graduate study at Yale, where course work with Richard Brodhead, George Hersey, R. W. B. Lewis, Margaretta Lovell, Jules Prown, Cynthia Russett, and Bryan Wolf and reading directed by Nancy Cott helped me rethink my whole approach to this book. Vincent J. Scully's advice during the rewriting of the book has been especially valuable to me.

Robert L. McNeil, Jr., the president of the Barra Foundation, Inc., provided the impetus for the book's complete rewriting during 1979 and 1980. Without his continuing enthusiasm and personal interest the manuscript drafted so long ago would perhaps never have seen print. The Barra Foundation, Inc., of Philadelphia has generously supported the last crucial phases of rewriting, additional research, editing, and preparation of the manuscript for publication. The foundation has assembled an able team headed by Regina Ryan, whose editing, production coordination, and attention to many details have been heroic. I thank her, and the talented people who have worked with her to turn my work into a book: Ulrich Ruchti, designer; Suzi Arensberg, copy editor and editor; Edmée Reit, indexer. I thank Scott Hyde for the many excellent photographs he has produced for this book, and Samuel A. Dornsife and Christian Rohlfing for reading and commenting on the manuscript.

When I returned to the Cooper-Hewitt to prepare the manuscript for publication, the entire staff assisted me in many ways, and I especially thank Robert Kaufmann and Margaret Luchars for their help, as well as Anne Hysa Dorfsman and Regina Rehkamp.

I also acknowledge with gratitude debts of longer standing: To Charles F. Montgomery, who introduced me to the study of American decorative arts; to Roy P. Frangiamore, who shared and refined my enthusiasm for nineteenth-century decorative arts and architecture, and to my parents, Catherine Mitchell and William Willis Lynn, who have always helped me. Finally, I thank the people who are listed below for kindnesses literally too many to specify:

Alice Davidson Abel, Edith Adams, Jean Adhémar, Christopher Allan, Don Anderle, Elaine Andrews, Kay Atwood, Bob Balay, William Nathaniel Banks, Jairus Barnes, Lu Bartlett, James A. Bear, Jr., Ellen Beasley, Alice Baldwin Beer, Helen Belknap, Sonja Bey, Norman Bielowicz, Eleanor Bishop, Mary Blackwelder, Edward Blair, Louise Blunt, Ruth Blunt, Louis W. Bowen, Mrs. J. Williamson Brown, Charles Brownell, Yvonne Brunhammer, Paul Buchanan, Mr. and Mrs. Richard Bumstead, Fred Burroughs, Janet Byrne, Xenia Cage, Marjorie Cahn, Richard Campbell, Jay Cantor, Orville W. Carroll, William L. Cawthon, Jr., Narcissa Chamberlain, Bettye Thomas Chambers, S. Allen Chambers, Jr., John Cherol, Helen Chillman, Oliver Conant, Barry Conduitt, Clement Conger, Maxine E. Cooper, Wendy Cooper, Phillip H. Curtis, Anne Daniels, Elaine Evans Dee, Francis Deransart, Jane desGranges, Davida Deutsch, Angelo Donghia, Murray Douglas, Philip H. Dunbar, Frances Duniway, Mary Dunn, Frances Edmondson, Anne Kneeland Ellis, Richard Ellis, Richard Eltzroth, L. V. Emmert, Nancy

Goyne Evans, Jonathan Fairbanks, Katharine Gross Farnham, Mme. M. de Fayet, J. Everitt Fauber, Wilson Faude, Susan G. Ferguson, E. McSherry Fowble, Charles Freeman, Nancy Freirich, Donald Friary, Elizabeth Donaghy Garrett, Wendell D. Garrett, Beatrice Lippincott Garvan, George Gassett, David Gebhard, Roger Gerry, Kristin Gibbons, Sue Gillies, Anne Castrodale Golovin, Mme. Andrei Gourlay, Douglas Gracie, Florence Griffin, Frank Grunewald, Elton W. Hall, Jean Hamilton, Jeanne B. Hamilton, Frances G. Hanahan, David Hanks, Henry J. Harlow, Thompson R. Harlow, Calvin Hathaway, Helena Hayward, Morrison H. Heckscher, Ambrose Hercules, Anne Herman, Constance Hershey, William Hershey, Elizabeth H. Hill, William B. Hill, Graham Hood, Daniel Hopping, Josephine Howell, Conover Hunt-Jones, Sibley Jennings, J. Stewart Johnson, Philip A. Johnson, Edward Kallop, Odile Kammerer, Patricia E. Kane, John Keefe, David W. Kiehl, Phyllis Kihn, Arthur Leibundguth, Martin Leifer, Marilynn J. Bordes Lissauer, Nina Fletcher Little, Calder Loth, John Lovari, Robert B. MacKay, Millie McGehee, Elizabeth McLane, Jean Mailley, Christine Meadows, Ernst Wolfgang Mick, Ellen G. Miles, Ronald W. Miller, Roger Mohovich, Christopher Monkhouse, Florence Montgomery, Roger Moss, Jack Murray, Mary Noon, Jane C. Nylander, Thomas W. Parker, W. J. Patterson, Ann Percy, William Pollard, William Pressly, Mrs. John W. Price, Maureen O'Brien Quimby, Robert L. Raley, Joe Rankin, Lisa Reynolds, Lee Roberts, Elizabeth Roth, Rodris Roth, Beatrix Rumford, Karol A. Schmiegel, Jean Pierre Seguin, George Shackelford, Wendy Shadwell, Alfred Simon, Marc Simpson, Margaret W. M. Shaeffer, Barclay Smith, Gaddis Smith, Larry Soloman, Milton Sonday, Joseph Peter Spang, Isobel Spencer, Lynn E. Springer, Mrs. John A. Sprouse, Margaret Stearn, Susan Stitt, Donald L. Stover, Mrs. Wilson L. Stratton, Judy Tankard, Elizabeth Taylor, Sandra Shaffer Tinkham, Joseph VanWhy, David B. Warren, Dorothy Waterhouse, Susan Finlay Watkins, Lucille Watson, Mrs. Edward K. Webb, Martin Weil, John J. Whitfield, Jr., Kenneth Wilson, Stefanie Munsing Winkelbauer, Fronia Wissman, Charles B. Wood, Elizabeth Ingerman Wood, Helena Wright, Mrs. Wesley Wright, Jr., Richard P. Wunder, and D. Lorraine Yerkes.

INDEX

Catherine Lynn is the acknowledged expert in the field of wallpapers used in America. She began her intensive study of the subject while serving as curator of the Cooper-Hewitt Museum's collection of wallpapers in New York City, a post she held for seven years. She went on to become curator of the Atlanta Historical Society for two and a half years and is now a doctoral candidate in American Studies at Yale University. She received her B.A. from Sweetbriar College and her M.A. degree from the University of Delaware, where she was a Fellow in the Winterthur Program. She is the author, under the name Catherine Lynn Frangiamore, of *Wallpapers in Historic Preservation,* a publication of the National Park Service. Ms. Lynn was born in Lynchburg, Virginia, and currently lives in New Haven, Connecticut.

This book was set in Times Roman. The composition was done by The Haddon Craftsmen; the complete manufacturing by R.R. Donnelley and Sons Co. The text paper is Quintessence Dull by Northwest Paper Co. The cloth binding is Natural Sail Cloth by Holliston.